69.95

MISSION COLLEGE
LIBRARY

The Essence of Logic Circuits

Also of Interest from IEEE Press ...

Digital Systems Testing and Testable Design, Revised Printing
Miron Abramovici, *AT&T Bell Laboratories*; Melvin A. Breuer, *University of Southern California, Los Angeles*; Arthur D. Friedman, *George Washington University*
1995   Hardcover   680pp   IEEE Order No. PC4168   ISBN 0-7803-1062-4

The Internet for Scientists and Engineers, Revised Edition
Online Tools and Resources
Brian J. Thomas
Co-published with SPIE Optical Engineering Press
1996   Softcover   480pp   IEEE Order No. PP5379   ISBN 0-7803-1194-9

Understanding Object-Oriented Software Engineering
Stefan Sigfried
Published in cooperation with the IEEE Computer Society Press
1995   Softcover   478pp   IEEE Order No. PP4507   ISBN 0-7803-1095-0

The Essence of Logic Circuits

Second Edition

Stephen H. Unger
Columbia University

**IEEE
PRESS**

The Institute of Electrical and Electronics Engineers, Inc., New York

This book may be purchased at a discount from the publisher when
ordered in bulk quantities. For more information, contact:

IEEE PRESS Marketing
Attn: Special Sales
P.O. Box 1331
445 Hoes Lane
Piscataway, NJ 08855-1331
Fax: (908) 981-9334

For more information about IEEE PRESS products, visit the
IEEE Home Page: http://www.ieee.org/

Printed in the United States of America

10   9   8   7   6   5   4   3   2

ISBN 0-7803-1126-4

IEEE Order Number: PC 5590

Library of Congress Cataloging-in-Publication Data

Unger, Stephen H. (date)
  The essence of logic circuits/Stephen H. Unger. — 2nd ed.
     p.   cm.
  Includes bibliographical references and index.
  ISBN 0-7803-1126-4
  1. Logic circuits.   I. Title.
TK7868.L6U54   1996
621.39′5—dc20   95-52708
                                                                               CIP

To My Sister

Muriel Krell

Contents

Appendix   284

Preface

As suggested by its title, this book is about what I consider to be the essentials of logic circuits. I have tried to present the material in a clear, concise form, striking a balance between theory and practice. Too much concern with theory makes a subject such as this unnecessarily difficult to learn and detracts from its usefulness to those who are going to put it to work. At the other extreme, presenting "useful" procedures without justifying them, and concentrating on details of current practice would also make for dull reading. Even worse, it would leave readers stranded when advances in technology make the material presented obsolete. I have relied heavily on the use of examples to clarify the general theory, and have tried to make these examples do double duty by using them to introduce useful devices.

Logic circuit design has gone through some interesting evolutions in the past few decades. Initially, logic gates were constructed from discrete components, such as vacuum tubes, diodes, transistors, resistors, etc. Each gate and the associated wiring entailed a cost in components, labor, bulk, and power requirements. A major design objective was to use the individual gates most efficiently. Next came the era of small- and medium-scale integration (SSI and MSI), in which the building blocks became sets of gates on a chip, or small functional units such as decoders. Now it became important to learn how to use these units efficiently. There was less value attached to minimizing logic expressions. This was also true when such devices as read-only memories (ROMs) and programmable logic arrays (PLAs) became available off the shelf. Either a logic function could be realized on a certain size PLA or it couldn't. Except in marginal cases, there was not much point working to minimize logic.

With the advent of large- and very-large-scale integration (LSI and VLSI) the situation has, to a surprising extent, reverted to the first stage. Now substantial systems and subsystems are realized on a single chip. Reductions in the number of elements required to realize each function may make it possible to use fewer chips or to do more on a given chip. In high-performance systems, eliminating redundant gates not only frees up chip area, it reduces power dissipation, often the limiting factor. The size of the PLA may be specified by the designer, and there are even special tricks for reducing the size of a personalized PLA. There are of course factors that were not present in the days of discrete logic elements, such as the value of regularity in the arrangement of elements on a chip. But the need for powerful methods for generating efficient logic circuits has indeed returned. A major consequence of the larger scale of integration is the enormous size and complexity of our systems, and the great importance of testing. An effort has been made in this book to focus on those problems that seem to be of greater importance for the technologies that are here now and that are coming in the foreseeable future. While there remains a place for SSI in "gluing" together larger chips, it must be recognized that the emphasis is now on large-scale integration.

This book is a suitable text for an introductory course in logic circuits for undergraduate students in computer science, computer engineering, or electrical engineering, as well as for some more advanced courses. It should also be of interest to practicing engineers and logic designers who wish to learn about the subject or update their knowledge. There are no prerequisites except perhaps for some sort of beginning course in programming.

Only a few topics covered here require some knowledge of electrical circuits and electronics. However, I believe that some such knowledge is helpful in acquiring a deeper appreciation of some important issues, such as trade-offs between power and speed, tristate devices, and wired logic. Since most computer science students lack this background, a simple presentation of the essentials has been added in an appendix.

Another relevant area, sometimes covered in other courses, pertains to manipulations of numbers in different number bases and some related matters such as the treatment of negative numbers. Since these are not central to logic circuits, they have also been relegated to an appendix.

Pointers to sources have been confined to a section at the end of each chapter, so as to avoid cluttering up the text. These all refer to a unified list of references at the end of the book. There is a summary at the end of each chapter, and a substantial number of problems dealing with virtually every topic. Solutions to a subset of the problems (those marked with asterisks) are presented at the end of the book.

Although it may be possible to cover all the topics in the book in an intensive one-semester course, many instructors might prefer to omit, or just skim lightly over, some of the more advanced topics. It was not possible to include detailed treatments of every interesting aspect of this field. Hence, I chose to do a rather thorough job of explaining the fundamentals, particularly the subject of combinational logic, treating other topics in varying degrees of detail. In some cases, e.g., asynchronous sequential circuits, I included complete treatments of the basics, selectively omitting material deemed less important relative to its complexity. In other cases, I presented some key points in detail, giving only overviews of the topics as a whole. This was done, for example, with respect to the vast subject of fault detection.

A number of topics included here generally receive little or no attention in introductory texts. These include iterative circuits, particularly the discussion of iterative circuits in tree form (culminating in a derivation of the full carry lookahead adder). The material on clocking schemes is new. Metastability has become an increasingly important problem in recent years; it is important that it be brought to the attention of logic designers. Some of the key concepts underlying self-timed systems are presented. The use of computers to assist logic designers, both in the design and simulation stages, is another subject introduced here that is not usually presented to beginning students. While many texts include some mention of duality, it is not usually given the attention it merits as a valuable conceptual tool that can sometimes be a real work saver.

In the introductory chapter, the basic elements of digital logic are introduced with discussions of switching circuits, gate circuits, and pass networks, and how they are related. The second chapter is about Boolean algebra and its correspondence to logic circuits. The simplification of expressions with the aid of the theorems is illustrated. Duality is explained and applied to the relations between positive and negative logic, relations between NAND- and NOR-gates, and to circuits with XOR-gates.

In Chapter 3, Karnaugh maps are introduced both as n-dimensional cubes and as Venn diagrams, and a rigorous procedure for using them to find minimal sum-of-products expressions is explained and illustrated. These ideas are extended to encompass variable-entered K-maps, a very useful concept. Then the Quine-McCluskey procedure for handling the same problem is presented. Next, these methods are extended to multioutput circuits. Some approaches to multistage logic are outlined. The special properties of unate (monotone) Boolean functions are pointed out. Efficient implementations of the XOR-gate are used to illustrate design with complex complementary metal-oxide silicon (CMOS) gates and with CMOS transmission gates. The treatment of the very important problem of finding diagnostic tests for combinational logic circuits is based on the idea of path sensitization.

In the first part of Chapter 4, the most useful combinational logic macros—decoders, multiplexers (MUXs), and encoders—are presented. This is followed by a section on how to realize irregular logic expressions with quasi-regular arrays; i.e., ROMs, programmable array logics (PALs), PLAs, and gate arrays. The incorporation of PLAs on LSI and VLSI chips is stressed; material on the use of decoders and folding is included. The handling of functions with inherent regularity is treated in Chapter 5. The flow table is introduced to describe iterative functions, and the analogy between spatial and temporal sequences is pointed out.

The realization of symmetric and iterative functions with linear iterative circuits is then explained. Next it is demonstrated that such functions can also be implemented with regular circuits in tree form, where complexity grows linearly with the number of inputs, while delay increases only logarithmically. Important examples are parity circuits, comparators, and the binary adder.

Sequential circuits are treated in Chapter 6. Since concepts involving synchronous and asynchronous circuits are intertwined, the discussion in the chapter alternates between them, beginning with an introduction to simple clocked circuits. Flow tables and state diagrams are presented along with procedures for merging equivalent states, and going from a flow table to a circuit. This is followed by a discussion of the basic elements of asynchronous sequential circuit theory, including types of functions, types of delays, flow table forms and manipulations, combinational hazards, sequential hazards, delay and data rate calculations, analysis, and the state assignment problem. No attempt is made to cover the last of these topics thoroughly. Rather, the problem of critical races is explained, state assignments of various types are illustrated with examples, and one simple general solution, the 1-hot code, is presented. The examples in this section generally involve useful devices such as the SR-FF, the latch, the edge-triggered D-FF, and the JK-FF.

In the following section, it is shown how the duality concept can be extended to sequential circuits. Next, some examples are given showing how informal design procedures are often adequate in important practical cases, such as registers with various capabilities. Formal design procedures for synchronous systems are then described. These entail problem specifications in various forms ranging from informal verbal descriptions, to register transfer sequences in structured form, state graphs, and ASM charts. Examples include a multiplier controller, a controller for a cache interface unit, and a synchronous arbiter. The generation of logic from these descriptions is illustrated, one novel approach being via a state map that leads to variable entered K-maps. Implementations are shown with latches, edge-triggered D-FFs, and JK-FFs. Next, it is shown how to choose the parameters of clocking systems so as to maximize operating speed while ensuring proper operation.

Where signals that may change at arbitrary times relative to one another must interact, there is a danger of indeterminate behavior arising through the mechanisms of metastability or runt pulses. A general method for identifying such situations is presented next, along with appropriate countermeasures. In the course of this discussion, arbiters and synchronizers are introduced. Chapter 6 continues with an introduction to pulse-mode circuits. A detailed treatment of the design of a pulse-mode circuit is used to illustrate several aspects of real-world design. Some basic ideas about self-timed circuits, including two- and four-phase handshakes, C-elements, and dual-rail coding are then presented. This material is applied to the design of a speed independent buffer that can be used independently or as a building block to construct more complex networks. The last section deals with the problem of testing sequential systems. Scanning (e.g., level-sensitive scan design, or LSSD) is described, and the elements of built-in testing, using linear feedback shift registers and the signature concept are outlined.

Chapter 7 concerns computer aids for digital system designers. Several different approaches to using computers to generate minimal combinational logic are discussed, and illustrated, including the Logic Synthesis System used at IBM. The less developed field of aids to sequential logic design is only briefly scanned. Next, the very important area of simulation is introduced, starting with brief discussions of high-level (or behavioral level) and logic-level simulations. The value of switch-level simulators for MOS-based technologies is illustrated with some examples motivating key features of the MOSSIM II simulator. The role of circuit-level simulators is then outlined. In the last section, a computer algorithm is described for timing analysis, i.e., finding bounds on propagation delays through a large logic network.

A brief postscript outlines what I feel are some important aspects of the applied science and engineering professions that all too often "fall between the cracks."

The first part of the Appendix is concerned with arithmetic operations in binary, and number conversions of both integers and fractions. Other topics are octal, hexadecimal, binary coded decimal (BCD), and American Standard Code for Information Interchange (ASCII). Gray codes are also introduced. The handling of negative numbers via two's complement encoding is explained in detail. On a higher level is the discussion of error-correcting codes and how to implement them with logic circuits. (Reading this should be postponed until after XOR gates are understood.) Section 2 of the Appendix is a short introduction to simple electrical circuits and electronics, beginning with a treatment of circuits with resistors, voltage sources, and switches. The voltage divider, a key configuration in digital elements, is given particular attention. Next comes a simple explanation of how an MOS transistor works. The section concludes with a discussion of capacitors and simple RC circuits. This is then related to digital circuitry, to show how rough estimates of propagation delays can be made.

The use of nMOS transistors to construct standard gates, complex gates, and pass transistor circuits is discussed in Section 3. Then the CMOS concept is introduced and the corresponding applications are explained. The current status of galium arsenide devices is sketched. There is a brief discussion of the role of bipolar transistors in BiCMOS, transistor-transistor logic (TTL), emitter-coupled logic (ECL), and integrated injection logic (I^2L) families.

In this second edition, some improvements have been made in the presentation of certain topics, new material has been added on CMOS (including transmission gates), JK-FFs, pulse-mode operation, two's complement arithmetic, galium arsenide technology, and self-timed systems (particularly handshaking and data bundling). A number of new examples and problems have been added, and various references have been updated.

1
Introduction

1.1. WHAT IS LOGIC DESIGN?

Logic circuits are what make things happen in digital computers. They govern the input of information, the fetching and execution of instructions, the conversion of data from one form to another, and the output of results. Logic design is about interconnecting large numbers of simple elements to make them perform intricate operations. The elements themselves are an amazing culmination of a vast array of engineering and scientific knowledge. But, using them entails a rather different kind of knowledge.

Logic designers do not employ the classic techniques of electrical engineering or physics or chemistry. Nor are they much concerned with the kinds of mathematics, largely based on differential equations, that underpins much of these disciplines. Corresponding to the simple physical elements used in logic design is a mathematics based on 1s and 0s rather than on real numbers. While some of this mathematics, such as Boolean algebra, is well developed, much of it is highly irregular, with bits and pieces invented as needed. Those who enjoy solving puzzles are likely to feel at home in this field.

Logic circuits and many sophisticated techniques for designing them antedate the digital computer by many years. The initial applications were to the design of telephone central office equipment. The key concept, which transformed the design process from an art to a science, was the idea of describing both the functions performed and the circuits themselves in terms of Boolean algebra. This was embodied in Claude Shannon's master's thesis in electrical engineering at MIT, surely one of the most brilliant master's degree theses ever written. It was published in 1938. A substantial body of knowledge has been generated since then, making logic design perhaps the most highly developed area of computer science. Yet many important and interesting problems remain to be solved.

There are two basic classes of logic circuits. *Combinational* logic circuits are those whose present outputs are functions only of the present inputs. *Sequential* logic circuits are those whose outputs depend on *past* as well as present inputs. The examples in this chapter all illustrate combinational logic; the more complex subject of sequential logic is not introduced until Chapter 6.

A major use of logic circuits is to perform arithmetic operations on numbers represented in terms of two-valued, or binary, signals. Circuits that count, add, and multiply in binary are basic applications of logic-circuit techniques, and are used in subsequent chapters to illustrate them. Hence, logic designers should have a good grasp of how numbers are represented and manipulated in binary. They should also understand how nonnumeric information can be represented in terms of two-valued signals, as in

American Standard Code for Information Interchange (ASCII). An introduction to this subject is in Appendix A.1.

An understanding of simple electrical circuits with resistors and batteries is sufficient to master the material in this book pertaining to logic-circuit components. A little further acquaintance with circuits and electronics is very helpful in enhancing one's appreciation of certain important issues in logic design. Strangers to these subjects are cordially invited to "break the ice" by perusing the introductory material in Sections A.2–A.4 of the Appendix.

The basic elements from which logic circuits are constructed are introduced in the following section. Some simple, but nontrivial, examples are then used to illustrate how these devices can be put to work.

1.2. Building Blocks For Logic Circuits

There are two principal types of logic circuits: those made of switches and those made of gates. These are related in that one can think of gates as being constructed from switching circuits. In some systems, both types are used.

1.2.1. Switching Circuits

It will be assumed here that when a switch is open, no circuit can pass through it (i.e., it has infinite resistance), and that when it is closed, current can pass through it without any voltage being developed across it (i.e., it has 0 resistance). Although real switches can only approximate such behavior, the approximations are adequate in most situations. Associated with each switch is a binary-valued control signal, usually represented by a voltage value. The switch is open or closed depending on whether the value of this signal is 0 or 1, respectively.

Such switches might be implemented as contacts on an electromechanical relay, in which case the control signal corresponds to the voltage across the coil or winding (the value of the control signal being 1 or 0, respectively, depending on whether this voltage is or is not sufficient to activate the relay). A modern implementation would be the region between source and drain of a metal-oxide silicon field-effect transistor (MOSFET), with the control signal corresponding to the voltage on the gate. While a MOSFET switch acts very much like an open circuit when the gate voltage is low, its resistance when the gate voltage is high is far from 0. This is a factor that must be taken seriously in electrical aspects of the design process. Those not acquainted with the terms and concepts just introduced can find more information in Sections A.2 and A.3 of the Appendix (or in some other source), or else they can accept on faith the idea that devices behaving as switches really do exist.

Our concern here is with two-terminal circuits as shown in Figure 1.1a. The voltages controlling the switches are $V_A, V_B, \ldots$. Between the two terminals is a circuit composed of switches. Some simple examples are shown in Figure 1.1b, 1.1c, and 1.1d. The issue of interest concerning a switching circuit is whether or not there is a path through it. This of course depends on the structure of the circuit *and* on the states of the variables controlling the switches in it. The *transmission* of a switching circuit, T, is defined as a variable with value 1 if there is a conducting path, and 0 if there is no path. In the very simple case of a single switch (Figure 1.1b), T is 0 if A is 0 and T = 1 if A = 1, i.e., T = A. For circuit (c), T is 1 if and only if A *and* B are both 1. (We often refer to a

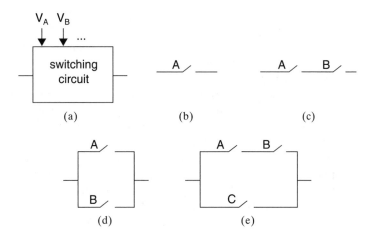

Fig. 1.1. Block diagram of switching circuit and some examples.

switch control variable or a circuit transmission that is equal to 1 or 0 as being *on* or *off*, respectively.) For circuit (*d*), T is on if A *or* B is on (or both—the word *or* here is being used in the *inclusive* sense). For circuit (*e*) T is on if C is on, or if both A and B are on.

Transmission variables are often converted to voltages, which can then be used as switch control variables. Two ways to do this are shown in Figure 1.2. In Figure 1.2*a*, if T = 0 for switching circuit S, the output voltage Z will also be 0 (we always assume that the voltage level of the ground connection is 0). Otherwise, if T = 1, Z will be pulled *up* to the value of the supply voltage V (which we assume is high enough to qualify for on status), and is therefore 1. We refer to a switching circuit in such a position (connecting the output to the supply voltage) as a *pullup* circuit. The Figure 1.2*b* circuit works in the opposite direction, with S connecting the output to the ground connection thus pulling *down* the output when T = 1, or allowing the output to rise, when T = 0. Hence this is referred to as a *pulldown* circuit. A pullup or pulldown circuit whose transmission can change between 0 and 1, i.e., a switching circuit, is said to be *active*, whereas a pullup or pulldown circuit consisting of a fixed resistor, is referred to as being *passive*.

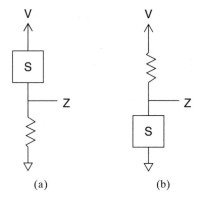

Fig. 1.2. Switching circuits as (*a*) pullups or (*b*) pulldowns.

Switching networks can also be used to route or *pass* logic signals from one terminal to another. Several such configurations are illustrated in Figure 1.3. In Figure 1.3*a*, the signal A is transmitted to Z if the transmission of the switching circuit S (an *active* pullup) is 1; otherwise the Z-signal is pulled down to 0 through the resistor connected to ground (a *passive* pulldown). The Figure 1.3*b* configuration differs in that, if S is off, the value of Z is pulled up to 1 by the resistor connected to the supply voltage.

A somewhat more complex situation is depicted in Figure 1.3*c*. Here, if only circuit P is on, then Z takes on the value of A. If only circuit Q is on, then the value of B is passed to Z. It is generally assumed for such arrangements that the signals controlling the transmissions of P and Q are so related that there will always be *exactly one signal* passed to Z, i.e., it is forbidden to have *both* P and Q on or *both* P and Q off, since that might leave Z undetermined. Clearly, this circuit could be generalized to include more than two switching circuits. Another variation would be to permit both P and Q to be simultaneously off, but to add passive pulldown or pullup circuits to determine the value of Z when neither switching circuit is on. The circuits shown in Figure 1.3*a* and Figure 1.3*b* illustrate such situations.

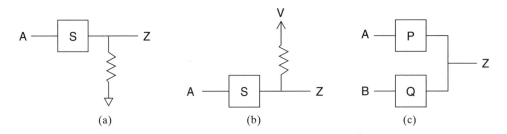

(a)   (b)   (c)

Fig. 1.3. Switching circuits as pass networks.

1.2.2. Logic Gates

The most common type of building block used in the design and analysis of logic circuits, and the one primarily used in this book, is the logic *gate*. These are devices with one or more inputs and one output. Both inputs and outputs are binary-valued logic signals. A basic set of logic gates is shown in Figure 1.4.

Part (*a*) shows a gate, called an *inverter*, with one input. The output Z takes on the value opposite that of the input. Inverters are often realized by circuits in the form of Figure 1.2*b*, in which the pulldown circuit consists of a single switch, controlled by the input to the inverter. The gate shown in Figure 1.4*b*, called an *OR-gate*, has two inputs, and its output is 1 if A *or* B or both are equal to 1. (The word *or* is used here in what is called the *inclusive* sense, as opposed to the *exclusive* sense that excludes the case where

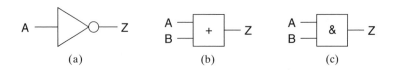

(a)   (b)   (c)

Fig. 1.4. Logic gates. (*a*) Inverter, (*b*) OR-gate, (*c*) AND-gate.

both variables involved are 1.) An OR-gate can be realized by a circuit in the form of Figure 1.2*a*, in which the pullup circuit consists of two switches in parallel (as in Figure 1.1*d*), one controlled by each input. The third type of gate, called an *AND-gate*, is shown in Figure 1.4*c*. Its output is 1 only if inputs A *and* B are equal to 1. It too can be realized by a circuit in the form of Figure 1.2*a*, with a pullup circuit consisting of two switches in *series* (as in Figure 1.1*c*). The output of an OR-gate is 1 if *at least one* of the inputs is 1, while the output of the AND-gate is 1 only if *all* of its inputs are equal to 1. This set of three types of gates is *complete*, in the sense that it is sufficient to realize any logic function.

However, for reasons pointed out in Appendix A.3, the commonly used *n*-channel MOS (nMOS) transistor performs poorly in pullup circuits. Therefore, the gates described next are more popular. Consider the circuit shown in Figure 1.5*a*. If switch A or switch B is closed, then output Z is connected to ground, so that the Z = signal will be 0. The Z-signal will be 1 only if neither A *nor* B is closed. Hence, we refer to this as a *NOR-gate*. The output of a NOR-gate is always precisely opposite to that of an OR-gate with the same inputs. We can also explain this circuit in terms of circuit transmissions. If the transmission of a pullup network is 1, then the gate output is 1. If the same network is used as a pull*down*, it has the opposite effect: when its transmission is 1, the gate output is 0. Thus, if the same switching network used as a pullup circuit in Figure 1.2*a* to create an OR-gate is used as a pulldown circuit, as in Figure 1.5*a*, the result is the opposite of an OR-gate, namely a NOR-gate. The circuit symbol for a NOR-gate (Figure 1.5*b*) is an OR-gate symbol with a "bubble" at the output. (Such bubbles are commonly used in logic-circuit diagrams to represent signal inversion.)

The other major type of logic gate is shown in Figure 1.5*c*. If A and B are *both* closed, then the output, Z, is connected to ground, so Z = 0. The value of Z is 1 only if it is *not* the case that A *and* B are both closed. Since this behavior is exactly the opposite of that of an AND-gate with the same inputs, we call this a *NAND-gate* (for "*not* and"). The argument about the inverting effect of moving a switching circuit from a pullup to a pulldown position applies here as in the case of the NOR-gate discussed earlier. The symbol for a NAND-gate is shown in Figure 1.5*d*. NAND-gates and NOR-gates are both very widely used in modern digital systems. (In complementary MOS [CMOS] technology

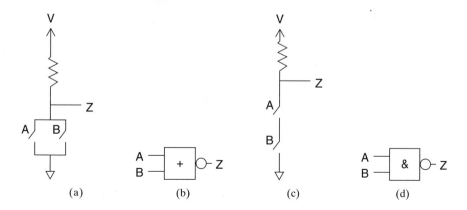

Fig. 1.5. NOR- and NAND-gates, implemented with pulldown networks.

—introduced in Appendix A.3.2—the pullup and pulldown circuits are *both* active. They consist of transistor networks realizing mutual complementary transmissions.)

NOR- and NAND-gates may have many more than two inputs (10 would not be uncommonly large in some technologies). However, in many important technologies, including CMOS, increasing the number of inputs beyond three or four may entail a significant penalty in terms of slower operation. The inverter, which uses a single switch as a pulldown, can be thought of as a 1-input NAND- or NOR-gate. As is shown in Section 2.6, NOR- and NAND-gates are each complete sets. The subject of the next section is a special way of interconnecting logic gates that is useful in some technologies. It can be skipped at first reading, as it is not a prerequisite for what follows.

1.3. Direct Connections of Gate Outputs

In certain common integrated circuit technologies it is possible to reduce the number of logic gates needed to realize a given function and also speed up circuit operation by connecting together the outputs of certain logic gates. In one approach, the gates involved must each have an additional special input that can make the gate appear to vanish from the circuit. In the other approach, discussed first, the only requirement is that the electrical parameters of the gates be chosen so as to accommodate such connections.

1.3.1. Wired (or Implied) Logic Gates

What happens if we connect together the outputs of two inverters (as shown in Figure 1.6a)? From the point of view of our formal definitions of logic operations, the signal so produced is undefined. But, if this were actually done with real devices the voltage could be measured and some sort of output would result. What might this be?

The answer depends on the resistances of the inverter transistors; see the equivalent circuit shown in Figure 1.6b, which is in effect a voltage divider. There are four cases to consider, corresponding to the four possible states of the two input signals. Clearly, the two cases where one of the inputs is on and the other is off are equivalent (we assume that the two inverter circuits are identical). If we consolidate these two cases, then there are three situations to analyze.

1. Both inputs are off; i.e., both switches in Figure 1.6b are open. Clearly, in this case, no current flows through the pullup resistors, and so $V_Z = V$.
2. Both inputs are on. This places both pulldown resistors in parallel (Figure 1.6c), so that, given that they are equal, the equivalent resistance of the combination is $R_{pd}/2$. Hence, considering this circuit as a voltage divider, we have

$$V_Z = V \frac{\dfrac{R_{pd}}{2}}{\dfrac{R_{pu}}{2} + \dfrac{R_{pd}}{2}} = V \frac{R_{pd}}{R_{pu} + R_{pd}}$$

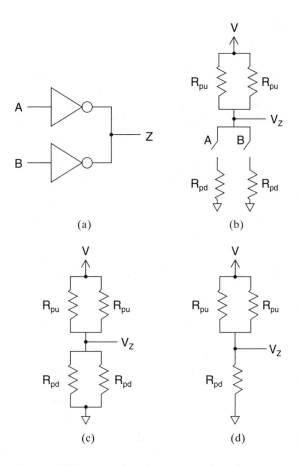

Fig. 1.6. Wiring together the outputs of two inverters.

3. One input is on and the other off. Only one of the pulldowns appears in the voltage divider (Figure 1.6d). So we have

$$V_Z = V \frac{R_{pd}}{R_{pd} + \dfrac{R_{pu}}{2}}$$

From the preceding calculations, we can draw some conclusions about the *logic* value of the output signal as a function of the input signals. It is evident from case 1 that if both inputs are logic 0s (which means both switches are open), then the output value is a logic 1. In analyzing the expressions derived for cases 2 and 3, we must bear in mind that, for an inverter to function properly, it is necessary that R_{pu} be much larger than R_{pd} (typically by a factor of about 6). It is therefore evident that V_Z is significantly less than V in both cases. Note that V_Z is larger for case 3 than for case 2. If the ratio of R_{pu} to R_{pd} is made somewhat larger than usual for an inverter, then the logic values for both case 2 and case 3 will fall safely in the range of logic 0s.

The net effect is that if the output of either inverter standing alone is 0, then if both outputs are connected together, the result will also be 0. Otherwise, if each output alone would be 1, the output signal resulting when the outputs are joined will also be 1. In other words, wiring the outputs together implements the AND operation and is referred to generally as a *wired*-AND (in some circles the term *dotted*-AND is used.) Thus the output of the circuit of Figure 1.6*a* is 1 if *neither* of the inputs is a 1. This corresponds to the NOR operation. In fact, NOR-gates are often implemented with circuits differing from Figure 1.6*b* only in that just *one* pullup resistor is used.

If the outputs of more than two inverters are wired together, the effect is to place *more* pullup resistors in parallel. This further reduces the value of the resultant pullup resistor. If the pullup resistor is too small, then the low value of the output signal voltage may not be low enough to qualify as a logic zero (the worst case is, as illustrated by case 3 of our example, where only one of the inverter outputs is driven toward 0). In any event, it is clear that NOR-gates constructed in this manner do not reduce chip area. They are also inferior to those built with a single pullup resistor, which can be carefully chosen to ensure an adequate safety margin for the low output voltage.

More generally, there are a number of logic circuit technologies, including nMOS and a subclass of transistor–transistor logic (TTL) (*without* totem pole outputs—see [McCluskey, 1986]), where wired-ANDs are acceptable and useful. The technique is usually applied to NAND-gate outputs. It is not hard to show that, for such technologies, wiring together the outputs of NOR-gates is not useful. It simply produces NOR-gates with more inputs that are also inferior due to the reduced value of the pullup resistor.

There are other technologies, principally emitter-coupled logic (ECL), where the result of connecting together the outputs of two or more gates implements the OR-function. Representations of wired logic gates, both AND and OR, are shown in Figure 1.7*a* and 1.7*b*.

In Section 2.6 (where applications of wired logic are discussed), it is shown that the circuits of Figure 1.7*a* and 1.7*b* are equivalent, respectively, to the circuits of Figure 1.7*c* and 1.7*d*. In the latter two circuits, input signals must pass through two gates to reach

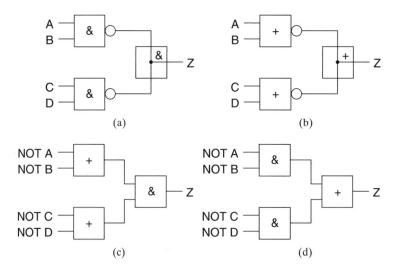

Fig. 1.7. Logic circuits with wired AND and wired OR.

the outputs, whereas in the wired logic circuits, there is only one real gate in each path. This reduction in path length, and, hence, in delay, is an important advantage of wired logic.

1.3.2. Tristate Outputs

There is another situation in which the outputs of logic elements may be wired together to produce a useful result. In this case the gates involved are all equipped with what are called *enable* inputs. When the signal to the enable input is 1, the gate is said to be *enabled*, i.e., it behaves normally, as an ordinary gate of the same type. When the signal to the enable input is 0, the gate is *disabled*: it acts as though the output wire has been cut, or disconnected. When a gate is disabled in this manner, it is said to be in the *third* state—sometimes referred to as the *high-impedance* state. This is of course different from having an output of 0 or 1, the usual two states to which output signals in logic circuits are restricted. Devices with this capability are called *tristate* devices. When two or more tristate devices have their outputs connected together, their operation must be restricted so that exactly one of them is enabled at all times. If none are enabled, then the output is said to *float*. If a node remains in a floating state for more than some time interval determined by circuit parameters, then the signal at that node becomes essentially random.

The principle involved is illustrated in Figure 1.8*a*, in which the enable (En) signals are shown to be controlling switches at the outputs of three NOR-gates. In some technologies, tristate devices are implemented essentially in this manner, while in other cases, the implementation is more complex. From the point of view of the logic designer, the Figure 1.8*a* model is quite adequate. A logic-circuit diagram depicting the same circuit with standard symbols is shown in Figure 1.8*b*. The triangular symbol is used to indicate that a particular output signal is a tristate signal.

Tristate devices are commonly used where it is necessary to select one of several signals. The number of logic stages necessary is reduced, reducing delay, and the logic is often simplified. The outputs of complex devices such as memory chips are often in tristate form to facilitate combining their outputs.

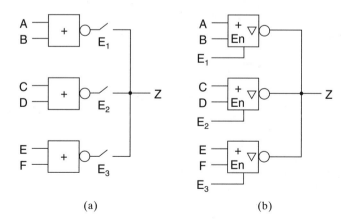

Fig. 1.8. Logic circuit using tristate devices.

1.4. Examples of Logic Circuits

The first example (Figure 1.9) is an input-selector circuit (commonly called a *multiplexer* or *MUX*, treated more fully in Subsection 4.1.2); it selects input A or input B as its output, depending on the value of the signal S. If S is on, then the output Z is set equal to A. Otherwise, Z is set to B. In subsequent chapters useful formal methods are developed for analyzing and synthesizing such circuits. For the present, readers should examine the circuits informally in the light of the previously introduced definitions to satisfy themselves that the circuits do indeed realize the stated goals.

In the switching circuit of Figure 1.9a, if S is on, then the corresponding switch is closed, and the switch below it (labeled NOT S) is open. (Note that the diagram is somewhat deceptive in showing both the switch labeled S and the switch labeled "not S" open. Actually, if one of these switches is open, the other *must* be closed, and it is not possible for *both* to be closed simultaneously.) Hence, there will be a path through the circuit if, and only if, A is on. Conversely, if S is off, then the overall circuit acts as a closed switch if, and only if, B is on. In the gate circuit of Figure 1.9b, when S is on, the upper input to the OR-gate is equal to A and the lower input is 0. When S is off, the lower input to the OR-gate is B and the upper input is 0. These conditions, in conjunction with the behavior of the OR-gate, ensure that the appropriate selection is indeed made.

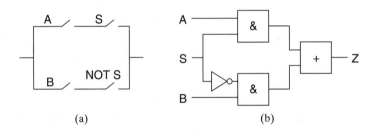

(a)   (b)

Fig. 1.9. Logic circuits that select A or B depending on S. (*a*) Switching circuit, (*b*) gate circuit.

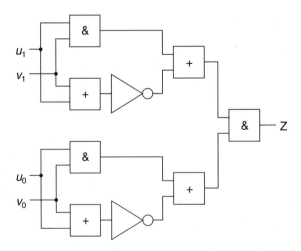

Fig. 1.10. Gate circuit that compares two 2-bit words.

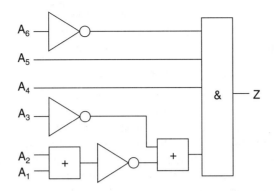

Fig. 1.11. Logic circuit that detects an ASCII-coded numeral.

The second example is the gate circuit shown in Figure 1.10. It compares two 2-bit words, u_1u_0 and v_1v_0. If their values are the same, then the output Z is 1. The subcircuit feeding the upper input to the output AND-gate matches u_1 and v_1; if they are equal, then that signal is 1. The lower input to the AND-gate similarly receives the results of a similar matching of u_0 and v_0. If *both* matches have positive results, then the AND-gate output is 1.

The third example (Figure 1.11) is a logic circuit that has as its input the seven bits of an ASCII character (Appendix A.1.10). Its output indicates whether this character is one of the 10 numerals. If we designate these bits as $A_6A_5A_4A_3A_2A_1A_0$, then the value of the $A_6A_5A_4$ prefix for an ASCII numeral is 011 ($A_6 = 0$, and $A_5 = A_4 = 1$). The upper three inputs to the AND-gate in Figure 1.11 will be 1 for precisely this condition. But only 10 of the 16 ASCII codes with this prefix are numerals; these are the ones for which the values of $A_3A_2A_1A_0$ correspond to a binary-coded decimal (BCD) code (Appendix A.1.9). This condition is satisfied if either A_3, the most significant bit (with weight 8) is 0, or if both of the next most significant bits, A_2 and A_1 (with weights 4 and 2, respectively) are 0. A careful examination of Figure 1.11 will show that the output of the OR-gate is 1 for this condition. Hence, the Z signal will indeed be 1 if the input corresponds to an ASCII numeral.

OVERVIEW AND SUMMARY

Although they are built out of relatively simple devices, and operate on binary variables, logic circuits can nevertheless realize arbitrarily complex functions. One kind of basic building block for logic circuits is a simple switch controlled by a binary variable. Such devices can be interconnected to form a switching circuit, the equivalent of another switch, but one that is on or off depending on the values of the set of binary variables that control the component switches. Switching circuits are sometimes used to connect a circuit node to the voltage supply, in which case they are referred to as pull*up* circuits. More often they are used to connect nodes to ground, in which case they are called pull*down* circuits. Alternatively, the building blocks might be logic gates, devices with one or more binary inputs and one binary output. When used appropriately with voltage sources and resistors, switching circuits can be used to construct logic gates. A basic set of gates sufficient to realize any logic functions consists of inverters, AND-gates, and OR-gates, but NOR- and NAND-gates, each of which is similarly sufficient, are the gates most used. Wired logic and devices with tristate inputs are economical in certain circumstances.

SOURCES

(References are to the alphabetized reference list at the end of the book.)

The seminal work by [Shannon, 1938] covers an amazing range of ideas in addition to the use of Boolean algebra (introduced in the next chapter). A good introduction to the various technologies used to implement modern logic circuits can be found in [McCluskey, 1986]. Additional discussion of wired logic and tristate devices can be found in [McCluskey, 1986], [Hodges, 1983], and [Kambiashi, 1986]. Other introductory texts include [Fletcher, 1980], [Hayes, 1993], [Friedman, 1975], [Hill, 1981], [Katz, 1994], [Kohavi, 1978], [Lee, 1976], [Mano, 1991], [Roth, 1992], [Wakerly, 1994], and [Winkel, 1980]. Introductory texts on computer organization, relevant to applications of some of the material presented here are [Ercekovac, 1985], [Mano, 1993], and [Hennessy, 1994]. A more advanced book on this subject is [Hennessy, 1990]. An interesting compendium of thoughts on the practical aspects of digital design is in [Seidensticker, 1986]. References appropriate to specific topics treated here are introduced at the end of each chapter.

*PROBLEMS

1.1. For each of the logic circuits shown in Figure 1.12, describe the function performed (i.e., state, in terms of the input variables, when the transmission of the switching circuit is 1, or when the output Z is 1).

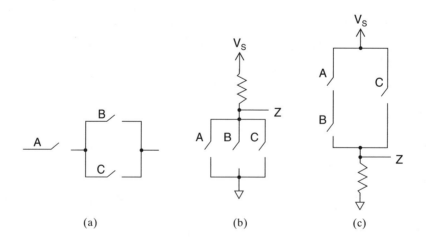

(a)   (b)   (c)

Fig. 1.12.

*Solutions to problems (or parts of problems) marked with asterisks are supplied at the end of the book.

1.2. For the logic circuits of Figure 1.13, describe, in terms of the input variables, the conditions under which the outputs Z are 1.

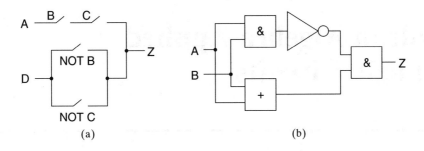

Fig. 1.13.

*1.3. (a) Design a switching circuit that has transmission 1 if, and only if, A or B is 1, and C or D is 1, where "or" is used in the inclusive sense.
(b) Design a logic gate circuit realizing the same function.

1.4. Design logic circuits that recognize the conditions A, C, and D all on, or B on with both C and D off. Use a switching circuit for part (*a*), a logic gate network for part (*b*), and the pass network concept for part (*c*) (this can be done with just one resistor and four switches, two controlled by "not X"-type signals).

1.5. Suppose we were operating with a technology in which the basic gates were NAND-gates. Would it be more useful to add a wired-OR capability or a wired-AND capability?

1.6. Design a circuit with inverters, AND-gates, and OR-gates that transmits an input signal X to output Z_1 if input S = 0 or to output Z_2 if S = 1. A 0 should appear at the output terminal not receiving X.

1.7. Design a pass network that realizes the same function as the circuit shown in Figure 1.9*b*.

1.8. Using NOR-gates with the wired-OR capability and an inverter, design a circuit with inputs A, B, and C, whose output, Z, is 1 if, and only if, A = B = 0 or A = 0 and C = 1.

*1.9. Using an inverter and two NOR-gates with tristate outputs, design a circuit whose output is 1 if, and only if, A = 1 while B = C = 0, or A = D = E = 0.

1.10. The input to a logic circuit is a 4-bit binary number, ABCD (where A is the most significant bit). Design an economical circuit using AND- and OR-gates that has a 1-output only if the input number is 0 or 5. Assume that, for each input variable X, NOT X is available as well as X.

2

Boolean Algebra Applied to Logic Circuits

In the introductory chapter, examples were presented illustrating the analysis and synthesis of some simple logic circuits. Especially if they have worked out some of the problems at the end of the chapter, readers will appreciate the importance of finding good ways to specify precisely what a logic circuit does, or is supposed to do. Thus far, only two "languages" have been employed here for this purpose: English and logic-circuit diagrams.

While natural languages such as English are very powerful in the sense that they can be used to express a vast variety of concepts, emotions, etc., they are also notoriously ambiguous and imprecise. Consider, for example, the sentence, "Your dinner will consist of fish or beans and rice." Does this mean that we get rice and either fish or beans, or that our choice is between: (1) fish, and (2) rice and beans? We normally resolve such ambiguities on the basis of the context of the situation, and our general fund of knowledge. In the present instance we might use the fact that beans and rice often go together, in the sense that together they constitute a complete protein. Hence we would probably interpret this sentence in the second sense. To convey in a natural language the information contained in a partial differential equation would be very difficult, but not nearly as difficult as it would be to understand the result. Describing really complex logic functions in a natural language would be about as awkward.

While a logic circuit diagram unambiguously specifies the function performed by any circuit corresponding to it, the form of the description is often not best suited to facilitate our understanding. It is obviously not the answer to the problem of how best to describe functions for which we have not yet designed logic circuits. In the following sections, a powerful and unambiguous language is presented for describing *both* logic functions *and* logic circuits. By manipulating the forms of the descriptions, it is possible to generate logic circuits satisfying various constraints. Techniques for designing and testing efficient combinational logic circuits are presented in the next chapter.

2.1. BOOLEAN ALGEBRA—DEFINITIONS AND POSTULATES

A basic property of Boolean algebra is that Boolean variables can range over just two values. Boolean algebra can be thought of as a mathematical system for dealing with statements that are either true or false. Simple statements of this type can be combined in various ways to form compound statements whose truth or falsity depend on the truth or falsity of the constituent statements. For example, elementary statements A and B, respectively, might be: "The disk drive has issued an interrupt to the central processor," and, "No other interrupts are currently in effect." A compound statement formed from these is: "The disk drive has issued an interrupt to the central processor and no other interrupts are currently in effect." This might be expressed as A *and* B. Clearly, this compound statement is true if *both* of the constituent statements are true.

In the Boolean algebra, precise formal meanings are assigned to such terms as *and*, *or*, and *not*. These definitions, which can be thought of as postulates, can be used to derive useful theorems. The theorems can then be used to transform compound statements into more desirable forms without altering their meanings.

The two-valued signals discussed in Chapter 1 can be thought of as representing the "truth values" of elementary statements. The common practice, followed here, is to use the values 1 and 0, respectively, to represent true and false. The symbols, such as A, B, A_1, etc., are referred to as *Boolean variables*, or, when no confusion will result, as simply *variables*. In the last example described in the Introduction, A_5 can be thought of as a Boolean variable that has the truth value 1 if bit number 5 of the American Standard Code for Information Interchange (ASCII) character being processed is equal to 1. It is important to understand that, in the context of Boolean algebra, 1s and 0s are not numerals (not even in the binary system), but *truth values*. We might just as easily have chosen to use T and F instead of 1 and 0.

The Boolean OR operation, also referred to as Boolean *addition*, is symbolically represented by a plus sign. Analogously, the Boolean AND operation, also referred to as Boolean *multiplication*, is treated symbolically the same way we treat arithmetic multiplication (i.e., we may use a $\times$ sign or a dot, or, most often, simply regard the juxtaposition of two symbols as implying multiplication). The Boolean NOT operation, or Boolean *complementation* or *inversion*, is represented by an overbar. What follows is a precise definition of the basic rules of Boolean algebra for two-valued logic:

1. For every variable X, $X = 0$, or $X = 1$. (We are dealing with a two-valued system.)
2. Complementation: $\bar{0} = 1$; $\bar{1} = 0$.
3. Multiplication: $0 \cdot 1 = 1 \cdot 0 = 0 \cdot 0 = 0$; $1 \cdot 1 = 1$.
4. Addition: $1 + 0 = 0 + 1 = 1 + 1 = 1$; $0 + 0 = 0$.

From these rules, a number of useful theorems are derived in the next section. But first, some variations in notation and terminology often found in the literature should be noted. Complements are sometimes denoted by the use of primes, e.g., X′, by a rather special "not" symbol as in $\neg X$; or by the use of a negation sign as in $-X$. In addition to referring to the OR operation as addition, the term *disjunction* is also sometimes used, generally along with *conjunction* for the AND operation. The symbols $\vee$ and $\wedge$ are sometimes found (usually in works by mathematicians or logicians) to represent OR and AND, respectively.

2.2. BOOLEAN EXPRESSIONS

Using these basic operations, complex expressions can be formed, analogous to the expressions that we construct in ordinary algebra. The syntax rules are actually the same as those for ordinary algebra, including the use of parentheses to group subexpressions. For example, $(AB) + C$ is an expression that would be evaluated by first finding the Boolean product of A and B, and then ORing the result to C. In ordinary algebra it would be similarly evaluated, but using arithmetic multiplication and addition.

In order to reduce the need for parentheses, we have a convention in ordinary algebra that gives multiplication "precedence" over addition, in the sense that, in ambiguous cases, we multiply before we add. Hence, if we had omitted the parentheses in the last expression, writing AB + C, it would still have been evaluated the same way. If we had wished to have the addition performed first, we would have forced this by using parentheses: A(B + C). Precisely the same precedence rule is used for Boolean expressions: multiplication takes precedence over addition. Thus the following examples of Boolean expressions are unambiguous:

$$AB + C; A(B + C); AB + CD; A(B + C)D; \overline{A}C + BD; ((AC + \overline{B})(B + CD)) + D.$$

Readers should satisfy themselves that they understand the meanings of these expressions by evaluating each of them for the case where $A = B = 0$ and $C = D = 1$. (The answers, in order, are $1, 0, 1, 0, 1, 1$.)

There is one other way in which the use of parentheses is minimized in ordinary algebraic expressions. In this realm we do *not* ordinarily write (A + B) + C; rather we write A + B + C. But how would we interpret the second expression if it were Boolean? Our definition of Boolean addition is in terms of just *two* operands. Does A + B + C mean (A + B) + C or does it mean A + (B + C)? In ordinary algebra these two expressions mean the same thing because arithmetic addition is an *associative* operation. Does Boolean addition share this property? One of a number of useful theorems proved in the following section is that Boolean addition is indeed an associative operation.

2.3. TRUTH TABLES AND THEOREMS

Consider a circuit with three inputs, A, B, C. Since each input variable can assume one of two values, 0 or 1, the possible input combinations are: $000, 001, 010, 011, 100, 101, 110, 111$. The total number of possible input combinations is $2 \times 2 \times 2 = 2^3 = 8$. We shall refer to each such combination as an *input state*. In general, if a logic circuit has k input variables, then, since each of them may independently be 0 or 1, the total number of input states cannot exceed 2^k. (In Chapter 3, situations are treated in which the number of input states actually expected is less than this number. Suppose, for example, that the input is a binary coded decimal (BCD) digit (see Appendix A.1.9). Then 6 of the 16 possible input states would never occur, since they do not correspond to numbers between 0 and 9.) Since the number of possible input states is finite, the desired behavior of the circuit can be specified simply by a *truth table*, which is a listing of all the input states with the desired output for each. An example of such a table is shown as Figure 2.1. There are three input variables, and hence eight rows; these are ordered by treating the input states as an increasing sequence of binary numbers, which they are *not*.

Truth tables constitute a third way of describing logic functions, which can be thought of, in an abstract sense, as mappings from input states (*binary vectors* such as 011) to the domain consisting of 0 and 1. The truth table is also a valuable tool for proving theorems, as is demonstrated below.

An equation is called an *identity* if it is valid for all possible values of its variables. For example, in ordinary algebra, $(X + Y)^2 = X^2 + 2XY + Y^2$ is an identity. It is true for all values of X and Y. The expression $(X + Y)^2 = 3XY$ is *not* an identity. In ordinary algebra, each variable can take on an infinite number of values. Therefore, we cannot prove that a given equation is an identity by testing its validity for each possible

ABC	Z
000	0
001	0
010	0
011	1
100	0
101	0
110	1
111	1

Fig. 2.1. Truth table for a function.

set of variable values. (We *can*, however, prove that an equation is *not* an identity if we can find a single set of values for its input variables for which it is false. Thus, $(X + Y)^2 = 3XY$ is clearly not an identity since it is false when $X = Y = 1$.)

The situation is much nicer for Boolean algebra. Each variable is restricted to one of two values. Therefore, it is possible to test a given equation for all possible values of its variables. If it is valid in every case, then it is an identity. (We generally refer to important identities as *theorems* or *laws*.) This method of proof is called *perfect induction*.

Consider $X = X + X$ as a Boolean expression. Is it an identity? There is only one variable, X, which must be either 0 or 1. If $X = 0$, the expression becomes $0 = 0 + 0$. Referring back to the list of postulates at the beginning of this section, we see that this corresponds to the second postulate under item 4, the definition of addition. (What actually appears there is $0 + 0 = 0$, which amounts to the same thing since, in general, $A = B$ implies $B = A$.) Hence, the statement is true for the case where $X = 0$. Next we set $X = 1$ to obtain $1 = 1 + 1$. This too appears as a postulate under item 4. Therefore the statement is valid for $X = 1$ as well. Since it is valid for *all* possible cases, we have established it, by perfect induction, as a valid identity. (It is important enough to be called a law, namely the *idempotent law for addition*.)

Another example involving one variable is $X + \overline{X} = 1$. For $X = 0$ this becomes $0 + \overline{0} = 1$. Using one of the postulates under item 2, which defines complements, we see that $\overline{0} = 1$. Substituting in the expression, yields $0 + 1 = 1$, which is valid, since it corresponds to one of the postulates defining addition. For $X = 1$, the original expression becomes $1 + \overline{1} = 1$. Referring again to the definition of the complement operation, we obtain $1 + 0 = 1$, again one of the postulates for addition. Hence, by the method of perfect induction, $X + \overline{X} = 1$ is a valid theorem. Readers should now apply the same method to establish the validity of three more one-variable theorems:

$$X + 0 = X, \quad X + 1 = 1, \quad \text{and} \quad \overline{\overline{X}} = X.$$

Now let us consider an important theorem involving two variables. This is the statement that the order of the operands in Boolean addition does not matter, that is, Boolean addition is *commutative*. This is equivalent to the validity of the identity: $A + B = B + A$.

The method of perfect induction is again applicable, but this time there are four states of the input variables to consider. The process can be organized neatly in the form of a truth table (see Figure 2.2). As usual, each row corresponds to a state of the input

AB	A + B	B + A	A + B = B + A
00	0	0	1
01	1	1	1
10	1	1	1
11	1	1	1

Fig. 2.2. Truth table establishing the commutative law for Boolean addition.

variables, listed in the first column. For each row, the second column specifies the value of the left side of the expression, the third column specifies the value of the right side of the expression, and the fourth column specifies whether the values of the second and third columns are equal. In other words, a 1 in column 4 indicates that the statement is valid for the corresponding input state. The identity is valid if all of the entries in column 4 are 1s.

Check out this method on another two-variable identity:

$$A + \overline{A}B = A + B.$$

Consider next the associative law for Boolean addition, discussed in the previous section. In its simplest form it can be expressed by the identity:

$$A + (B + C) = (A + B) + C.$$

The truth table method could of course be applied here (it would be an eight-row table). But a shorter approach would be to exploit what has already been established. The strategy (which might be called *perfect induction on A*) is to show that the theorem is valid when $A = 0$, regardless of the values of the other variables, and then do the same for $A = 1$. This amounts to showing that it is true for all possible inputs.

Assuming, then, that $A = 0$, and letting $X = B + C$, we can use the theorem $0 + X = X$, that some readers proved, to show that the left side becomes $B + C$. On the right side, the subexpression within the parentheses becomes, applying the same theorem, simply B. Hence, the right side also reduces to $B + C$, and so the theorem is established for $A = 0$. Now set A to 1 and apply the theorem: $1 + X = 1$ to show that the left side becomes 1. Using the same theorem twice—once within the parentheses to show that $A + B$ becomes 1, and then a second time with the C term—it is clear that the right side is also 1 when $A = 1$. Therefore the theorem is also valid when $A = 1$, and is therefore true in general. Hence, we are free to use subexpressions such as $A + B + C$, knowing that they are well defined. Next it is shown that *any* number of variables can be included in such a sum. The result is independent of the way in which the terms are grouped for addition, and the ordering of the variables is also irrelevant.

More precisely, the theorem to be proved is:

THEOREM.   For any $k > 1$, the Boolean sum of k variables is 0 if all of the variables are 0, and is otherwise 1. This is true regardless of the way the variables are grouped, or of the order in which the variables appear.

PROOF.   The proof is sketched informally. First assume all k variables have the value 0. Then regardless of which two variables are chosen to start the summation process, the result of the first summation is $0 + 0 = 0$ (according to the corresponding postulate for addition). Each time another variable is added to the partial sum, the result is again $0 + 0 = 0$. Hence, the overall sum must be 0, as specified in the theorem statement. Now suppose that not all of the variables are equal to 0. Let X be an instance of a variable not 0, and therefore, by the first postulate, equal to 1. Let S be the cumulative sum of the variables added just prior to the addition of X. Then, when X is added to S, we have $S + X = S + 1 = 1$ (according to the previously proved theorem, $X + 1 = X$). Subsequently, each time another variable is added, we are adding a 1 (from the partial sum) to some variable Y, so that the same theorem comes into play again. Hence, the final sum must be 1, and the theorem is proved.

Note that this theorem subsumes and generalizes the theorem on the commutative law as well as the theorem on the associative law. The next theorem states that multiplication distributes over addition, i.e., that $A(B + C) = AB + AC$.

It is easily proved by perfect induction on A. But now let us see how this can be generalized to permit the number of variables in the sum to be 3 or 4 or, in fact, *any* finite number k. The theorem to be proved is:

For any finite number $k > 1$,

$$A \sum_{i=1}^{k} X_i = \sum_{i=1}^{k} (AX_i).$$

The proof employs conventional finite induction.

PROOF. Assume the theorem is true for $k \le j - 1$ (it has already been proved for $k = 2$), then we shall show that it must also be true for $k = j$. That is, we assume

$$A \sum_{i=1}^{j-1} X_i = \sum_{i=1}^{j-1} (AX_i). \tag{1}$$

Decomposing the summation of j terms into two parts yields

$$A \sum_{i=1}^{j} Xi = A\left(\sum_{i=1}^{j-1} X_i + X_j \right). \tag{2}$$

Applying the distributive laws for two terms to the right side of (2) produces

$$A \sum_{i=1}^{j} X_i = A \sum_{i=1}^{j-1} X_i + AX_j. \tag{3}$$

Applying (1) to the first term on the right side of (3) and combining yields

$$A \sum_{i=1}^{j} X_i = \sum_{i=1}^{j-1} (AX_i) + AX_j = \sum_{i=1}^{j} (AX_i). \tag{4}$$

The validity of the theorem now follows by induction. Note that when applied from right to left, the distributive law theorem corresponds to "factoring out" the A, a very useful technique in manipulating logic expressions.

A key theorem used in the simplification of logic circuits is that $AB + A\overline{B} = A$. Since its effect is to combine, or *unite* two terms to produce one term, it is referred to here as the *uniting* theorem. The proof, by perfect induction, is left to the reader. As is the case with all of the theorems, subexpressions can be substituted for any of the variables. For example, if we have the expression

$$(P + Q)RV + (P + Q)R\overline{V},$$

the preceding identity can be applied by considering $(P + Q)R$ as replacing the variable A. Thus, the expression is reduced to $(P + Q)R$.

Complementing complex expressions can be carried out with the aid of a pair of theorems, known as the DeMorgan laws. The first of these is

(DeMorgan Law 1).

$$\overline{A + B} = \overline{A} + \overline{B}.$$

Again, the proof, by perfect induction, is left to the reader. Given this theorem, we can get the following generalization to a sum of an arbitrary number of variables with the aid of finite induction:

$$\left(\overline{X_1 + X_2 + \cdots + X_n}\right) = \overline{X}_1\overline{X}_2 \cdots \overline{X}_n.$$

As an exercise in the use of the theorems just developed, let us apply some of them to the derivation of another interesting theorem:

$AB + \overline{A}C + BC$

$\quad = AB + \overline{A}C + (A + \overline{A})BC \qquad (A + \overline{A} = 1, and\ 1 \cdot BC = BC) \qquad (1)$

$\quad = AB + \overline{A}C + ABC + \overline{A}BC \qquad (multiply\ out;\ i.e.,\ use\ the\ distributive\ law) \qquad (2)$

$\quad = AB + \overline{A}C \qquad (X + XY = X;\ use\ this\ twice). \qquad (3)$

We have established, then, that $AB + \overline{A}C + BC = AB + \overline{A}C$. (We could, of course, have proved this theorem with perfect induction.) This is called the *consensus* theorem, and the BC term is called the *consensus* of AB and $\overline{A}C$. In general, two product terms have a consensus if there is *exactly one* variable that appears uncomplemented in one of them and complemented in the other. Defining a *literal* as a variable or a complemented variable (e.g., X and $\overline{X}$ are both literals), the consensus is the product of the union of all other literals appearing in two terms. For example, the product terms $A\overline{B}CD$ and $AB\overline{C}EF$ have the consensus $A\overline{B}DEF$, while the pair of terms $ABCDE$ and $AC\overline{E}F$ does *not* have a consensus (because they differ in *two* variables, C and E). Nor do the terms $AC\overline{E}F$ and AB have a consensus (because there is *no* variable in which they differ). The consensus theorem is another useful tool in simplifying logic expressions.

The next theorem is one that enables us to focus attention on any one variable in an expression. Consider, for example, the following function of six variables, expressed as

$$f(A, B, C, D, E, F) = A\overline{B}C + \overline{A}BD + \overline{B}DE + A\overline{E}F.$$

The theorem will allow us to expand the function about any chosen variable, i.e., to produce an equivalent expression for the function in which the chosen variable appears once in uncomplemented form and once in complemented form.

The statement of the theorem, known as the *Shannon expansion theorem*, is

(Shannon Expansion Theorem).

$$f(X_1, X_2, \ldots, X_i, \ldots, X_n) = X_i f(X_1, X_2, \ldots, 1, \ldots, X_n) + \overline{X}_i f(X_1, X_2, \ldots, 0, \ldots, X_n).$$

This theorem is easily proved by perfect induction on X_i. Applying it to the preceding example, with B playing the role of X_i, we obtain

$$f = B(\overline{A}D + A\overline{E}F) + \overline{B}(AC + \overline{D}E + A\overline{E}F).$$

A somewhat similar theorem, also easily proved, is: $X + f(X) = X + f(0)$. An example of its use in simplifying expressions is: $A + B(\overline{C}D + \overline{A}E + \overline{A}C) = A + B(\overline{C}D + E)$, where $X = A$.

Now we introduce a powerful concept that allows us to double the number of theorems at our disposal without having to generate any new proofs. Turn back to Section 16 and examine the set of postulates for Boolean algebra. Choose any one of them and replace each occurrence of 0 with 1, each occurrence of 1 with 0, each occurrence of $+$ with $\cdot$, and each occurrence of $\cdot$ with $+$. These operations would transform the postulate $0 + 1 = 1$ to $1 \cdot 0 = 0$. Observe now that the result of the transformation is itself one of the postulates! This is true of *all* the Boolean algebra postulates. Some other examples are $\overline{1} = 0$ maps into $\overline{0} = 1$, and $1 \cdot 1 = 1$ maps into $0 + 0 = 0$. Of course, if postulate P_i maps into postulate P_j, then postulate P_j maps into postulate P_i. The two postulates for inversion map into one another. The postulates for addition map into the postulates for multiplication, and the postulates for multiplication map into the postulates for addition. The corresponding pairs of postulates are referred to as *duals* of one another.

Consider next any of the theorems derived earlier. All are based on the postulates. Suppose we transform the statement of any theorem in the same way that we obtain the dual of a postulate, i.e., by interchanging 0 and 1, and interchanging $+$ and $\cdot$. (Note that the complement operation is *not* affected.) Then we could prove that the resulting statement is also a valid theorem by simply transforming each step of the proof in the same manner. In particular the duals of the postulates used for the original proof would be used. A proof based on perfect induction implemented with a truth table would be transformed into a proof of the dual by replacing each column heading by the dual expression, and each row entry by its dual (i.e., 1s and 0s would be swapped).

The result is that we have a whole new set of theorems. An example is the dual of the consensus theorem. This is obtained by replacing the identity corresponding to the consensus theorem by the dual expression. Thus, the dual of $AB + \overline{A}C + BC = AB + \overline{A}C$ yields

$$(A + B)(\overline{A} + C)(B + C) = (A + B)(\overline{A} + C).$$

Another example that is particularly interesting is the dual of the law that states that multiplication distributes over addition. This law applies to ordinary algebra as well. But the dual law in Boolean algebra states that addition also distributes over multiplication, i.e., that

$$A + BC = (A + B)(A + C).$$

This of course has no validity in ordinary algebra. Skeptical readers might try proving some of these new theorems by paralleling the previously generated proofs of the dual theorems.

An important addition to our set of theorems is the dual of the DeMorgan theorem for complementing expressions (introduced earlier). The aforementioned theorem and its dual are

$$\overline{A + B} = \overline{A}\,\overline{B} \quad \text{and} \quad \overline{AB} = \overline{A} + \overline{B}.$$

The two DeMorgan laws can be used in concert to find the complement of any logic expression. As an example, let us complement

$$Z = A(B\overline{C} + \overline{B}(C + D)).$$

Our goal is an expression, with no complemented subexpressions, that corresponds to the complement of the given expression. Observing that Z can be expressed as the product of two subexpressions, A and $(B\overline{C} + \overline{B}(C + D))$, we apply the second DeMorgan law to obtain

$$\overline{Z} = \overline{A} + \left(\overline{B\overline{C} + \overline{B}(C + D)}\right).$$

We now focus attention on the complemented second component of the sum on the right-hand side, and observe that it can be expressed as the *sum* of two subexpressions. We can therefore apply the first DeMorgan law to convert that sum into the product of two complemented subexpressions, yielding

$$\overline{Z} = \overline{A} + \left(\overline{B\overline{C}}\right)\left(\overline{\overline{B}(C + D)}\right).$$

Next we apply the second law to the first complemented subexpression, and the first law to the second complemented subexpression, to obtain

$$\overline{Z} = \overline{A} + (\overline{B} + C)\left(B + (\overline{C + D})\right).$$

Finally, the first law can be applied to the complemented sum to yield the result:

$$\overline{Z} = \overline{A} + (\overline{B} + C)(B + \overline{C}\overline{D}).$$

Any expression can be partitioned as either the sum or the product of two (or perhaps more) subexpressions. Thus, it is always possible to apply one of the DeMorgan laws. Therefore, the preceding process is generally applicable. Note that there is a one-to-one correspondence between the literals of the original expression and those of the complemented expression, each individual literal being complemented in the course of the process. Furthermore, each AND operation of the original expression maps into an OR operation of the complemented expression, and vice versa. These effects are not peculiar to this particular example; they hold true in general. Let us see why.

Each time the first DeMorgan law is applied to a sum of literals or subexpressions, the connecting pluses are converted to multiplication operators. Furthermore, if any of the operands of the sum are literals, they are complemented. A dual statement applies to the second DeMorgan law; it changes connecting AND operators to OR operators, and also complements any literal operands. Once a literal has been complemented, or an operand converted, they are not subjected to any further applications of the laws. Since each operator eventually does become the subject of a DeMorgan law application, as does each literal, we see that each operator undergoes a net transformation (AND to OR, or OR to AND), and each literal is complemented exactly once. If we define the *dual of an expression* as the result of swapping AND and OR operations, as well as swapping any 1s or 0s that it may contain (we don't usually see 1s or 0s in logic expressions), then the process of complementing an expression can be carried out by finding the dual of the expression and then complementing all of the literals. This is known as the *generalized DeMorgan theorem*. It can be symbolically described by the following expression (in which we introduce the abbreviation of a set of variables, $X_1, X_2, \ldots, X_n$ by the "vector" $\underline{X}$):

(Generalized DeMorgan Theorem).

$$\bar{f}(\underline{X}, 0, 1, +, \cdot) = f(\overline{\underline{X}}, 1, 0, \cdot, +).$$

More is said in the sequel about the duals of expressions and the relations between duals and complements.

The principal theorems useful in logic design are summarized in Figure 2.3 (arranged in dual pairs). All but a few simple ones have been introduced earlier in this chapter. Since these theorems are used as tools in subsequent sections and chapters, readers should begin familiarizing themselves with them.

2.4. RELATING LOGIC EXPRESSIONS TO LOGIC CIRCUITS

2.4.1. Gate-Type Circuits

There is considerable variation in the symbols used to represent logic gates. The symbols introduced in the previous chapter are consistent with current Institute of Electrical and Electronics Engineers (IEEE) standards; they are repeated in Figure 2.4, along with the most commonly used alternatives.

Given any logic circuit, it is a straightforward process to determine the logic function realized at any output. A preliminary definition is necessary; we define a *primary* input as a signal that is an input to the overall circuit under consideration. In analyzing a circuit to determine the output function, we start at any gate that has as its input only primary inputs, and write down at the output end of that gate a logic expression corresponding to the function of the gate (which we shall refer to as a *primary level* gate). Refer now to the circuit of Figure 2.5, used as a running example.

The inverter receiving A as input is a primary-level gate, and so its output has been labeled $\overline{A}$. Another such gate is the OR-gate at the left end, with inputs C and D. Its

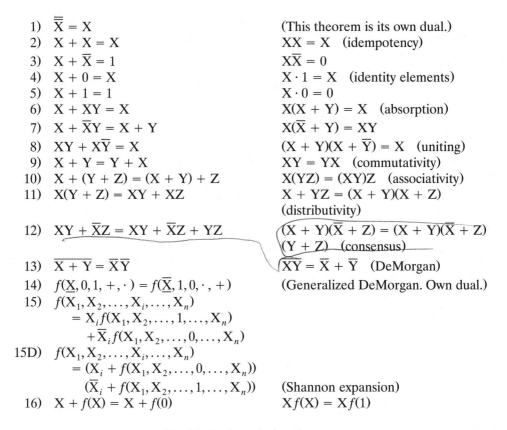

1) $\overline{\overline{X}} = X$ (This theorem is its own dual.)

2) $X + X = X$ $XX = X$ (idempotency)

3) $X + \overline{X} = 1$ $X\overline{X} = 0$

4) $X + 0 = X$ $X \cdot 1 = X$ (identity elements)

5) $X + 1 = 1$ $X \cdot 0 = 0$

6) $X + XY = X$ $X(X + Y) = X$ (absorption)

7) $X + \overline{X}Y = X + Y$ $X(\overline{X} + Y) = XY$

8) $XY + X\overline{Y} = X$ $(X + Y)(X + \overline{Y}) = X$ (uniting)

9) $X + Y = Y + X$ $XY = YX$ (commutativity)

10) $X + (Y + Z) = (X + Y) + Z$ $X(YZ) = (XY)Z$ (associativity)

11) $X(Y + Z) = XY + XZ$ $X + YZ = (X + Y)(X + Z)$
                                       (distributivity)

12) $XY + \overline{X}Z = XY + \overline{X}Z + YZ$ $(X + Y)(\overline{X} + Z) = (X + Y)(\overline{X} + Z)$
                                        $(Y + Z)$ (consensus)

13) $\overline{X + Y} = \overline{X}\,\overline{Y}$ $\overline{XY} = \overline{X} + \overline{Y}$ (DeMorgan)

14) $f(\underline{X}, 0, 1, +, \cdot\,) = f(\overline{\underline{X}}, 1, 0, \cdot\,, +)$ (Generalized DeMorgan. Own dual.)

15) $f(X_1, X_2, \ldots, X_i, \ldots, X_n)$
       $= X_i f(X_1, X_2, \ldots, 1, \ldots, X_n)$
       $+ \overline{X}_i f(X_1, X_2, \ldots, 0, \ldots, X_n)$

15D) $f(X_1, X_2, \ldots, X_i, \ldots, X_n)$
       $= (X_i + f(X_1, X_2, \ldots, 0, \ldots, X_n))$
       $(\overline{X}_i + f(X_1, X_2, \ldots, 1, \ldots, X_n))$ (Shannon expansion)

16) $X + f(X) = X + f(0)$ $Xf(X) = Xf(1)$

Fig. 2.3. Boolean algebra theorems.

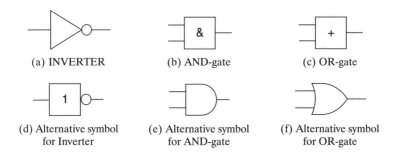

(a) INVERTER   (b) AND-gate   (c) OR-gate

(d) Alternative symbol   (e) Alternative symbol   (f) Alternative symbol
for Inverter   for AND-gate   for OR-gate

Fig. 2.4. Logic gate symbols (top row shows those used in this book).

output has been labeled C + D. When all primary-level gates have been so processed, we proceed to determine the output functions at other gates, whose inputs are either primary inputs or outputs of previously processed gates. In our example, we could at this point determine the output of the lower left AND-gate, which has as inputs primary input B and the output of the primary-level OR-gate, which has been found to be C + D. Thus, the output of the AND-gate must be the Boolean product of these or, as shown on the diagram, B(C + D). Readers can see the completion of this process on the diagram. The result is the output (labeled Z, the letter conventionally used to connote

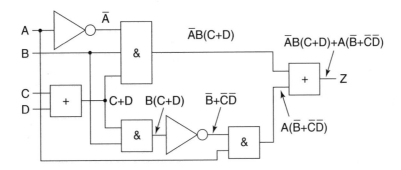

Fig. 2.5. Analysis of a gate-type logic circuit.

outputs) of the rightmost AND-gate, which is

$$\overline{A}B(C + D) + A(\overline{B} + \overline{C}\overline{D}).$$

Apart from possible trivial reorderings of terms in sums or products, the result of this analysis is unique: the structure of the logic expression directly mirrors the structure of the circuit.

Some additional important terms are illustrated in this diagram. Primary input signals A and B each *fan out* to two circuit nodes. This is very common. In fact, most nontrivial functions cannot be realized without some fanout. There is also fanout from the output of the leftmost OR-gate. Such *internal* fanout is *not* necessary to realize any function, but, as is the case here, it can lead to a more economical circuit. In this case, the function corresponding to C + D is generated once and then used in two places in the circuit. The alternative would be to generate it once for each node where it was needed. In this case, an additional OR-gate would have been needed. Readers familiar with computer programming will recognize the similarity with the concept of subroutines.

The topological opposite of fanout is fan*in*, which refers to the number of signals that converge at any gate, i.e., the number of inputs to a gate. In Figure 2.5, apart from the inverters, which of course are restricted to one input, the fanin to each gate is 2, except for the upper AND-gate, which has a fanin of 3.

The inverse of the problem just considered is that of generating a logic circuit corresponding to a given logic expression. This is no more difficult than the analysis problem. Again, a running example is used; refer to Figure 2.6. The expression to be realized by a gate circuit is

$$\left(\overline{B + C + \overline{A}D}\right) + B(C + D).$$

There are several ways to proceed. The method described here can be thought of as a "top-down" approach, in that the outermost operations of the expression are taken care of first, and then we work through the layers of parentheses from the outside in. The circuit is generated as a tree, from the right (output) end to the left (input) end. It resembles to some extent the procedure discussed earlier in connection with the use of the DeMorgan theorems to generate complements (p. 21).

The outermost operation (outermost in the sense that it is not contained in any parentheses) for the given expression is addition, with two subexpressions as operands. Therefore, we start with an OR-gate, with its output labeled Z (the *primary* output of

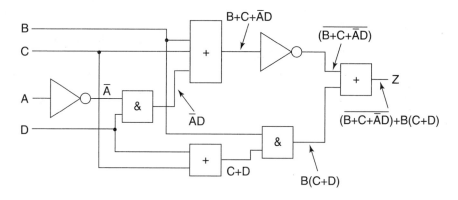

Fig. 2.6. Construction of a gate-type logic circuit.

our circuit), and with two inputs, one for the subexpression $\overline{(B + C + \overline{A}D)}$, and the other for the subexpression $B(C + D)$. We must now construct a circuit for each of these (observe the recursive nature of the process). The outer operation for the first of these is inversion, so we add an inverter with its output going to the OR-gate, and its input coming from the yet-to-be-constructed realization of $B + C + \overline{A}D$. We continue the process along this path by noting that the outermost operation of this expression is addition with three operands, which leads to our adding to the circuit a 3-input OR-gate. Following the other path to generate the subcircuit for $B(C + D)$ would lead us first to add a 2-input AND-gate at the other input to the output OR-gate. At this point the process should be clear, and readers can see from Figure 2.6 how it is completed. An alternative approach is to work from the "bottom up," i.e., from inside the innermost parentheses, and generate the circuit from the input end to the output end. Both methods, or a combination, are satisfactory—it is a matter of personal preference.

2.4.2. Switching Circuits

It is shown in this subsection that, just as there is a one-to-one correspondence between logic expressions and logic gate circuits, there is a similar correspondence between logic expressions and switching-circuit transmissions. That is, for any switching circuit, a logic expression can be found that mirrors its structure, and that indicates the conditions under which a conducting path exists through it. Furthermore, for any logic expression, a switching circuit can be constructed with matching structure and with its transmission corresponding to the conditions under which the expression is equal to 1. There are, however, some qualifications to the preceding assertion.

First, switching circuits cannot implement the complementing of subexpressions more complex than a single literal. For example, the term

$$\left(\overline{B + C + \overline{A}D}\right)$$

in the expression implemented by Figure 2.6 cannot be handled directly by a switching circuit. In such cases, the expression must be transformed, usually through the application of the DeMorgan theorems, to a form in which only input variables are complemented. Even then there may still be a problem. While electromechanical relays have built-in ways of dealing with complemented variables (normally open contacts), no corresponding feature exists for the modern implementations in the form of MOS transistors. If an expression includes a complemented literal, and if no corresponding signal is available at the input to the system, then it must be produced *outside* of the switching circuit with an inverter. There are also limitations on the sizes of switching circuits implemented electronically. These are due to engineering considerations, such as limits on the amount of stray capacitance that can be tolerated without making the time constants excessively large.

It should be understood that, while we shall discuss switching circuits directly in terms of their transmissions, they are almost invariably used today to process voltage levels, just as gate circuits do. In practice, switching circuits are generally used as pulldown circuits, or as pass-transistor networks.

Consider now the switching circuit shown in Figure 2.7. (Note a change in notation. Rather than showing a picture of a switch, with the associated symbol next to it, as is done in Figure 1.1, a switch is shown simply as a gap in a line, with the variable in it.) The circuit can be considered as consisting of two branches in parallel, one containing A and $\overline{C}$, the other consisting of $\overline{A}$ and a subcircuit, itself consisting of two parallel branches. If the transmission of either of two parallel branches is 1, then clearly the transmission of the combined circuit is 1. Hence, again thinking in terms of a top-down, recursive process, we can express the overall transmission as $T_1 + T_2$, where the T_i represent the (yet to be determined) transmissions of the parallel branches.

When there are switches in series, as in the top branch, it is clear that *both* must be on in order for there to be a path. Hence, the transmission of the top path is $A\overline{C}$. The bottom branch is more complex. We can think of it as consisting of two branches in series, the first of which is $\overline{A}$. Then the transmission of the lower branch is of the form $\overline{A}T_3$, where T_3 is the transmission of the network in series with $\overline{A}$. Thus far our expression consists of $A\overline{C} + \overline{A}T_3$. Continuing in this way, T_3 is found to be $BC + D$, so that the expression for the overall transmission is $A\overline{C} + \overline{A}(BC + D)$.

This process is valid for all circuits that can be decomposed, step by step, into subcircuits that are either in series or in parallel with one another. It is not difficult to carry out.

Constructing a switching circuit corresponding to a logic expression closely parallels the corresponding procedure for gate circuits. The expression is progressively decomposed from the outside; each time an OR-operation is encountered the operand subfunctions are relegated to parallel branches (analogous to the insertion of OR-gates).

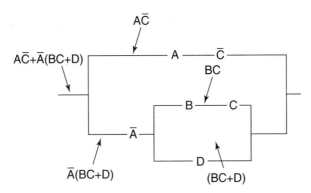

Fig. 2.7. Analysis of a switching circuit.

AND-operations generate series branches (analogous to the insertion of AND-gates). The same expression used in connection with Figure 2.6 to illustrate the process for generating a gate circuit from a logic expression, will be used again here. Since it includes a complemented subexpression (corresponding to an inverter in the interior of the circuit, the generalized DeMorgan theorem is applied first to eliminate that feature. This leads to

$$\overline{\left(B + C + \overline{A}D\right)} + B(C + D) = \overline{B}\,\overline{C}(A + \overline{D}) + B(C + D).$$

In the first step, the expression is decomposed into the operands of the outermost sum, and each is assigned to correspond to the transmission of a parallel branch. Thus, we have two parallel branches, with transmissions $\overline{B}\,\overline{C}(A + \overline{D})$ and $B(C + D)$. Readers should have no trouble in following the remainder of the process by studying Figure 2.8.

In n-channel MOS (nMOS) and complementary MOS (CMOS) technologies, and in some technologies employing bipolar transistors, switching circuits are sometimes used as pulldown and/or pullup networks to realize simple functions. Such configurations, often referred to as *complex gates*, may be used along with standard logic gates to realize more complicated logic functions. Complex gates are generally limited to functions realizable with no more than four switches in series.

2.5. USING BOOLEAN ALGEBRA TO MANIPULATE LOGIC CIRCUITS

In the preceding section, it was demonstrated that there is a one-to-one correspondence between Boolean algebra expressions and logic circuits. Thus, Boolean algebra may be regarded as a fourth language for describing and manipulating logic functions (recall that the first three, introduced earlier in the book, are natural language, logic-circuit diagrams, and truth tables). Next, it is shown that logic expressions can be derived directly from truth tables in a straightforward manner.

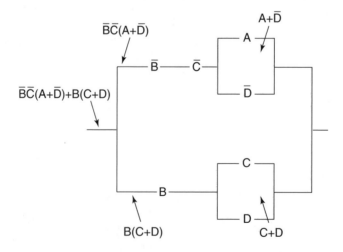

Fig. 2.8. Constructing a switching circuit corresponding to a logic expression.

2.5.1. Truth Tables and Minsums

Consider the truth table of Figure 2.1 (page 17). Each row for which the output Z is 1 corresponds to exactly one state of the input variables. Each such input state (which is called a *1-point* of the function) can also be described by a Boolean product of all of the input variables or their complements. For example, the row for which A = 0, B = 1, C = 1 (such a state will be abbreviated as ABC = 011) corresponds to the Boolean product $\overline{A}BC$. This product term (the abbreviation *p-term* will be used) has the value 1 only for ABC = 011. The other rows of the truth table for which Z = 1 correspond to 1-points represented by the *p*-terms $AB\overline{C}$ and ABC. Consider the Boolean sum of these product terms

$$Z = \overline{A}BC + AB\overline{C} + ABC.$$

For each state of the input variables for which the output Z is to be 1 (i.e., for each 1-point of the function) exactly one of the *p*-terms of the expression is 1, and hence the Boolean sum is 1. For any input state corresponding to a truth table row in which Z = 0, *none* of the *p*-terms is 1, and hence the sum is 0. Therefore, the sum of all the *p*-terms corresponding to the truth table rows in which Z = 1 is a logic expression that describes the function. Some additional definitions are necessary to facilitate discussion of this important topic. A *p*-term that includes exactly one literal for each variable of the function, such as the *p*-terms under consideration here, is called a *minterm* (alternate names are *complete product term* or *standard product term*). An expression for a function that consists of a sum of minterms is called a *minsum* (*canonic sum-of-products, canonic sum, standard sum*, and *disjunctive normal form* are some of the other ways in which these expressions are referred to in the literature). The minsum is often used as the starting point for the synthesis of logic circuits. (We could, of course, immediately specify a logic circuit corresponding directly to the minsum, but this does not usually correspond to a desirable realization.)

A less frequently used alternative to the minsum form can also be derived from the truth table. This is a product-of-sums form, called a *maxproduct*, in which each sum, defined as a *maxterm*, contains a literal corresponding to each variable. Each maxterm in this product is obtained from a row of the truth table in which Z = 0 (e.g., corresponding to row 100 of Figure 2.1, the maxterm is $\overline{A} + B + C$). The literals of the maxterm are chosen to be the *complements* of the values of the input variables in that row. Thus, a particular maxterm has the value 0 *only* for the one row (or input state) that it corresponds to, since for all other rows, at least one of the literals in it will have the value 1. Then the product of all of these maxterms must be 0 for precisely those input states for which the value of the function is 0. For all other states (the 1-points of the function), the maxproduct is 1. In our example, the maxproduct is

$$Z = (A + B + C)(A + B + \overline{C})(A + \overline{B} + C)(\overline{A} + B + C)(\overline{A} + B + \overline{C}).$$

Although there is a sense in which the *form* of the maxproduct is the dual of the *form* of the minsum, they do not, of course, represent dual *functions*, since they represent the *same* function.

2.5.2. Goals of Logic Design

An important goal in logic design is to minimize the cost of the resulting circuits. How the cost of an actual logic circuit translates into terms relevant to the abstract logic circuit diagrams under discussion here, is not a simple matter. All other things being equal, the fewer gates, the less the cost. Not only are fewer gates likely to translate into less area on a chip, but the power requirement is also reduced. The reduced area requirement may lead to the need for fewer, or less costly chips. The reduced power requirement may mean a less costly power supply can be used, and that the cost of dissipating heat generated on the chip is cut. In some high-performance systems, heat dissipation is the principal limiting factor. Fewer gates also means less opportunity for failure, and, hence, better reliability.

But "all other things" are not always equal. A characteristic of modern integrated circuit technology is that, as the sizes of the basic elements (usually the transistors) get smaller, the importance of on-chip wiring grows, both with respect to area and effects on speed. Thus, there are cases where a logic circuit in which the elements can be arranged in a very orderly manner, with minimum wire lengths between elements, may be more economical than a less orderly circuit that performs the same function with fewer transistors. In general, conceptual simplicity and regularity usually are helpful in reducing manufacturing and testing costs. Both regularity and minimization of component counts tend to increase reliability and reduce maintenance problems.

With respect to minimizing logic, there are several parameters that can be used. Instead of counting gates, we might count the total number of gate inputs, since the cost of a gate and the associated wiring in any technology increases with the number of inputs. Or, if we are designing a switching circuit, we might count the total number of literals in the logic expression to be implemented, since this corresponds to the number of switches (transistors in MOS technology). But if it is necessary to generate complements of input variables, the cost of inverters must be added in.

Fortunately, we seldom encounter functions where the implementation with the smallest number of gate inputs, for example, is not also the implementation with the smallest number of gates. In fact, this may not be possible at all for sum-of-product realizations of functions of fewer than eight variables. We shall use the total number of gate inputs as the measure of logic circuit cost. In specific situations, other factors should be taken into account.

Speed is also often an important consideration. With respect to logic design, it is enhanced by minimizing maximum logic path lengths and minimizing fan-in and fanout. Physically, speed increases can generally be obtained (within limits) by using more power.

Time is a factor in at least two other ways. One is the elapsed time to completion of a design project. Where this is important, it may be necessary to use simpler design procedures that take less time to carry out and less time for verification and testing. Simplicity is, of course, always a virtue in any piece of engineering work. Related to this is the time of the designers. If the product being designed is to be used for important purposes over an extended period of time, and if other costs, such as manufacturing or distribution, are very high, then it is worthwhile to invest substantial designer time to find the best possible solution. But, at the other extreme, if only one copy of the product is to be built, and if there is no inherent need for a tight design, it may be wiser to use a very straightforward approach to get the job done as quickly and cleanly as possible.

2.5.3. Using Algebra to Simplify Logic

Let us see how a good circuit can be generated to implement the function specified in the truth table of Figure 2.1. Begin with the minsum for the function, which was found in Subsection 2.5.1 to be $Z = \overline{A}BC + AB\overline{C} + ABC$.

Observe that the first and third minterms fit the conditions of Theorem 2.14 (the uniting theorem) of Figure 2.3, where A plays the role of Y, and BC plays the role of X. Hence, we can replace these minterms by the single p-term, BC, reducing the expression to: $Z = BC + AB\overline{C}$. Next we might apply the distributive law, Theorem 11, to factor out the B from the two p-terms, yielding $Z = B(C + A\overline{C})$. Applying Theorem 7 to the subexpression within the parentheses, reduces the expression further to $Z = B(C + A)$. Using the techniques introduced in Subsection 2.4.1, this expression can be translated into a logic circuit consisting of a 2-input OR-gate and a 2-input AND-gate.

Thus, it is apparent that Boolean algebra theorems can be useful in converting minsums into more economical equivalent expressions that can, in turn, lead to correspondingly economical logic circuits. Another example, illustrating other techniques, is presented in this section. All this is preliminary to the development, in the next chapter, of more powerful, systematic methods for accomplishing similar ends. There are situations, however, where the less structured use of the theorems, as demonstrated in this subsection, is more appropriate.

The next example serves to illustrate the use of some other theorems, the problems encountered in relying only on the direct use of the theorems, and some of the choices that a designer must make. Readers are strongly advised to carry out the detailed steps themselves with pencil and paper. It is difficult to appreciate the processes involved by simply reading.

Rather than specify the function to be realized by means of a truth table, we shall introduce and then use a useful shorthand notation. A function may be specified by listing the input states (equivalently, the rows of the truth table) for which its value is 1. Thus, we could list a set of binary vectors to convey the same information as the truth table. But, as a further step in the direction of compactness, we shall use the technique mentioned in Appendix A.1.7 of using octal numbers as abbreviations for binary strings. (We use octal rather than decimal because it is so easy to convert between octal and binary.) Individual minterms are specified as m_i, where the subscript is the octal abbreviation. Thus the minterm $AB\overline{C}D$ is abbreviated as m_{15}. Similarly, a function can be specified by listing a set of octal numbers representing the 1-points (or minterms) of the function. In particular, the example to be considered here is a function specified as $f(A, B, C, D) = \Sigma m(2, 3, 4, 5, 13, 14, 15, 17)$.

We begin by writing the minsum expression corresponding to the preceding description; to each octal number, there corresponds one minterm. The expression is

$$Z = \overline{A}\overline{B}C\overline{D} + \overline{A}\overline{B}CD + \overline{A}B\overline{C}\overline{D} + \overline{A}B\overline{C}D + A\overline{B}CD + AB\overline{C}\overline{D} + AB\overline{C}D + ABCD.$$

The cost in gate inputs of realizing this expression is 4 for inverters (1 for each input variable), $8 \cdot 4 = 32$ for AND-gates, and 8 for the final OR-gate, a total of 44; the number of gates is 13. There are many ways that we can apply our theorems to simplify this expression. As will soon be evident, the *order* in which we apply them to various subexpressions can make a difference in the outcome.

For our first try, consider the sum of the first and second minterms,

$$m_2 + m_3 = \overline{A}\overline{B}C\overline{D} + \overline{A}\overline{B}CD.$$

Applying the uniting theorem, $XY + X\overline{Y} = X$, with $X = \overline{A}\overline{B}C$ and $D = Y$, we can replace these two terms with $\overline{A}\overline{B}C$. Similarly, we can replace $\overline{A}B\overline{C}\overline{D} + AB\overline{C}\overline{D}$ with $B\overline{C}\overline{D}$, and $AB\overline{C}D + ABCD$ with ABD. This reduces the expression to

$$Z = \overline{A}\overline{B}C + B\overline{C}\overline{D} + ABD + \overline{A}B\overline{C}D + A\overline{B}CD.$$

Now, using the distributive law, we factor $B\overline{C}$ out of the second and fourth terms, and factor AD out of the third and fifth terms to obtain

$$Z = \overline{A}\overline{B}C + B\overline{C}(\overline{D} + \overline{A}D) + AD(B + \overline{B}C).$$

Applying the theorem $X + \overline{X}Y = X + Y$ to each of the parenthesized subexpressions further reduces the expression to

$$Z = \overline{A}\overline{B}C + B\overline{C}(\overline{D} + \overline{A}) + AD(B + C).$$

Next we apply the first DeMorgan law to $\overline{D} + \overline{A}$ (also the commutative law for addition) to obtain

$$Z = \overline{A}\overline{B}C + B\overline{C}(\overline{AD}) + (AD)(B + C).$$

Note that AD can be generated once and used twice (i.e., we can fanout from an AND-gate generating AD). The cost of realizing this expression can be calculated as four inverters @ 1 each, two 2-input AND-gates @ 2, a 2-input OR-gate @ 2, two 3-input AND-gates @ 3 each, and a 3-input OR-gate @ 3, totaling 10 gates and 19 gate inputs. (It would be a useful exercise for readers to draw the logic circuit and verify the calculation.) Although this is a substantial reduction from the cost of 44 for a circuit corresponding directly to the minsum form, we have no reason to believe that this is the best that can be done. There are no obvious further simplifications of the last expression obtained, so we shall return to the original minsum form to take a different path.

This time we apply the uniting theorem to $m_2 + m_3$ (again), to $m_4 + m_{14}$ (again), to $m_5 + m_{15}$, and to $m_{13} + m_{17}$ to obtain

$$Z = \overline{A}\overline{B}C + B\overline{C}\overline{D} + B\overline{C}D + ACD.$$

Now observe that the uniting theorem can be applied again to $B\overline{C}\overline{D} + B\overline{C}D$ to reduce the expression further to

$$Z = \overline{A}\overline{B}C + B\overline{C} + ACD.$$

The cost for this second expression is only 14 gate inputs; there are seven gates. But perhaps we can do better. Factoring out the C from the first and third terms and applying the second DeMorgan law to $\overline{A}\overline{B}$ (to save an inverter) yields

$$Z = C(\overline{A + B} + AD) + B\overline{C}.$$

Although this expression has one fewer literal, the cost remains at 14 gate inputs, and the number of gates is eight. A possible advantage of the third solution is that the maximum fan-in is two (the second solution involves two 3-input AND-gates and a 3-input OR-gate), and there is no internal fanout. However, although these factors tend to increase speed somewhat, a comparison of the logic circuits (which readers should draw) indicates that, for the third solution, signals A and B must pass through a chain of five gates to reach the output, whereas for the second solution, the longest signal path includes only three gates. Since a signal is delayed by each gate through which it passes, the third circuit is likely to be slower. The second solution is thus probably better in most situations.

2.6. NAND- AND NOR-GATES

As pointed out in Subsection 1.2.2, NOR- and NAND-gates are the most commonly used gates for all of the integrated circuit technologies in use today, including transistor–transistor logic (TTL), emitter-coupled logic (ECL), nMOS, and CMOS (terms discussed in the Appendix). Because these technologies are so prevalent, it is important that logic designers be thoroughly familiar with techniques for implementing logic circuits with NOR- and NAND-gates. The approach here is to treat the NAND-gate in depth and then show how, using the concept of duality, essentially the same methods are applicable to NOR-gates.

First it will be shown that NAND-gates alone constitute a *complete set* of logic elements in that any logic function can be realized using only NAND-gates. We already know that any logic function is realizable using inverters, AND-gates, and OR-gates. To show that any other set of elements is complete, it is sufficient to show that inverters, AND-gates, and OR-gates can all be built out of members of this set. It has already been pointed out that an inverter *is* a NAND-gate (with 1 input). Clearly, an AND-gate can be constructed by attaching an inverter to the output of a NAND-gate. It remains only to show that the OR-function is realizable with NAND-gates. This follows from an examination of the expression $Z = (\overline{\overline{X}\,\overline{Y}})$. Using DeMorgan's second theorem, this becomes $Z = \overline{\overline{X}} + \overline{\overline{Y}} = X + Y$. The corresponding circuit, showing the actual implementation of an OR-gate with NAND-gates is shown in Figure 2.9a. (Figure 2.9b shows the dual circuit in which an AND-gate is implemented with NOR-gates.)

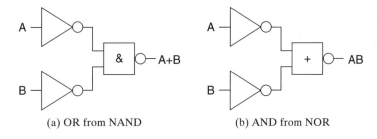

(a) OR from NAND   (b) AND from NOR

Fig. 2.9. Implementing OR and AND with NAND and NOR.

This argument is a constructive proof that NAND-gates constitute a complete set of logic elements. (Readers should have no trouble in completing the dual argument for NOR-gates.) Since it takes three NAND-gates to build an OR-gate, and two NAND-gates to build an AND-gate, one might conclude that logic circuits constructed from NAND- (or NOR-) gates must be rather cumbersome. Fortunately, this is not the case, since, as is shown below, we do not generally use such constructions to design NAND-gate circuits.

Before proceeding, a few more terms must be defined and a simple circuit equivalence pointed out. An *n-stage logic circuit* is one in which there are *n* gates in the longest signal path. We generally refer to logic circuits corresponding to sum-of-product (SOP) expressions, or to product-of-sums (POS) expressions as *two-stage* logic circuits. Actually, these are not always two-stage circuits, since it may be necessary to include inverters to generate complements of some of the inputs. In some cases we may have *double-rail inputs*, meaning that both complemented and uncomplemented input signals (both *polarities*) are available. (Where this is not the case the inputs are referred to as being *single*-rail.) The circuit equivalences shown in Figure 2.10a and b are direct reflections of the two DeMorgan theorems. These alternative ways of depicting NAND- and NOR-gates are useful in what follows. (The use of bubbles at the inputs as well as at the outputs of gates to represent inversion is a useful common practice.)

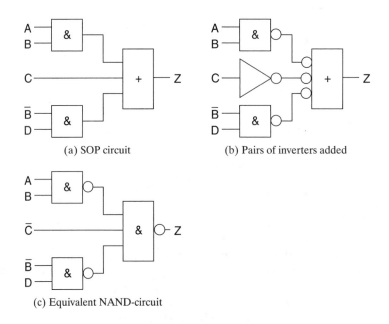

Now let us consider implementing SOP expressions with NAND-gates, assuming double-rail inputs. The expression $Z = AB + C + \overline{B}D$ will serve to illustrate the method. Figure 2.11a shows the direct implementation of this expression with a two-stage AND-to-OR circuit. In Figure 2.11b bubbles, representing inversions, are inserted in pairs between the logic stages, and an inverter and bubble are inserted between the C-input and the OR-gate. Since pairs of inverters cancel out, the function performed by Figure 2.11b is the same as that performed by Figure 2.11a.

But now we can think of the first stage, consisting of AND-gates followed by inverters, as NAND-gates. Since the second-stage OR-gate now has inverters at each of its inputs,

Fig. 2.11. Deriving a two-stage NAND-gate circuit.

it too, according to the equivalence shown in Figure 2.10a can be considered to be a NAND-gate. Instead of the inverter following the C-input, we can use the "other rail" of C, the $\overline{C}$-input, which is assumed to be available. This gives us the circuit of Figure 2.11c, a two-stage NAND-gate circuit that realizes the given function. It could have been obtained directly from the AND-to-OR-circuit of Figure 2.11a simply by replacing *all* of the gates with NAND-gates and taking the complement of any input that goes directly to the second-stage gate. This is a perfectly general procedure applicable to any SOP expression.

If the inputs are single-rail, the only change is that inverters must be added or removed between the last-stage gate and any inputs that feed it directly. Readers should go through the dual process to show that POS expressions can be realized by replacing all of the gates in a two-stage OR-to-AND implementation with NOR-gates. This is illustrated in Figure 2.12 for the expression $Z = (A + B)C(\overline{B} + D)$.

The same principles apply to the conversion of multistage AND-to-OR circuits to NAND logic. Each gate is converted to a NAND-gate by the addition of pairs of bubbles in the appropriate connecting lines (between the outputs of AND-gates and the inputs of OR-gates). Where necessary, inverters are inserted (either at inputs or where a gate feeds another gate of the same type). The process is illustrated in Figure 2.13.

The process for generating multistage NOR logic is similar, except that the bubble pairs are inserted between OR-gate outputs and AND-gate inputs, as illustrated in Figure 2.14 (in which the circuit of Figure 2.13a is transformed into a NOR circuit). Note that if the original circuit terminates with an OR-gate, as in the example, then an inverter must be added at the output of this gate in order to compensate for the bubble necessary to convert it to a NOR-gate. Comparing the NAND and NOR circuits corresponding to the same AND/OR/NOT circuit (i.e., Figure 2.13c and Figure 2.14b), we see that the NOR circuit has six inverters, as compared to three for the NAND

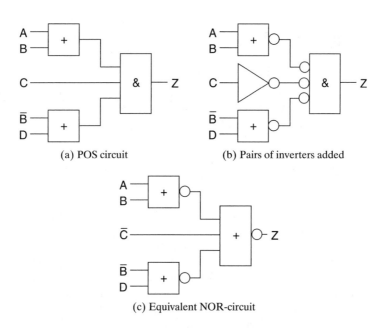

(a) POS circuit   (b) Pairs of inverters added

(c) Equivalent NOR-circuit

Fig. 2.12. Deriving a two-stage NOR-gate circuit.

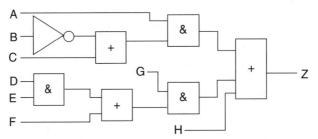

(a) AND/OR/Inverter realization of $Z = A(\bar{B} + C) + (DE + F)G + H$.

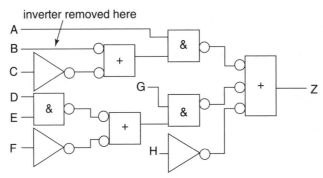

(b) Same circuit with inverter pairs added

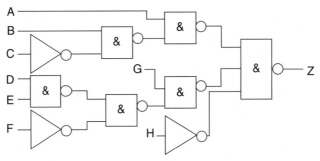

(c) NAND-gate realization of same function

Fig. 2.13. Generating a multistage NAND-gate circuit.

circuit. This reflects the fact that, for this particular circuit, the sequence of AND- and OR-gates terminates with an OR, and has four uncomplemented primary inputs feeding AND-gates. The last example (see Figure 2.15) illustrates what happens when, in the original AND/OR/NOT circuit, the AND- and OR-gates do not alternate. As indicated earlier, each such instance leads to the need for an inverter after some gate.

The possibility of directly connecting together the outputs of logic gates to perform an additional logic operation without using an actual gate, i.e., the concept of *wired logic gates*, is introduced in Subsection 1.3.1. It is pointed out there, without proof, that if the outputs of a set of NAND-gates with a wired-AND capability are joined, the result (Figure 1.7*a* and *c*) is equivalent to a two-stage OR-to-AND circuit, with all inputs complemented. The validity of that observation is evident if we simply replace each

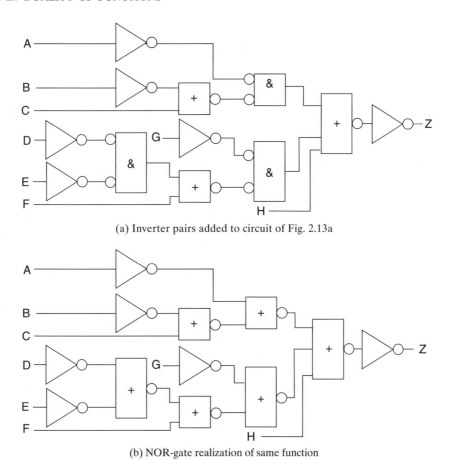

(a) Inverter pairs added to circuit of Fig. 2.13a

(b) NOR-gate realization of same function

Fig. 2.14. Generating a multistage NOR-circuit.

NAND-gate in that configuration with the alternative representation of a NAND-gate shown in Figure 2.10a, i.e., the OR-gate with complemented inputs. Similarly, if the outputs of NOR-gates with wired-OR capabilities are connected, the resulting circuit is equivalent to a two-stage AND-to-OR circuit with complemented inputs.

Exploiting wired logic is thus a simple matter. Given an arbitrary function to realize, suppose the available technology is NOR-gates and wired-ORs. We need only find a minimal SOP expression, complement all of the literals, connect the variables associated with each resulting p-term to the inputs of a NOR-gate, and wire together the outputs of all these gates to obtain the desired output. An analogous process applies to NAND-gates and POS expressions.

2.7. DUALITY OF FUNCTIONS

Closely related to the concept of duality of Boolean algebra theorems is that of duality of logic *functions*. The dual of a logic function is obtained by interchanging the AND and OR operations and the 0s and 1s in any description of the function. Thus, if the

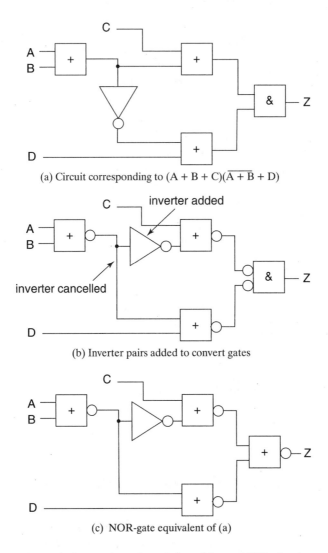

(a) Circuit corresponding to $(A + B + C)(\overline{A + B} + D)$

(b) Inverter pairs added to convert gates

(c) NOR-gate equivalent of (a)

Fig. 2.15. Generation of another multistage NOR-circuit.

description is in the form of a truth table, all of the entries in the table, the input-state descriptions, and the output values are complemented. If the description is in the form of an algebraic expression, then the AND and OR operations are all swapped; if any 0s or 1s appear in the expression, they are also changed. Descriptions in the form of gate circuit diagrams are converted by swapping AND- and OR-gates (also, of course, NORs become NANDs and vice versa). Finding the dual of a function expressed as a switching circuit can be thought of in topological terms, but rather than go into that area, one can find a corresponding logic expression, get its dual, and then, if desired, convert this to a switching circuit.

Duals and complements are different, but related. Let the dual of a function f be represented as f^D. Then, from the generalized DeMorgan theorem, we have: $\bar{f}(\underline{X}) = f^D(\underline{\overline{X}})$. Complementing the input variables (denoted by $\underline{X}$) on both sides and transposing gives us the very useful equation

$$f^D(\underline{X}) = \bar{f}(\underline{\bar{X}}).$$

This means, for example, that we can get the dual of a gate circuit by either of two methods: (1) the swapping process described earlier, or (2) complementing the output signal and all of the input signals.

Depending on whether inverters are inserted at the output, the inputs *and* the output, or just at the inputs, a function can be transformed in any of three ways. We get, respectively, the complement, the dual, or the dual of the complement (which can be shown to be the same as the complement of the dual—see Problem 2.23). For some important functions, treated in later sections, a function may be equal to its own dual. Note that if there is an inverter at the output of a logic circuit, we can eliminate it, if we insert inverters at the inputs and convert the circuit to its dual (by changing ANDs to ORs, etc.). The same operation may be carried out in the opposite direction, pushing inverters from the input to the output end.

2.8. POSITIVE AND NEGATIVE LOGIC

Consider now the way electrical signals represent logic values. It was assumed in our discussions as to how logic gates are implemented, that a logic 0 is represented by a low voltage—say close to 0 volts—and that a 1 is represented by a relatively high voltage (in integrated circuits, this may typically be in the range from 2 to 5 volts). The operation of the various gates we have treated in the Appendix, the previous chapter, and in the discussion of NAND- and NOR-gates of the previous section, all depended on this assumption. For example, we presented a circuit that generates a high output voltage if both of the input voltages are high. Under the assumptions just pointed out, such a circuit can serve as an AND-gate. But there is no natural law that compels us to associate the higher voltage with 1 and the lower voltage with 0. What would be the consequences of reversing that association?

If a particular logic circuit had been built with the higher voltage equated to 1, and if we then decided to consider the higher voltage as representing 0, then the meanings of all the signals, including the inputs and outputs, would be reversed. The physical behavior of the circuit would, of course, be unchanged by our change in viewpoint. Since 0-inputs have been transformed into 1-inputs and vice versa, and since the same is occurring at the output, it is precisely as though inverters were placed in all of the input and output terminals. But we have just seen that the effect of complementing both the output and all of the inputs in a circuit is to cause it to generate the dual of the original function. Hence, if we reverse our assumption about the interpretation of voltage levels, our logic circuits are transformed to perform the duals of the original functions. Thus, in the preceding example, the AND-gate becomes an OR-gate under the reversed assumptions.

Logic circuits designed under the first assumption, that high means 1, are said to operate under the *positive logic* assumption. If the higher voltage is used to represent logic 0, then the system is said to operate under the *negative logic* assumption. In practice, both assumptions are made, although positive logic is probably more common. Sometimes, within one system different assumptions are made in different parts in order to eliminate the need for some inverters. Whether we are talking about the behavior of a single gate or of a complex circuit, the effect of a change between positive and negative logic is simply to convert the functions involved to their duals.

XY	$X \oplus Y$
00	0
01	1
10	1
11	0

Fig. 2.16. Truth table for the XOR function.

2.9. EXCLUSIVE-OR-GATES

In Chapter 1, when the OR function was introduced, it was defined as having the value 1 if at least one of its inputs has this value. This is often referred to as the *inclusive*-OR. In common usage, this is the meaning that would be assigned in the sentence, "Don't buy it if it is too expensive *or* if it doesn't fit properly." But, in our everyday language, we sometimes use the word "or" in another sense, as in the sentence, "You get soup *or* a salad with today's special." Here the word is being used to mean one or the other, but not both. This is referred to as the *exclusive*-OR.

The exclusive-OR (the common abbreviation XOR will be used here) is a very useful and interesting operation, and so it merits careful study. It is formally defined as a binary operation having two operands, and written as $X \oplus Y$. Then it is generalized to allow any number of operands. In terms of our basic Boolean algebra operations, the definition is

$$X \oplus Y = X\overline{Y} + \overline{X}Y.$$

It can also be thought of as "X not equal to Y," or as having the value 1 if the number of its operands having the value 1 is odd. The equivalence of these definitions, and their equivalence to the truth table definition of Figure 2.16, should be evident.

The principal properties of the XOR function are summarized in the set of identities listed in Figure 2.17. Proofs, which are very simple using perfect induction, are left to the reader.

Since the XOR function has the associative property, expressions of the form: $X_1 \oplus X_2 \oplus \cdots \oplus X_n$ are meaningful (the situation is analogous to that for the inclusive-OR, which, as has been shown, also has the associative property). Such expressions can be evaluated by grouping the elements in pairs, as convenient, replacing each pair with the result of the binary operation, and then repeating the process until all of the elements have been accounted for. The order in which this is done does not affect the result.

1. $0 \oplus 0 = 1 \oplus 1 = 0; 0 \oplus 1 = 1 \oplus 0 = 1$
2. $X \oplus 0 = X$ (identity element is 0)
3. $X \oplus X = 0$ ($\oplus$ behaves like subtraction)
4. $X \oplus \overline{X} = 1$
5. $X \oplus 1 = \overline{X}$
6. $\overline{X \oplus Y} = \overline{X} \oplus Y = X \oplus \overline{Y} = 1 \oplus X \oplus Y = \overline{X}\overline{Y} + XY$ (complementing the XOR function)
7. $X \oplus Y = Y \oplus X$ (commutative property)
8. $(X \oplus Y) \oplus Z = X \oplus (Y \oplus Z)$ (associative property)
9. $X(Y \oplus Z) = XY \oplus XZ$ (distributive law)
10. $X \oplus Y = X \oplus Z$ implies $Y = Z$ (cancellation property)

Fig. 2.17. Properties of the XOR function.

Suppose that, in the preceding expression, $n = 2$. Then the value S of the result is 1 if, and only if (we shall abbreviate this stock mathematical phrase by iff) exactly *one* of the operands is 1 (see Property 1 in Figure 2.17). Assume now that for every value of n up to $j - 1$, the value of S is 1 iff the number of 1-values among the X_i is odd. We have just shown this to be true for $j = 3$—it is trivially true for $j = 2$. Then if an additional variable X_j is added to the sequence, and is XORed with S, then S changes value only if $X_j = 1$. In other words, the value of S changes only if the total number of 1s in the sequence changes from odd to even, or from even to odd. It follows, then, that for $n = j$, the value of S is 1 iff the number of 1-values among the X_i is odd. Hence, by finite induction, we see that, for *all* values of n, S is 1 iff there is an odd number of 1s in the sequence.

This justifies calling the generalized (to two or more inputs) XOR operation *modulo-2-addition*, since we can get the value of S by arithmetically adding the X_i, dividing by 2, and taking the remainder. (In general, the modulo-k value of any number is the remainder after division by k; e.g., 17 modulo-5 is 2.) Another name for the XOR operation is *ring sum*. In a context considered in a later chapter, yet another designation is *parity check function*. Circuit symbols for XOR are shown in Figure 2.18, along with an efficient NAND-gate realization.

From the SOP expression for XOR, $X\overline{Y} + \overline{X}Y$, we can obtain the POS form by first writing the dual expression, $(X + \overline{Y})(\overline{X} + Y)$, then multiplying out (and deleting the $X\overline{X}$ and $\overline{Y}Y$ terms) to obtain $XY + \overline{X}\overline{Y}$, the SOP expression for the dual. If we take the dual again, we obtain $(X + Y)(\overline{X} + \overline{Y})$, the desired POS expression for XOR. We could also have obtained this result by using complements instead of duals—a very similar process.

Still another way is to use the dual of the distributive law (the second part of Theorem 11 of Figure 2.3) to "add out" $X\overline{Y} + \overline{X}Y$ to obtain $(X\overline{Y} + \overline{X})(X\overline{Y} + Y)$. Adding out each of the parenthesized terms yields $(X + \overline{X})(\overline{Y} + \overline{X})(X + Y)(\overline{Y} + Y)$. The final result is generated by applying the identity $Q + \overline{Q} = 1$ twice. All these methods are applicable in general to generate POS expressions from SOP expressions.

In the preceding development, an ancillary result is that the dual of XOR is $XY + \overline{X}\overline{Y}$. If we complement the literals of this expression, we obtain the *complement* of XOR, $\overline{X}Y + X\overline{Y}$. This is equal to the previous expression (just reverse the order of the two p-terms in the sum). Thus, the complement and the dual of the 2-input XOR function are the same!

Consider again the n-input XOR function

$$Z = X_1 \oplus X_2 \oplus \ldots \oplus X_n$$

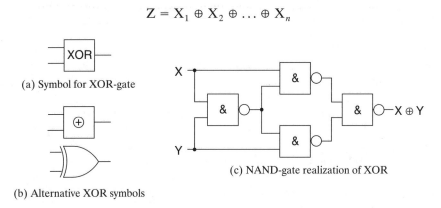

(a) Symbol for XOR-gate

(b) Alternative XOR symbols

(c) NAND-gate realization of XOR

Fig. 2.18. XOR-gate symbols and implementation.

Suppose we find the dual by complementing the output and all of the inputs (according to the result obtained in Section 2.7). We can complement the output by adding (modulo-2) a 1 to the right side (property 5 of Figure 2.17), and we can complement each input X_i by replacing it by $X_i \oplus 1$. The result is that we have added, modulo-2, to the right side of the expression a total of $n + 1$ 1s. If n is even, then this number is odd, and, hence, equal to 1, modulo-2. The result is that the original expression is complemented. Hence, the result obtained in the previous paragraph for $n = 2$ generalizes for all *even* values of n. That is, the dual of an XOR function of an even number of variables is equal to its complement.

Now suppose n is odd. Then the total number of 1s added is even, and so equal to 0 modulo-2. Therefore the function remains unchanged. Hence, for odd values of n, an n-input XOR function is equal to its own dual. A consequence of this result is that if we were to design a circuit implementing an XOR function with, say, seven inputs, using NAND-gates as components, the NANDs could all be replaced by NORs (the dual components) without altering the function performed. The general question of synthesizing XOR functions with many inputs, a matter of some practical importance, is treated in a later chapter.

2.10. Multivalued Logic

Allowing logic signals to take on more than two values would have several advantages. It might lead to fewer gates being required to perform certain operations (particularly arithmetic operations), speed advantages might result, and, most clearly, less wiring would be required to transmit information. A considerable amount of thought has, therefore, been given to the subject of multivalued logic, from both a theoretical point of view and with respect to implementations in hardware. It is the latter aspect that has proved to be the major obstacle. The electrical engineering advantages of two-valued as opposed to multivalued logic circuits with respect to reliability, cost, and performance, have thus far proved to be overwhelming. This is due to the great difficulty of controlling voltage levels on chips to within the close tolerances that would be necessary to discriminate among three or more values. Thus, there are only very limited situations in which multivalued signals are used; principally at the boundaries of chips where the bottlenecks involved in getting data into and out of chips sometimes justifies significant trade-offs in terms of circuit complexity, etc. No basic change in this situation seems to be on the horizon.

Overview and Summary

A simple listing of input states and the corresponding output states, called a *truth table*, is commonly used as a starting point for specifying what a logic circuit is supposed to do. Truth tables are easily translated into Boolean expressions, resembling ordinary algebraic expressions. The two-valued Boolean algebra, a very useful tool in dealing with logic circuits, can be characterized succinctly by specifying that all variables have the values 0 or 1 and by defining the basic operations of inversion (or complementation), AND (or Boolean multiplication), and inclusive-OR (Boolean addition). Using basic operations, expressions similar to ordinary algebraic expressions can be constructed that precisely describe functions to be realized. The expressions can be manipulated in accordance with various theorems, while preserving the functions described. The DeMorgan theorems are useful for finding the complements of logic expressions. Other theorems, such as the uniting theorem, the absorption theorem, and the consensus theorem are useful in simplifying logic expressions. Initial expressions obtained from truth tables are in SOP form. Each particular expression can be mapped into either a gate- or a switching-type logic circuit. SOP expressions map into two-stage AND-to-OR logic. The factoring theorem is the principal tool used to derive multilevel logic.

NOR- and NAND-gates are the most commonly used logic elements. It is easy to transform circuits built with AND-OR-inverter logic into either NAND- or NOR-type circuits. If, in any description of a logic function, we replace ANDs with ORs, ORs with ANDs, 0s with 1s, and 1s with 0s, then the new function is called the *dual* of the original function. In particular, replacing all of the gates in a NAND circuit with NOR-gates realizes the dual of the original function. An alternative way to get the dual of a circuit is to complement the output and all inputs. Circuits designed under the *negative logic* assumption represent logic 1s and 0s by low and high voltages, respectively, the opposite of the usual assumption. Changing the assumption for a particular circuit converts the function realized to the dual. The *exclusive-OR*-gate (XOR-gate), also called a *modulo-2-adder*, is another useful component in logic circuits.

Sources

Somewhat different introductions to Boolean algebra can be found in some of the texts mentioned in the last chapter, e.g., [Mano, 1991]. Discussions of relay contact circuits can be found in [Keister et al., 1951] and in [Caldwell, 1958]. Although the technology itself is obsolete, some of the ideas may be useful in connection with switching circuits based on MOS transistors. Special techniques applicable to the design of NAND- or NOR-gate circuits can be found in [McCluskey, 1963], in [Gimpel, 1967], and in [Maley, 1971].

Problems

2.1. Use a truth table to establish whether or not, "$A + B = C + B$ implies $A = C$" is a valid Boolean algebra theorem. (It is clearly a valid theorem in ordinary algebra.)

2.2. Using basic theorems, or a counterexample, prove the truth or falsity of the identity $(A + B)(\overline{A} + C) = AC + \overline{A}B$.

2.3. Prove that $(X + Y)(X + \overline{Y}) = X$ using perfect induction on Y.

2.4. Prove directly (i.e., without resorting to a duality argument) the validity of the dual of the Shannon expansion theorem (15D of Figure 2.3 on p. 25).

2.5. Prove the validity of the dual form of the consensus theorem *without* using duality or perfect induction.

2.6. Convert $A\overline{B} + CD$ to an equivalent POS expression by using the dual of ordinary "multiplying out" (second part of Theorem 11 of Figure 2.3) three times.

2.7. Simplify each of the following expressions, using *one* of the basic theorems (state which) in Figure 2.3:

*(a) $(AB + \overline{C}D)(AB + \overline{C}E)$

(b) $A + \overline{B}C + DE(A + \overline{B}C)$

(c) $A + BC + (DE + F)(\overline{A + BC})$

2.8. Simplify the expressions:

(a) $ABC + A\overline{C} + A\overline{B}$

(b) $AC(B + D) + \overline{B}C\overline{D}$

2.9. Simplify each of the following expressions using basic theorems:
  (a) $AD + BCD + \overline{A}C$
  (b) $AB(\overline{A}CD + AE + EG)$

2.10. Using the basic theorems, find minimal-sum expressions for each of the following:
  (a) $ABC + A\overline{B}CE + ACDE + \overline{A}\overline{B}D + \overline{B}CDE$
  (b) $\overline{A}\overline{C}\overline{D} + A\overline{C} + BCD + \overline{A}C\overline{D} + \overline{A}BC + AB\overline{C}$

2.11. Find SOP equivalents for each of the following:
  (a) $(A + \overline{B}C + \overline{D})(\overline{B}C + \overline{D} + E)(A + \overline{E})(AD + \overline{E})$
  (b) $(\overline{A} + B\overline{E})(B\overline{E} + C + D)(\overline{C} + E)$

2.12. Find POS expressions equivalent to each of the following:
  *(a) $AB + CD + A\overline{C} + D\overline{E}$
   (b) $AB(\overline{C} + D)E + F$

2.13. Draw gate circuits corresponding *directly* to:
  *(a) $Z = A(B + \overline{C}) + (\overline{B + D})E$
   (b) $Z = \overline{A} + (C + DE)(A + B)$

2.14. Find logic expressions corresponding *directly* to each of the circuits of Figure 2.19:

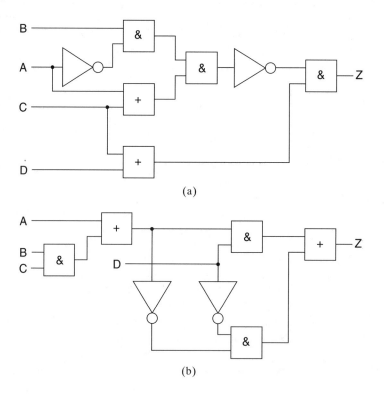

(a)

(b)

Fig. 2.19.

2.15. Convert each of the logic circuits of Problem 2.14 to an equivalent NAND-gate circuit.

2.16. Convert each of the logic circuits of Problem 2.14 to an equivalent NOR-gate circuit.

2.17. Assuming double-rail inputs are available, specify
(a) a NAND-gate
(b) a NOR-gate
realization of the function $Z = A(B + \overline{C}D)(B + E)$.

2.18. Memory banks M_1, M_2, M_3, M_4 are arranged in an ordered row. $X_i = 1$ when M_i is busy. Write a minsum expression for a function that has the value 1 if at least two adjacent memory banks are busy.

2.19. How many different functions of three variables are there?

2.20. Convert the following expression into an equivalent expression in which A and $\overline{A}$ each appears atmost once: $\overline{B}(D\overline{E} + A\overline{C}D + A\overline{E} + C\overline{D}) + B(ACE + \overline{A}\overline{C}D + \overline{A}DE)$.

2.21. Write dual expressions for each of the expressions of Problem 2.13.

2.22. Show that $f = \overline{A}\overline{C} + BC + A\overline{B}$ is the same as the dual of its complement.

2.23. Prove that the complement of the dual is always identical to the dual of the complement.

2.24. Find a three-variable function that is its own dual.

2.25. The table below shows the input and output voltages for a 2-input logic circuit with one output. Write a logic expression for the function realized if the logic is
(a) Positive.
(b) Negative.

V_A	V_B	V_Z
-3	-3	0
-3	0	-3
0	-3	-3
0	0	-3

2.26. Write a SOP expression corresponding to the function realized by the circuit of Figure 2.20.

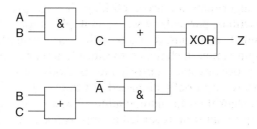

Fig. 2.20.

2.27. (a) Prove that $AB \oplus \overline{A}C = AB + \overline{A}C$.
(b) Prove that $P \oplus Q = P + Q$ iff $PQ = 0$.

2.28. Does the XOR operation distribute over Boolean multiplication (i.e., is $A \oplus BC = (A \oplus B)(A \oplus C)$)?

2.29. Using as components *only* an inverter and two 2-input XOR-gates, realize the function described by $Z = AB\overline{C} + \overline{A}BC + A\overline{B}C + \overline{A}B\overline{C}$.

2.30. For the function of Problem 2.29, draw logic circuits for minimum realizations using:
(a) NOR-gates with wired-OR capability.
(b) NAND-gates with wired-AND capability.

3
Designing Combinational Logic Circuits

3.1. GENERAL CONCEPTS

Good logic designs satisfy the functional specifications, and can be used to produce reliable circuits at minimal cost. Provided that it is not done at the expense of complicating matters in other respects (e.g., chip layout), minimization of the number of gate inputs tends to enhance reliability and to reduce costs. Methods for accomplishing this are the principal subject of this chapter.

The first step in the combinational logic design process is to generate a minimal sum-of-products (SOP) expression. The criterion for minimality (Subsection 2.5.2) is the total number of gate inputs. For some situations, such as where programmed logic arrays (PLAs) or programmable array logic (PALs) are used (Subsections 4.2.2 and 4.2.3), the logic design process terminates at this point (assuming limits on fanin are not violated). In other cases, where multistage logic is necessary to reduce the fanin or number of gates or gate inputs, further processing of the minimal-sum expressions is needed, factoring being one of the prime tools.

For functions involving no more than six or eight variables, and for many relatively simple functions of a larger number of variables, it is feasible to find truly minimal SOP expressions. Practical algorithms for accomplishing this are presented in this chapter. For more complex functions, particularly where multiple outputs are involved, we must settle for "nearly minimal" solutions, found by either the incomplete application of the algorithms, or by certain simple techniques specifically intended for this purpose. Both precise minimization processes and approximate methods have been implemented by computer programs. These are briefly discussed in Chapter 7.

There are no known algorithms for finding minimal *multi*stage logic circuits. All of the techniques for this purpose must be regarded as approximate. Minimal SOP (or minimal product-of-sum [POS]) expressions are usually good starting points for finding good multistage logic circuits. There are indeed examples of cases where a minimal SOP or POS form does *not* lead to the best multistage logic circuit, but we don't have any general way of identifying such cases in advance.

Rather than mastering the dual process for generating POS expressions directly, it is recommended that designers find minimal-sum solutions of the complements of the functions and then apply the DeMorgan theorem to the result. This yields the desired POS form.

In place of the direct use of Boolean algebra for finding minimal-sum expressions, two alternative approaches are developed here. The first, based on the Karnaugh map, is graphical in nature. It is the most efficient technique for hand computation, and facilitates the development of insight into the problem and its solution. The other approach, the Quine-McCluskey method, is rather tedious when executed by hand, but lends itself better to implementation on a computer. (There are, however, important competing methods in this realm.) In order to simplify the discussion, it is assumed, unless otherwise stated, that double-rail inputs are available.

In many real-world situations, logic functions are encountered that are incompletely specified, in the sense that, for some input states, the value of the output is optional. This can come about in one of several ways. Perhaps the most common case is where certain input states never occur. An example of this would be where the input to a circuit consists of a binary-coded decimal (BCD) character, i.e., a 4-bit word representing a decimal digit (Appendix A.1.8). Since only 10 of the 16 possible input states represent decimal digits, the other 6 states never occur. Hence it does not matter what output values the circuit is designed to produce for these inputs.

Another situation in which incompletely specified functions arise is where, for some reason, the outputs for certain input states do not matter, even though those states may indeed occur. An example is a circuit for determining the larger of two numbers. Suppose that the output Z is supposed to be 1 if the 2-bit binary number $A_1 A_2$ is greater than the number $B_1 B_2$ and 0 if it is smaller. In many cases it may not matter whether Z is set to 0 or 1 for the case where the two numbers are equal.

Regardless of why the output for a particular input state is unspecified, this is frequently referred to as a *don't care* condition. These can be exploited by designers explicitly or implicitly to assign values to the output in such a way as to optimize the resulting logic. Methods for accomplishing this are discussed in later sections.

3.2. PRIME IMPLICANTS

Some concepts important in explaining minimization procedures must be understood at the outset. First is the notion of an *irredundant form*. This is a logic expression that is "locally" minimum in the sense that no part of it can be deleted without changing the function that corresponds to it. A SOP expression, for example, is irredundant if the deletion of any product term (*p*-term) or the deletion of any literal in any *p*-term, changes the function. Conversely, if it is possible to delete any literal or any term from a SOP expression without changing the corresponding function, then that expression is said to be *redundant*.

Consider the expression on the right side of

$$f(A, B, C) = \overline{A}\,\overline{C} + \overline{A}BC + AC + A\overline{B} + \overline{B}\,\overline{C}. \tag{1}$$

Applying the consensus theorem we see that $\overline{A}\,\overline{C} + A\overline{B} + \overline{B}\,\overline{C} = \overline{A}\,\overline{C} + A\overline{B}$. Hence the last *p*-term can be deleted without changing the value of f, and so the expression is redundant. Removing this term gives us

$$f(A, B, C) = \overline{A}\,\overline{C} + \overline{A}BC + AC + A\overline{B}. \tag{2}$$

Consider now the first two *p*-terms. Factoring out $\overline{A}$ gives us $\overline{A}(\overline{C} + BC)$. Applying Theorem 2.13 (Figure 2.3) to the expression inside the parentheses eliminates the C to produce $\overline{A}(\overline{C} + B)$. Multiplying out gives us $\overline{A}\,\overline{C} + \overline{A}B$. The net result is that a single literal has been eliminated from the expression (2) without changing the value of f. It too, therefore, is redundant. The reduced expression

$$f(A, B, C) = \overline{A}\,\overline{C} + \overline{A}B + AC + A\overline{B} \tag{3}$$

cannot be pared down any further. Deleting any literal or p-term changes the value of f. Hence, expression (3) *is* irredundant.

Does this mean that (3) is a minimal SOP expression for the function it denotes? In general, while every minimal SOP expression is obviously irredundant, the converse is not always true. In fact, expression (3) illustrates this, since the expression

$$f(A, B, C) = \overline{A}\,\overline{C} + BC + A\overline{B}, \tag{4}$$

which is simpler than (3) can be shown to be equivalent to it. But it can *not* be obtained from (3) by simply deleting literals or terms. *Both* (3) and (4) are irredundant; only the latter is minimal.

An expression is said to *imply* a function if, whenever that expression has the value 1, the function also has the value 1. Such an expression is said to be an *implicant* of the function. What we have earlier called a product term of a function is a particular kind of implicant of the function. For example, the term BC is an implicant of the function f denoted by (3) or (4) (even though it does not appear in (3)). Another implicant of f is ABC. This follows since, if ABC = 1, then certainly BC = 1, and since BC is an implicant of f, then $f = 1$. Note further that ABC implies BC, or, ABC is an implicant of BC as well as an implicant of f.

Now comes the definition of a very important term. A p-term is a *prime* implicant of a function if it is an implicant of that function and if it does not imply any other p-term that implies that function. Thus, in the previous example, ABC, while a p-term implying f, and, hence, an implicant of f is *not* a prime implicant of f because it implies BC, which is also an implicant of f. What about BC? It is an implicant of f. Does it imply any other p-term that implies F? We might first ask how to identify p-terms implied by another p-term. Informally, the answer is that if, as in the case of ABC and BC, a p-term P_1 (e.g., ABC) includes all of the literals of another p-term P_2 (e.g., BC), then P_1 implies P_2. Clearly if $P_1 = 1$, then all of its literals must equal 1. Then all of the literals of P_2, a subset of these, must also equal 1, and, hence, $P_2 = 1$. Suppose P_3 is some p-term (e.g., BCX) that includes a literal X *not* in P_1. Then, for some input state for which X = 0 (e.g., ABCX = 1110) it is possible for P_1 to equal 1, while $P_3 = 0$. Hence, P_1 does *not* imply P_3.

The question as to what p-terms are implied by BC is thus easy to answer: any p-term that can be obtained by deleting one or more literals from BC. The only such p-terms are, therefore, B and C. Since by inspection of (4) it is evident that neither B nor C imply f, it follows that BC is indeed a prime implicant of f. Other prime implicants (abbreviated as pi) of f include *all* of the p-terms of expressions (3) and (4). A pi of f not present in (3) or (4) is $\overline{A}B$. Examples of p-terms that are *not* pi's of f include $A\overline{C}$ (which does not imply f) and $A\overline{B}C$ (which implies f, but which also implies $A\overline{B}$, another implicant of f). A definition of a pi in terms of Karnaugh maps is introduced in a later section.

The importance of pi's lies in the fact that an irredundant sum expression, and, hence, any minimal-sum expression, for any function must consist entirely of pi's. Why? The statement can be proved by demonstrating that if an expression for a function f includes a p-term P_1 that is not a pi, then P_1 can be replaced by some pi, P_2, that it implies. The result will be a simpler expression that still represents f. P_1 *must* imply some pi or else it would *be* a pi. The resulting expression must be simpler, since P_2 is obtained from P_1 by deleting one or more literals. (In fact, we can define a pi as a p-term of a function that ceases to be a p-term of that function if any literal is deleted.) Finally, the resulting expression still represents f, since whenever P_1 is 1, P_2 = 1, by definition. Hence, the absence of P_1 is compensated for by the presence of P_2. Since P_2 is an implicant of f, it does not introduce any false 1-values for f (i.e., f = 1 whenever P_1 = 1).

Every p-term is equal to 1 for a specific set of input states (called 1-points of the function) for which the function that it implies is equal to 1. We say that the 1-points for which the p-term is equal to 1 are *covered* by the p-term. Finding a SOP expression for a function f consists of selecting a set of p-terms that, collectively, cover all of the 1-points of f and that do not cover any points that are *not* 1-points of f. A necessary condition for such an expression to be minimal is that all of the p-terms be pi's of the function. In general, as illustrated by the fact that expression (3) and (4) of the previous section both consist entirely of pi's, not all such expressions are minimal. Thus, our design procedure must enable us to select the pi's judiciously so that we can cover the 1-points of the function at minimal cost. This is, in general, a difficult problem. In the next section, a powerful tool is introduced that greatly facilitates the process.

3.3. KARNAUGH MAPS AND THEIR USE IN FINDING MINIMAL-SUM EXPRESSIONS

When dealing with functions of real numbers, graphical representations are often useful. Many ideas, such as the concept of a derivative in calculus, become clearer and easier to visualize when viewed graphically. Graphical displays of functions and other abstract entities in the world of 1s and 0s are similarly useful. The techniques presented next are not only valuable in finding minimal logic expressions, they are also useful for clarifying a number of other topics in both combinational and sequential logic circuits.

3.3.1. Cubes in n-Dimensions

Suppose two 1-points of a function differ in the values of only one variable. For example, the points ABCD = 0110 and 0100. These correspond to the sum of minterms: $\overline{A}BC\overline{D} + \overline{A}B\overline{C}\overline{D}$. We can apply the uniting theorem to replace this pair of terms by a single term $\overline{A}B\overline{D}$. Obviously, this is a very good deal; two p-terms costing 4 gate inputs (4 gi) each (assuming double-rail inputs), are replaced by a single p-term costing 3 gi that covers the same set of 1-points. The repeated use of the uniting theorem is the general basis for generating prime implicants from minterms. Hence, it is important to be able to recognize sets of 1-points that differ in the values of only one or a few variables. A truth table is a simple linear list of input states. The ordering of the states in a truth table is of no consequence. We now consider a way of organizing the placement of input states so as to facilitate the minimization process.

The key is to associate logic variables with spatial dimensions. Figure 3.1a is a simple line segment connecting two nodes. At the first node, X_1 = 0, and at the second node, X_1 = 1. There is only one dimension; motion along this dimension is associated with changes in X_1. Figure 3.1b depicts a two-dimensional figure—a square. Motion in the horizontal dimension is associated with changes in X_1, and motion in the vertical

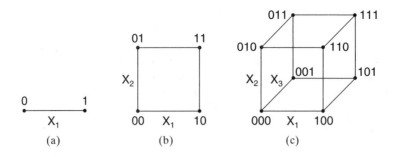

Fig. 3.1. Associating Boolean variables with dimensions of cubes.

dimension is associated with changes in X_2. Thus the lower left corner is associated with
the state $X_1X_2 = 00$, the lower right corner is associated with state 10, and the upper
right corner with state 11. A cube, or three-dimensional figure, is shown in Figure 3.1c,
with the third dimension, pictorially indicated by an oblique line, associated with X_3.
The states of the eight vertices or nodes are indicated on the figure.

Before proceeding to higher dimensions, let us examine some important properties of
the three-dimensional cube (3-cube). (It is useful to think of such figures as the line
segment, the square, and the cube, as all being cubes, in a generic sense, of various
dimensions. Dimensions exceeding three are considered later in this section.)

The 3-cube has as components cubes of smaller dimensionality. These can be identified in terms of the variables associated with the dimensions. Thus, all nodes for which
$X_1 = 1$ constitute a 2-cube (i.e., a square) that is one face of the 3-cube of Figure 3.1c.
The nodes for which $X_2 = 0$ form the 2-cube that is another face of the cube. The pair
of nodes for which $X_1 = 1$ and $X_3 = 0$ define the 1-cube (line segment) that is a vertical edge part of the first 2-cube described earlier. We can also think of nodes as zero-
dimensional cubes. Thus, if we specialize the values of any subset of the variables, a
component cube, or *sub*cube, is defined. Specializing one variable value defines a 2-cube,
specializing two variables defines a 1-cube, and specializing all three variables defines a
0-cube. (It is sometimes more fruitful to consider the number of variables *not* specialized. If k variables are *not* specialized, then the dimensionality of the subcube defined is
k.) We could indicate the values of the specialized variables associated with a subcube by
writing a product term equal to 1 for those values. For example, corresponding to the
line segment for which $X_1 = 1$ and $X_3 = 0$, we could associate the *p*-term $X_1\overline{X}_3$. This
term is equal to 1 precisely for those nodes comprising the line segment. The *p*-term $\overline{X}_1$
defines the 2-cube for which $X_1 = 0$. Thus each *p*-term covers the 1-points in the
subcube it describes, and does not cover any other 1-points. (It is sometimes useful to
describe a subcube of an *n*-cube by a vector of 0s, 1s, and −'s, where the 0s and 1s
indicate the values of the variables in the corresponding positions, and the −'s indicate
missing variables. Thus, corresponding to the *p*-term $X_1\overline{X}_3$, we would write 1–0.)

These properties can be exploited by distributing over the cube the information about
a function contained in a truth table. Suppose we consider each node to represent an
input state of a function. We could then mark those nodes that correspond to 1-points by
circling them. For example, the function $f(A, B, C) = \Sigma m(1, 4, 5, 6, 7)$ is plotted in this
manner on the cube shown in Figure 3.2.

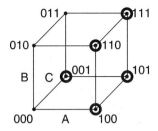

Fig. 3.2. Logic function $f(A, B, C) = \Sigma_m(1, 4, 5, 6, 7)$ plotted on a 3-cube.

Observe that the 2-cube defined by $A = 1$ (i.e., the p-term A) consists entirely of 1-points of f. In addition, all of the nodes of the 1-cube -01, defined by $B = 0$, $C = 1$ (i.e., the p-term $\overline{B}C$) are 1-points of f. There are no 1-points of f outside of both of these cubes. Then every 1-point of f is covered by at least one of the cubes corresponding to a p-term of the expression $A + \overline{B}C$ (and no points other than 1-points of f are covered). It follows that the expression is a valid description of f. If either of the p-terms is deleted, then some 1-points of f will no longer be covered. Suppose that some literal of a p-term in the expression is deleted. Then, since fewer variables would be specialized, the corresponding cube would be of a higher dimensionality. Inspection of Figure 3.2 indicates that none of the 2-cubes containing $\overline{B}C$ (i.e., $\overline{B}$ and C) is filled with 1-points of f. Hence the expression is irredundant. (Techniques developed further on in this section make it clear that the expression is also minimal.) Before discussing methods for exploiting this graphical approach to logic design, the question of depicting cubes of higher dimensionality must be considered. This problem inspires consideration of alternative ways of representing cubes.

Representing cubes of dimensionality not exceeding 2 on a two-dimensional surface poses no problem at all, since the necessary dimensions can be directly represented. When it comes to adding a third dimension, we are accustomed to representing it along an oblique line, as in Figure 3.1c and Figure 3.2. We do not usually find it necessary to depict four-dimensional objects, and so we do not have a convention for the fourth dimension. However, at least for a simple object, such as a cube, it is feasible to represent the fourth dimension by distance along oblique lines at an angle different from that used for the third dimension. Thus, we can construct a four-dimensional cube as shown in Figure 3.3. Such a diagram is sometimes referred to as a *four-dimensional hypercube*. (We often shorten this to *four-dimensional cube*, or to 4-*cube*.)

Figure 3.3 is very interesting, and is worth studying to gain some appreciation of what a hypercube is like. (One might, for example, try to identify the various subcubes of lower dimensionality; all of the 3-cubes and 2-cubes are undistorted and plainly identifiable). It could be used, as was Figure 3.2, to map a logic function (this time a four-variable function) and identify various p-terms. However, Figure 3.3 is a rather complex diagram, not easy to sketch rapidly, and its generalizations to higher dimensions are still more complex and difficult to deal with. There are simpler, less cluttered, ways to represent cubes of three, four, or more dimensions. We now introduce and discuss the most common of these, the Karnaugh map (abbreviated K-map).

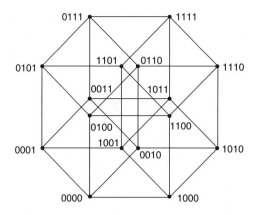

Fig. 3.3. A four-dimensional cube (or hypercube).

3.3.2. Structure of Karnaugh Maps

The states of the variables are represented on a K-map by cells (boxes) rather than by nodes as in the previous diagrams. Otherwise, the K-maps for 1- and 2-cubes, as shown in Figure 3.4a and b, respectively, correspond closely to parts (a) and (b) of Figure 3.1, discussed earlier. In the 1-cube, the cell in the right half, the region labeled A, represents the state where A = 1. The left half cell, outside this region, is the state where A = 0. In Figure 3.4b, representing the 2-cube, the right and left halves again represent the regions where A is 1 and 0, respectively. The upper half is the region where B = 0, and the lower half is the region where B = 1. Then the lower right cell is where both variables are 1, etc.

The 3-cube shown in Figure 3.4c can be thought of as being generated from the two-variable K-map (we shall call it a 2-map) as follows. Using the right edge of the 2-map as an axis of symmetry, reproduce the 2-map to the right of that edge, so that the pair of rows adjacent to the axis is copied over immediately to the right of the axis, and the pair of rows on the left edge of the original 2-map is reproduced as the second column to the right of the axis. As shown in Figure 3.4c, the right half of the 8 cell map so produced is labeled as the A (i.e., A = 1) region, the bottom half is the C-region, and the middle two columns constitute the B-region. Note now that all pairs of adjacent cells differ in exactly one variable, and that the leftmost and rightmost columns differ only in

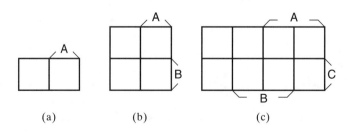

Fig. 3.4. K-maps for (a) one, (b) two, and (c) three variables.

the value of A, so that the cells in the upper left and upper right corners differ only in A, as do the lower leftmost and lower rightmost cells. (We might imagine that the map is wrapped around a cylinder so that the leftmost and rightmost columns are adjacent to one another.)

The function mapped on the 3-cube of Figure 3.2 is mapped again on the 3-map of Figure 3.5 with 1s in the cells that correspond to 1-points of the functions, and the other cells left empty. The filled subcubes are enclosed in loops (the loop enclosing the subcube corresponding to $\overline{B}C$ is broken, since the cells comprising it are not geometrically adjacent). More is said about recognizing and utilizing the filled subcubes after the 4-map is introduced and another way of thinking about K-maps is pointed out.

The four-variable K-map is constructed from the 3-map in the same general way that the 3-map was generated from the 2-map. In this case, the axis of symmetry used is the bottom edge of the 3-map. The result, shown in Figure 3.6, is a 16-cell map. As in the case of the 3-map, the right and left edges are considered to be adjacent (differing only in the value of A). Similarly, the top and bottom edges are *also* considered as connected (they differ only in C).

Before discussing 4-maps in more detail, it will be helpful to consider K-maps as *Venn diagrams*, a tool developed in the nineteenth century for analyzing problems in set theory (a topic closely related to Boolean algebra). An example of a classic Venn diagram is shown in Figure 3.7a. In Boolean algebra terms, each of the three ovals corresponds to a variable, which has the value 1 within that oval and 0 outside it. Thus, within the regions in which two or more ovals overlap, several variables have 1-values. In the figure, the regions are labeled with octal numbers that indicate the states of the variables A, B, and C. For example, in region 4, within only the A-oval, only A = 1. In region 2, only B = 1, and in region 6, contained within *both* of these ovals, A and B are both 1.

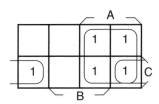

Fig. 3.5. $f(A, B, C) = \Sigma m(1, 4, 5, 6, 7)$ mapped on a 3-map.

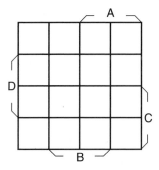

Fig. 3.6. Four-variable K-map.

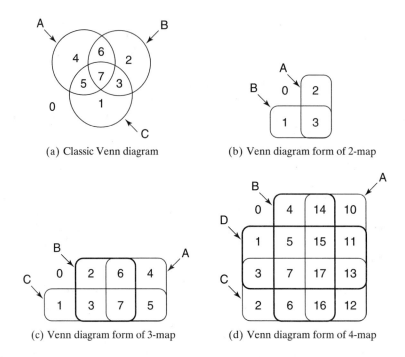

(a) Classic Venn diagram   (b) Venn diagram form of 2-map

(c) Venn diagram form of 3-map   (d) Venn diagram form of 4-map

Fig. 3.7. Venn diagrams.

Now consider Figure 3.7*b*. It is a simple Venn diagram for variables A and B. It has only four regions, including the region outside of both ovals (labeled 0). Observe how it resembles the two-variable K-map of Figure 3.4*b*, which has precisely the same four regions, interpreted in the same way. In the same manner observe the correspondences between the three- and four-variable Venn diagrams of Figure 3.7*c* and *d* and the three- and four-variable K-maps of Figure 3.4*c* and Figure 3.6.

Look at the area covered by regions 4 and 5 (which we denote by {4, 5}) of the four-variable Venn diagram of Figure 3.7*d*. Since it constitutes the entire area within the B-oval and outside the A- and C-ovals, it can be described by the *p*-term $\overline{A}B\overline{C}$, a term that is equal to 1 in this area and equal to 0 everywhere outside it. The area {2, 3, 12, 13} can similarly be described by the *p*-term $\overline{B}C$. On the other hand, given a *p*-term such as $A\overline{C}$, we can identify the complete set of regions in which it has the value 1. This is the area within the A-oval and outside the C-oval, or {10, 11, 14, 15}.

While for every *p*-term there is a corresponding set of regions, the converse is not true. For example (still referring to Figure 3.7*d*), there is no *p*-term corresponding to the area {1, 7}. Both are contained within the area within which A = 0 and D = 1, but this area also includes 3 and 5. No smaller area described by a *p*-term includes both 1 and 7 because both of the other two variables, B and C, have different values in regions 1 and 7. Observe that when moving from any region to any immediately neighboring region, we must cross the boundary of exactly one oval, so that the value of the corresponding variable changes, and no other variable changes value. (Immediate neighbors include regions in the same row at the left and right extremes, or regions in the same column at top and bottom extremes.)

As an exercise in finding our way around K-maps, let us consider the problem of constructing sequences of binary words such that consecutive members differ in exactly one variable (see Appendix A.1.11 on Gray codes). To find such a sequence using a K-map, one need only trace a continuous path of the desired length (if possible) between the desired endpoints.

Two examples are shown in Figure 3.8. In Figure 3.8*a*, a path consisting of 11 states labeled a, b, . . . , k is shown between 0000 and 1111. (The states are 0000, 0100, 0101, 0001, 0011, 0111, 0110, 0010, 1010, 1011, 1111.) No particular algorithm was used to find this path; it was done simply by moving from cell to cell and adjusting the path length by trial and error. But since there are many possible solutions to these problems, it is usually very easy to find one. In Figure 3.8*b*, a closed path is shown consisting of 10 states labeled a, b, . . . , j. (The states are 0000, 0100, 1100, 1000, 1001, 1101, 0101, 0001, 0011, 0010.)

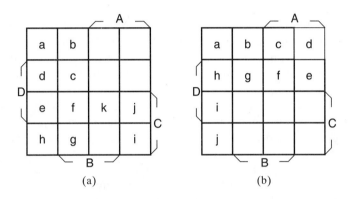

Fig. 3.8. Gray-code-type paths on 4-cubes.

3.3.3. *Identifying Subcubes on K-Maps*

Now we are in good position to consider the problem of identifying subcubes on the four-variable K-map. These correspond to the areas for which *p*-terms are equal to 1. The various types of patterns that subcubes fall into are illustrated in Figure 3.9. In Figure 3.9*a*, we see that a complete column, typified by {4, 5, 6, 7}, described by $\overline{A}B$ is one type. Complete rows constitute another. Squares, such as {4, 5, 14, 15}, the area where $B\overline{C} = 1$, is still another. Discontinuous 2-cubes are typified by {4, 14, 6, 16}. The least obvious form of 2-cube is that made up of the four corners of the map, {0, 2, 10, 12}, which can be defined as the area outside of the B-region and outside of the D-region, i.e., the area where $\overline{B}\,\overline{D} = 1$. The two types of patterns corresponding to 1-cubes are also illustrated in Figure 3.9*a*. They are the geometrically adjacent pair of cells, such as {13, 17}, described by ACD, and pairs at opposite ends of the map, such as {1, 11}, corresponding to $\overline{B}\,\overline{C}D$. In Figure 3.9*b* we see 3-cubes. Each occupies half of the map. The examples shown correspond to D (the middle two rows), $\overline{A}$ (the two left columns), and $\overline{D}$ (the top and bottom rows).

In the next section, these concepts are applied to the problem of finding simple SOP expressions. But first let us consider how to extend the K-map to higher dimensions. This

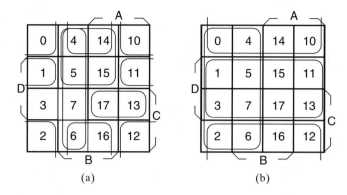

(a)   (b)

Fig. 3.9. Four-variable K-maps with some subcubes identified.

is accomplished by following the same principle used in going from the 2-map to the 3-map and then from the 3-map to the 4-map. In each case we double the size of the map by reproducing it symmetrically with respect to one of the edges. To get the 5-map from the 4-map, we carry out this process with respect to the rightmost edge. This is shown in Figure 3.10

The heavy vertical line down the middle marks the axis of symmetry. The pair of cells labeled p are immediate neighbors in the sense that they differ in just one variable (A). Other pairs of neighbors (also differing in A) are labeled q, r, s, and t. Since the t's and p's are neighbors, the set of t's and p's constitutes a 2-cube corresponding to $\overline{C}\overline{D}\overline{E}$. Another 2-cube, perhaps the least obvious, is made up of the four outside corners, i.e., the cells labeled s and t. They constitute the area for which the p-term $\overline{B}\,\overline{C}\,\overline{E} = 1$.

By using the bottom edge of the 5-map as a new axis of symmetry, a six-variable K-map can be formed. Such a diagram is shown as Figure 3.11. The adjacencies between cells in the upper and lower halves of this map are of the same type as those between cells in the left and right halves of the 5-map (or in the 6-map itself).

The less obvious adjacencies are again indicated by letters in the cells. (Of course, within any subcube, the same rules apply for adjacencies as if that subcube were the entire map.) The four cells labeled p are located symmetrically with respect to the center of the map. The left pair of p's are neighbors in that they differ only in the value of D.

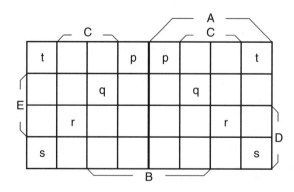

Fig. 3.10. Five-variable K-map.

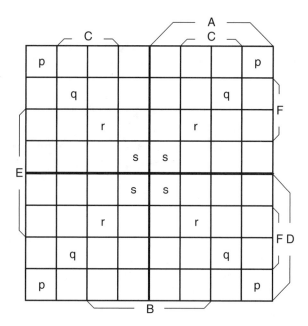

Fig. 3.11. Six-variable K-map.

The same is true of the right pair of p's. From the discussion of the 5-map, it follows that the top pair of p's are also neighbors, differing only in the value of A, and the same is true of the bottom pair of p's. The four p's form a 2-cube, corresponding to $\overline{B}\,\overline{C}\,E\,\overline{F}$. In a similar manner, the four q's form a 2-cube corresponding to $\overline{B}CEF$. The r and s cells each form 2-cubes, corresponding to BCEF and $B\overline{C}E\overline{F}$, respectively.

Sometimes it is useful to construct K-maps of a more "rectangular" form. For example, instead of a 4-map being represented as a 4×4 matrix, it may be convenient to represent it as a 2×8 matrix. (See Problem 3.21.)

3.3.4. Using K-Maps to Find Minimal SOP Expressions

In terms of K-maps, pi's can be thought of as maximal filled subcubes. That is, if a subcube S is filled with 1s and if it is not wholly contained within a larger filled subcube, then S corresponds to a pi. This statement is justified as follows: If S does *not* correspond to a pi, then it must be possible to delete some literal from the corresponding p-term so that the resulting p-term is a p-term of the function. This reduced p-term corresponds to a subcube that contains S, since it must cover all of the 1-points that S covers. Thus, in K-map terms, finding a minimal-sum expression corresponds to finding a set of subcubes that covers all of the 1-points of the map in the most economical manner.

The strategy for accomplishing this will be developed by means of a series of examples. (As a matter of convenience, subcubes are often referred to by the expression for the corresponding p-term; e.g., in Figure 3.11, we might have referred to the 2-cube filled with p's as $\overline{B}\,\overline{C}\,E\,\overline{F}$.)

Example 3.1

Consider the function mapped in Figure 3.12a. Our goal is to find a minimal set of pi's covering all of the 1-points on the map. Each 1-point must be covered by at least one pi. Suppose we start with the 1-point at 000 (upper left corner). Two pi's cover this point: $\overline{A}\,\overline{C}$ and $\overline{A}\,\overline{B}$. Both cover points other than 000, but not the *same* points. It is not obvious which to choose. Hence, we postpone the decision and examine another 1-point on the map, 001. Only one pi, $\overline{A}\,\overline{B}$, covers this point. It follows then that *any* minimal solution *must* include $\overline{A}\,\overline{B}$. A pi that *uniquely* covers a 1-point of a function is said to be an *essential* pi for that function. The subcube corresponding to this term is shown "looped" in Figure 3.12b and it is labeled as the first selection by the placement of a "1" in the cell corresponding to the 1-point that caused it to be chosen. Since this term also covers 000, the previously unresolved issue as to how to cover that point is now settled. Only two 1-points remain uncovered (010 and 110). This pair is uniquely covered by $B\overline{C}$, and so the solution (see Figure 3.12c), $Z = \overline{A}\,\overline{B} + B\overline{C}$ is clearly minimal (in this case it is a unique minimum solution). This was an easy case since both pi's were essential. Note that the two pi's might have been chosen in reverse order.

Example 3.2

Suppose now that our object is to find a minimal POS expression for the function mapped in Figure 3.13a. We shall employ the time-honored approach of the intelligent, labor-minimizing mathematician (or engineer): reduce the problem to one that we already know how to solve (or rather that we will soon know how to solve). In this case, our first step is to complement the function, by simply deleting all the 1s from the map and putting 1s in all the blank spaces (that of course represent 0-points), thereby obtaining Figure 3.13b. After finding a minimal SOP expression for this function (which happens to be no more difficult to handle than was the function of the previous example), we shall use the DeMorgan theorem to complement it, thereby obtaining the desired minimal POS expression.

The SOP solution, illustrated in Figure 3.13c, is $\overline{Z} = \overline{A}B\overline{C} + A\overline{C}D + ABC + \overline{A}CD$. Again, the four pi's, all essential, might have been chosen in any order. Note that the 2-cube corresponding to BD is not used at all. If it had been hastily chosen simply because it seemed to be a "bargain" (covering four 1-points at a cost of only two literals in the *p*-term) the same four essential pi's appearing in the minimal solution would also have been needed.

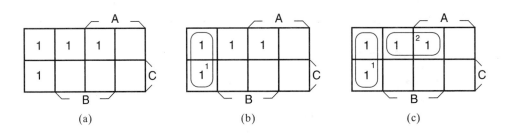

Fig. 3.12. Selection of pi's for a function.

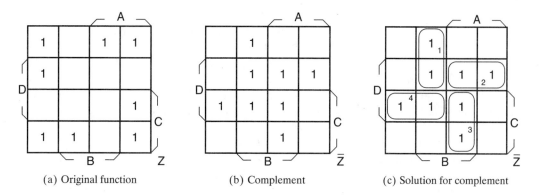

(a) Original function   (b) Complement   (c) Solution for complement

Fig. 3.13. Finding a minimal POS expression.

The POS expression is $Z = (A + \overline{B} + C)(\overline{A} + C + \overline{D})(\overline{A} + \overline{B} + \overline{C})(A + \overline{C} + \overline{D})$. (A minor variation of this approach is to find the SOP form with the blank cells treated as 1s—and the 1s as blanks—and then complement the result as above.)

Example 3.3

The function of Figure 3.14*a* is a little more difficult. For most 1-points there are two covering pi's. Even where one of these is a 2-cube and the other is a 1-cube (for instance, the point 0100), it would be risky to choose the 2-cube, as was illustrated in the previous example. Fortunately there are two points that are uniquely covered. These are 0111 and 1100, which we use to choose the essential pi's BD and $B\overline{C}$, respectively. Having made these choices (see Figure 3.14*b*), we reexamine the as yet uncovered 1-points.

It is now clear how to cover the point 0000. The competing pi's are $\overline{A}\,\overline{C}\,\overline{D}$ and $\overline{A}\,\overline{B}\,\overline{D}$. The first of these covers 0100 as well as 0000, but 0100 has already been covered by a previously chosen pi. Since, as is evident in the map, the subcube corresponding to $\overline{A}\,\overline{B}\,\overline{D}$ covers the currently uncovered point 0010, and since it costs no more than $\overline{A}\,\overline{C}\,\overline{D}$, we cannot possibly miss finding a minimal solution by choosing $\overline{A}\,\overline{B}\,\overline{D}$. (If the other pi had

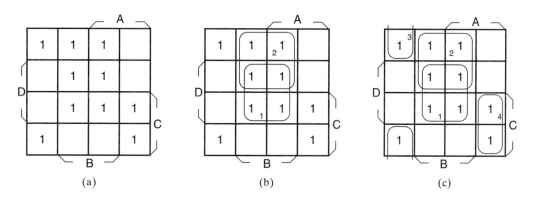

(a)   (b)   (c)

Fig. 3.14. Introducing good pi's.

been chosen, and if it appeared in some minimal solution, we could replace it in that solution by $\overline{A}\overline{B}\overline{D}$ and the result would still be a solution with the same cost in gi's [and number of gates].) Such a pi, which, though not essential, is guaranteed to be part of *some* minimal solution, is referred to as a *good pi*. A good pi is one that *dominates* the set of pi's that could be used to cover some uncovered 1-point in that it costs no more than any of them, and it covers the union of all uncovered 1-points covered by members of the set. Note that an essential pi is a special case of a good pi.

The 1-point 1011 also defines a good pi, namely $A\overline{B}C$, which dominates the other pi covering this point. The resulting solution is shown in Figure 3.14c. It happens to be a unique minimum, but this is often not the case when some of the chosen pi's are good, but not essential. Although in this discussion, the 1-points were identified by the corresponding states and the subcubes identified by specifying the corresponding *p*-term, it is not necessary to identify either type of entity during the process of finding a minimal cover. Only after a satisfactory set of cubes has been marked on the K-map is it necessary to determine the *p*-terms corresponding to the solution. In this case, the solution is $Z = BD + B\overline{C} + \overline{A}\overline{B}\overline{D} + A\overline{B}C$.

Example 3.4

The function displayed in Figure 3.15a is of the same type as in the previous example. All but one of the 1-points are covered by at least two pi's. The point 0101 is covered by three pi's. Only 1101 is uniquely covered, making $B\overline{C}D$ essential. After this choice is made, we have the situation shown in Figure 3.15b.

Now, the "log jam" is broken. Both 0100 and 0111 now define good pi's $\overline{A}\overline{C}\overline{D}$ and $\overline{A}CD$, respectively. These are the choices labeled 2 and 3 in Figure 3.15c. The remaining uncovered 1-point 0010 defines, in a trivial sense, two good pi's. The choice shown in Figure 3.15c, labeled 4, is $\overline{A}\overline{B}C$. This gives us the minimal-sum solution: $Z = B\overline{C}D + \overline{A}\overline{C}\overline{D} + \overline{A}CD + \overline{A}\overline{B}C$. Except for the first step, choosing the essential pi, the order of choices could have been different. For example, the order of steps 2 and 3 could have been reversed. More important, after step 2, we might have noted that $\overline{A}\overline{B}C$ became good (due to 0010), and made that choice next. In that case, 0111 could have been covered by $\overline{A}BD$ instead of by $\overline{A}CD$. The important concept to be noted here is

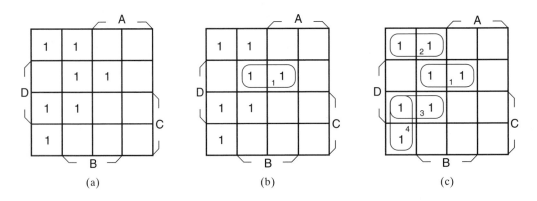

Fig. 3.15. Another function with good pi's.

that "goodness" of a pi is not an absolute property, but is dependent on the set of choices previously made. In fact, *every one* of the pi's in this function could have been found to be good in some valid sequence of choices. Hence, each of them appears in some minimal-sum solution for this function.

Before proceeding to the next level of complexity, let us consider how to deal with unspecified, or "don't care" entries (these were introduced at the end of Section 3.1). In a sense, we have already been handling these in the course of generating minimal-sum expressions from K-maps. Once any subcube has been selected, the 1-points that it covers become, in effect, don't care entries. They need not be covered again, but are available for use to complete subcubes needed to cover other 1-points. If the function has don't care entries to start with, then we simply ignore them except to recognize that they give us opportunities to complete subcubes containing uncovered 1-points. This concept is illustrated in the next example, which is of the same order of difficulty as the last two examples.

Example 3.5

The function shown in Figure 3.16*a* has five don't-care entries, represented by dashes "–". (In some of the literature, x's are used for this purpose.) Readers should verify that there are no essential pi's. The 1-point 0110 is covered by the subcube corresponding to

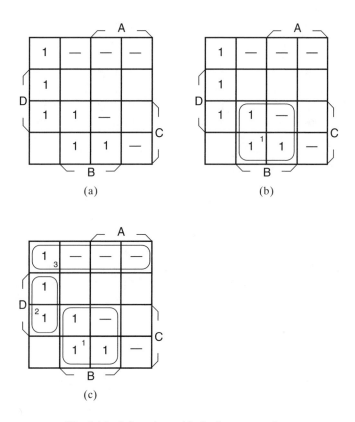

Fig. 3.16. A function with don't-care entries.

the pi BC, and by the subcube corresponding to $B\overline{D}$. The latter subcube includes two of the don't-care entries. Since they could, at our option, be made into 1-points, we can regard $B\overline{D}$ as a pi (clearly it cannot be enlarged without encompassing 0-points). But BC dominates $B\overline{D}$ in that it costs no more and covers all of the previously uncovered 1-points covered by it. Hence, BC is a good pi, and we choose it, as shown in Figure 3.16*b*.

Consider next the 1-point 0000. It is covered by $\overline{A}\,\overline{B}\,\overline{C}$, which covers all of the 1-points covered by $\overline{C}\overline{D}$, the other cube covering it. But since $\overline{C}\overline{D}$ is less costly than $\overline{A}\,\overline{B}\,\overline{C}$ (two literals as opposed to three), $\overline{A}\,\overline{B}\,\overline{C}$ does not dominate it and is therefore not a good pi (which is not to say that it is a *bad* pi; in fact, if our measure of cost were only the number of gates, then we could safely choose $\overline{A}\,\overline{B}\,\overline{C}$ at this point, or even before the previous choice had been made). Nor, of course, is $\overline{C}\overline{D}$ a good pi at this point, because of $\overline{A}\,\overline{B}\,\overline{C}$. But the 1-point 0011 now makes $\overline{A}\,\overline{B}D$ a good pi, so it is our second choice. The only uncovered 1-point remaining is 0000; it makes $\overline{C}\overline{D}$ good, and hence the third chosen pi. The solution then is $Z = BC + \overline{A}\,\overline{B}D + \overline{C}\overline{D}$. We have implicitly chosen 1 as the value assigned to the three don't-cares in the top row, and 0 as the value assigned to the don't-care entry in the bottom row.

Example 3.6

What do we do if there are no essential or good pi's, either at the outset of the problem or at some intermediate point while some 1-points remain to be covered? This is the situation shown in Figure 3.17*a*. The basic strategy is systematic trial and error, or *branching*. We choose a 1-point covered by a minimal number of pi's, arbitrarily select one of them to cover it, and proceed to identify, we hope, some pi's that become good as a result of this choice. If we are fortunate, we are able to find a sequence of good pi's that cover all of the remaining 1-points. If not, then we must branch again, i.e., make another such arbitrary choice, etc., until we arrive at a complete cover. We then return to the last branch point, make a different choice, and continue forward to find a second solution. This process is repeated, each time with a return to the last branch point for which there remain unexplored choices until none remain. We then choose the best of the solutions found, and can be sure that no less costly solution exists. Let us see how this works for the current problem.

Depending on where we choose to branch, the amount of work necessary may vary, and even the particular solutions found may be different. We can only be sure that *some* minimal solution will result. In this case, let us choose the 1-point 0010, which is covered by pi's $\overline{A}\overline{D}$ and $\overline{B}\overline{D}$. It makes no difference which pi we choose first. Let us start with $\overline{A}\overline{D}$, marking the 0010 cell with a star to indicate a branch on it, and with a 1 to indicate that the first covering term is being chosen to cover this point. This leaves us at the point shown in Figure 3.17*b*. From here we proceed as in the last few examples, choosing, in order (see Figure 3.17*c*), the pi's 2, 3, and 4, all of which are good when chosen. Thus we have one solution consisting of the sum of three pi's corresponding to 2-cubes, and one corresponding to a 1-cube. (There is no need to determine the actual *p*-terms at this point, since this may not be the best solution.)

Returning to the last branch point (there is only one in this problem), we start again from Figure 3.17*b*, at the starred 1-point labeled 1, and choose the other subcube that covers it (the one corresponding to pi $\overline{B}\overline{D}$). We then go forward again, observing that the 1-points labeled 2 and 3 (see Figure 3.17*d*) both define good pi's, $\overline{A}B$ and ACD,

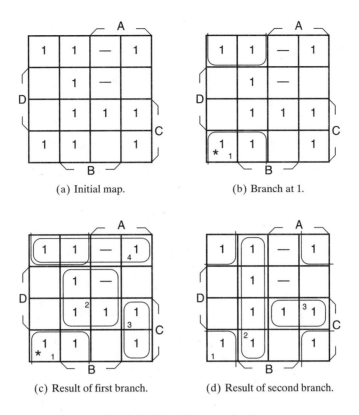

Fig. 3.17. Branching required.

respectively. These can thus be chosen in either order. The result is a covering requiring only three pi's, clearly better than the result of the first path. Our solution is therefore $Z = \overline{B}\overline{D} + \overline{A}B + ACD$.

Example 3.7

The principal method for using K-maps to find minimal-sum expressions has at this point been fully described. In this example, a refinement is presented that is not always easy to exploit, but which can sometimes obviate the need for branching.

For the five-variable function plotted on the map of Figure 3.18a, there are no essential or good pi's at the start, so that branching appears necessary. (This can only be determined after a careful examination of each 1-point.) The basic idea of the technique to be introduced here to find a 1-point that is guaranteed to be covered in any solution found, even if it is converted to a don't care. If this could be done, it *might* make one or more pi's good and thereby enable us to begin selecting pi's without branching.

Consider the 1-point labeled q in Figure 3.18a. It is included in subcubes $\overline{B}\overline{D}\overline{E}$ and $\overline{A}\overline{B}\overline{E}$. Both of these subcubes also cover the 1-point labeled p (which is also covered by the subcube $\overline{A}\overline{B}\overline{C}\overline{D}$). It follows then that we cannot cover q without also covering p. We can exploit this fact by changing the p-entry to a don't care, an action that builds

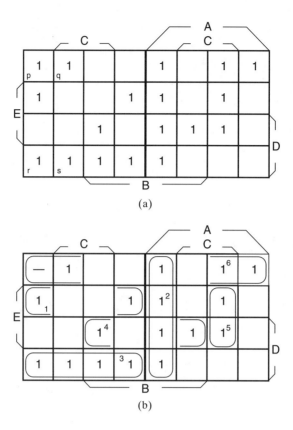

Fig. 3.18. Finding 1-point dominance.

into the K-map description the idea that we need not explicitly concern ourselves with covering p; p will be covered automatically when we cover q. The effect of this on our selection process is to reduce the "dominating power" of any pi that covers p but not q. In this example, the result is to identify the pi that should be chosen to cover the 1-point 00001. This point is covered by $\overline{A}\,\overline{B}\,\overline{C}\,\overline{D}$ and by $\overline{A}\,\overline{C}\,\overline{D}E$. As is evident from Figure 3.18b, after the p-entry has been made a don't care, $\overline{A}\,\overline{B}\,\overline{C}\,\overline{D}$ is dominated by $\overline{A}\,\overline{C}\,\overline{D}E$, which was *not* the case when p was a 1-point. Hence, $\overline{A}\,\overline{C}\,\overline{D}E$ becomes a good pi and is selected as shown in the figure. The result of this choice makes the pi marked as 2 in Figure 3.18b good, and the chain of pi's being made good by the preceding choices continues in this case to yield the solution shown on the map, $Z = \overline{A}\,\overline{C}\,\overline{D}E + AB\overline{C} + \overline{A}D\overline{E} + BCDE + A\overline{B}CE + \overline{B}\,\overline{D}E$.

In general, if p and q are 1-points of a function, we say that p *dominates* q if every pi covering q also covers p. The *dominating* 1-point may be changed to a don't care entry. Note that in the current example, 1-points r and s (see Figure 3.18a) are covered by precisely the *same* set of pi's ($\overline{A}\,\overline{B}\,\overline{E}$ and $\overline{A}D\overline{E}$). Hence, they dominate one another. We could therefore have changed either one, *but not both*, to a don't care. This would not, however, have had any beneficial effect because no pi would thereby be rendered more likely to be dominated by another pi (since no pi covers one but not the other).

Example 3.8

Our last example is a six-variable function requiring care and patience for its solution. A useful shortcut for determining that a 1-point does *not* define any pi as good may be appreciated at this point. A sufficient condition for a 1-point to be in this category is that it be covered by two different pi's, each of which covers a 1-point not covered by the other. Such 1-points often, but not always, have the property that they are adjacent to two other 1-points, but that the third member of the 2-cube defined by these three points is a 0. That is, a *sufficient*, but not *necessary*, condition for a 1-point to *not* define a good pi, is that it be the corner of a square with other 1-points at the adjacent vertices and a 0 at the opposite vertex. Examples of such points in Figure 3.19 are 101111, 101101, 101100, 111100, and 001100.

A point-by-point examination of Figure 3.19 reveals that there are no good pi's, and hence that branching is necessary. The starred point in Figure 3.19 was chosen as the branch point, simply because it is covered by two large pi's. The initial choice made (as indicated by the discontinuous rounded square in the figure) is $\overline{B}CF$. The sequence of further choices, each of a pi that was good at the point it was chosen, is shown in Figure 3.19.

Returning to the branch point, the other choice for covering the same 1-point, $\overline{C}D\overline{F}$, is made in Figure 3.20. Following this, the 1-point 111000 determines $BC\overline{F}$ to be good, and is marked on the map (with four "corner" markers and a pair of partly broken line segments) as choice number 2. Next, the 1-point labeled 3 determines $\overline{B}CD\overline{E}$ to be good; it is the third choice. At this stage of the process, it can be seen that the 1-point 110010 (labeled 4) can be covered by $A\overline{C}E\overline{F}$, which also covers the as yet uncovered 1-point 100010. The competing pi, $B\overline{E}\overline{F}$ does not cover 100010 or any other uncovered 1-point,

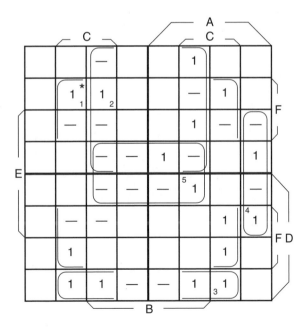

Fig. 3.19. First branch of a six-variable problem with branching.

but since it is cheaper than $A\overline{C}E\overline{F}$, the latter does not meet the requirements for being good. If we were to follow the procedure in a strict manner, we would find that there are no good pi's at this stage, so that another branching is called for. But, since only four 1-points remain uncovered, it is not difficult to look ahead, trying out the various possibilities in our heads. This leads quickly to the conclusion that we can do no better than to choose $A\overline{C}E\overline{F}$, as shown in Figure 3.20, and then the choice of the 2-cube labeled 5 concludes the process.

Each of the branches produces a solution with five *p*-terms, but the Figure 3.19 solution is cheaper by 1 gi, and so we choose it. Hence our solution is

$$Z = \overline{B}CF + BC\overline{D} + CD\overline{E}\,\overline{F} + A\overline{B}\,\overline{C}E + BE\overline{F}.$$

Summary of the basic procedure for finding minimal SOP expressions:

1. Examine the 1-points of the function, one by one, to identify good pi's and add each one found to the list of chosen pi's as it is identified. (Mark these on the K-map with ovals, etc., as they are found, identifying each with a number placed in the cell that led to its selection. The number should correspond to the order of selection. Do *not* determine the corresponding *p*-terms at this point.)
2. If, after one pass through all of the 1-points, there remain uncovered 1-points then, if at least one subcube has been chosen, repeat step 1. If, at any point, there remain uncovered 1-points and a pass through all of them has not generated any additional good pi's, then branching is necessary; go to step 3. (The advanced technique demonstrated in Example 3.7 for converting 1-points to don't cares can be employed here before resorting to branching.) If no

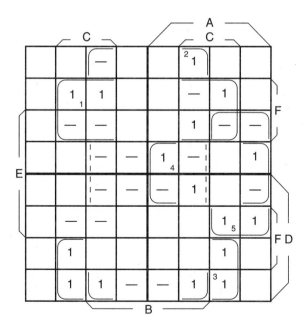

Fig. 3.20. Second branch of a six-variable problem with branching.

uncovered 1-points remain, then, if no branching has taken place, the list of chosen pi's is the solution. Else, go to step 4.

3. Choose a 1-point covered by as few pi's as possible, mark it with a star, choose any of the covering pi's, and go to step 1.

4. If there are no starred cells (indicating a branch point with at least one unexplored branch remaining), then compare all of the solutions found from the various branches and choose the best one as the solution. If starred cells remain, then return to the most recent one, i.e., the one with the highest numbered label, repeat all of the choices made prior to that point, choose an as yet unchosen pi covering that point, and go back to step 1.

The technique treated in the next subsection, useful for some important special cases (see Subsection 6.5.2), makes it possible to handle problems involving more than n variables on an n-variable K-map. It does not contain material essential for understanding subsequent topics and so can be skipped altogether or left for later study.

3.3.5. Variable Entered K-Maps

It sometimes happens that the description of a function to be realized by a logic circuit involves one basic set of input variables and one or more additional variables that play a special role in relation to the basic set. It then becomes possible to use a K-map with coordinates corresponding to the basic variables, with the secondary variables appearing as *entries* in the map. Provided that we know how to deal with these *map-entered variables*, this will be easier to handle than would be a map of higher dimensionality in which all of the variables are treated in the usual way. As in the previous subsection, the ideas involved are introduced below incrementally via a series of examples.

Example 3.9

Consider the following function description: "The output Z is a function of the five variables, A, B, C, D, and E. The first four variables represent the bits of a BCD number (with A the most significant bit). Z is to be 1 if the BCD number exceeds 6 or if it exceeds 3 and E = 1". We could plot this on a 4-map as shown in Figure 3.21a. (The 1s are in the cells corresponding to binary numbers 7, 8, and 9, the E-entries are in the cells corresponding to binary numbers 4, 5, and 6, and the don't care entries are in the cells corresponding to binary numbers 10 through 15.)

The philosophy involved in finding a minimal SOP expression corresponding to variable-entered maps such as that shown in Figure 3.21a is the same used for ordinary K-maps. Each 1-point and each point with a variable entry is examined to see if it defines a good pi. First consider the 1-points, treating the Es as 0s. The 1-point 1000 defines A as an essential pi, so that is our first selection (see Figure 3.21b). Next, the 1-point 0111 defines BCD as essential, so this is our second selection. Having covered all of the 1-points, we turn our attention to the E-points. Consider the E-point 0100. The smallest p-term covering it corresponds to $\overline{A}B\overline{C}\overline{D}E$, the product of E and the 0-cube corresponding to 0100 on the map. But we could certainly enlarge the cube to include the E-point below 0100, and the don't care points to the right of this pair, which would give us $B\overline{C}E$ as a dominating term. But this is not all. The 2-cube below the cube $B\overline{C}$

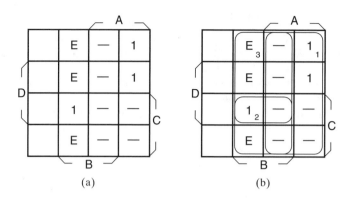

Fig. 3.21. Map-entered variable.

consists of an E-point, two don't cares, and a 1-point. Since all of these are or could be 1-points, when E = 1, they can be included in a larger cube covering $B\overline{C}$, i.e., B. Thus, the pi covering 0100 is BE. This gives us the solution Z = A + BCD + BE. (Readers may verify this by working out the same problem on a conventional five-variable K-map.) A key point is that 1-points may be regarded as don't-cares when we are considering coverings of variable entries.

Example 3.10

Variable entries can include complements, as indicated in the function mapped in Figure 3.22a. As in the previous example, we first take care of the 1-points. The 0-cube labeled 1 in Figure 3.22b accomplishes this. Considering the 1-point as a don't care entry, we can then cover the D-points with the 2-cube labeled 2. The corresponding pi is $\overline{C}D$. Finally, the $\overline{D}$ term is accounted for by the 1-cube labeled 3, which generates pi $\overline{A}B\overline{D}$. The solution is thus Z = $AB\overline{C}$ + $\overline{C}D$ + $\overline{A}B\overline{D}$.

Example 3.11

More than one variable can be entered on a map, as shown in Figure 3.23a. Referring to Figure 3.23b, the 1-point is covered by the pi $A\overline{C}$ ($B\overline{C}$ would have been equally good). The 2-cube $\overline{C}$ covers the D-point, yielding the pi $\overline{C}D$. Essential pi's $\overline{A}BE$ and $A\overline{B}\overline{E}$ cover the E and $\overline{E}$ points, respectively. The resulting minimal SOP expression is Z = $A\overline{C}$ + $\overline{C}D$ + $\overline{A}BE$ + $A\overline{B}\overline{E}$.

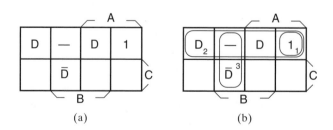

Fig. 3.22. Map with complemented variable entered.

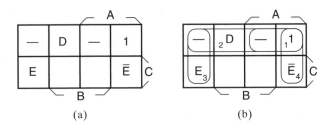

Fig. 3.23. Map with two different variables entered.

Example 3.12

In addition to entering simple literals on a K-map, *functions* of variables can be entered. In the map shown as Figure 3.24a, simple sums and products appear as entries. First, we dispose of the 1-points and literal entries, as discussed earlier. Selections 1 and 2 shown in Figure 3.24b take care of the 1-points with pi's BC and AB, respectively.

Next consider the D-point. Assume that D = 1 and E = 0. This makes the D + E cell a 1-point and the DE-cell a 0-point. Then the D-point is covered by the maximal 2-cube A (labeled 3 in Figure 3.24b), which gives us the pi AD. Note that this term also *partially* covers the D + E point in that it takes care of that cell when D = 1. In order to see how best to complete the coverage of the D + E point, we proceed by assuming that the value of the function D + E is 1. This occurs if D = 1 or if E = 1. We have already taken care of the D = 1 situation (for both values of E) with AD. Now assume E = 1 and D = 0. Then the largest cube covering the D + E point under this condition is $A\bar{C}$. To account for the E = 1 constraint we must multiply by E, to get the pi $A\bar{C}E$.

For the DE-point, we set DE = 1, which converts the map into one for which the 2-cube C is an essential cover of the target point. This gives us the pi CDE. The resulting minimal SOP expression is thus Z = BC + AB + AD + $A\bar{C}E$ + CDE.

Example 3.13

An example of a situation requiring special attention is illustrated by the function mapped in Figure 3.25. The application of the concepts discussed earlier would lead to the three choices shown in Figure 3.25b. First, AB (cube-1) is chosen to cover the 1-point (an equally good choice would have been $A\bar{C}$). Next, the D-entry is covered by $B\bar{C}D$ (cube-2), and finally the $\bar{D}$-entry is covered by $A\bar{D}$ (cube-3). The resulting solution,

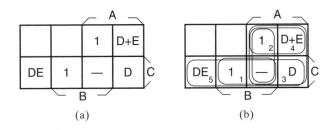

Fig. 3.24. Map with functions entered.

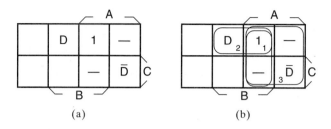

Fig. 3.25. Example involving consensus theorem.

$Z = AB + B\overline{C}D + A\overline{D}$, is valid, but it is *not* minimal. The key point is that the 1-point is covered by two overlapping cubes, one with a D-variable entered and the other with a $\overline{D}$-variable entered. For either value of D, exactly one of the cubes is activated at this point. Thus it is not necessary to cover the 1-point with a special cube (such as AB). More formally, we can see that the point $AB\overline{C}$ is the consensus of the other two terms, $B\overline{C}D$ and $A\overline{D}$. In general, if a 1-point is covered by overlapping cubes with complementary entered variables (or functions), then it need not be covered by a special cube.

Although no detailed algorithm has been presented, the previous examples should give readers a good idea as to how to proceed to find minimal SOP expressions from variable-entered maps. The approach may be summarized as follows: First cover all of the 1-points by assuming that the variable entries are all 0s. From then on treat the 1-points as don't cares. Then, considering each variable entry in turn, specify the variables involved for each entry in all ways that make that entry a 1 and again find maximal covering cubes in the usual way. Check to see if any of the cubes chosen to cover 1-points are unnecessary due to consensus terms. Obtain the pi's by multiplying the *p*-terms describing the selected cubes by appropriate products of the variables involved in the entry.

3.4. QUINE-MCCLUSKEY METHOD FOR FINDING MINIMAL SOP EXPRESSIONS

The subject of this section is an alternative method for solving the same problem treated in the preceding section, namely that of finding minimal SOP expressions. It consists of two basic parts; an algorithm for generating *all* of the pi's for the given function, and a procedure for then selecting from this set a minimum-cost subset. Neither of these procedures is very difficult to understand, but executing them is a tedious process. Hence, the method is generally recommended only as the basis for minimization procedures to be executed by computers. It is also worth learning in order to enhance our insight into the nature of the overall problem. As is pointed out later, the selection algorithm, by itself, is applicable to a number of other interesting problems.

3.4.1. Finding Prime Implicants

The basic tool in generating the pi's is the uniting theorem ($XY + \overline{X}Y = Y$; see Figure 2.3). Given the set of all complete *p*-terms of a function, i.e., a *p*-term for each 1-point, we could generate all pi's by applying the uniting theorem to all possible pairs to obtain all of the *p*-terms corresponding to 1-cubes. Then, given all of the *p*-terms corresponding to 1-cubes, we could apply the uniting theorem to all possible pairs of these to obtain all of the 2-cubes, etc. At each stage, we would delete all of the terms that were actually used to generate terms corresponding to larger cubes. The result is the set of all pi's. In order to make this process practical, it is necessary to organize it carefully, and a scheme for doing this is the subject of the following discussion.

First, instead of writing products of literals, we shall represent p-terms by words composed of 0s, 1s, and dashes (this scheme was referred to in Subsection 3.3.1 as an alternative way of designating cubes). Assuming n variables, each word consists of n-elements, representing the variables in order starting from the left with A. If an element appears complemented in the product, then a 0 is placed in its position; if it is uncomplemented, a 1 appears; and if it is missing, a dash appears. Thus, for a five-variable case, the term $A\overline{C}D$ would be represented by 1-01-, and the term $\overline{B}\,\overline{C}E$ would be written -00-1.

The uniting theorem combines two terms in which the same variables are represented, with all but one of the literals being the same. For example, $A\overline{D}E$ and $A\overline{D}\,\overline{E}$ can be united to produce $A\overline{D}$. In the new notation, this is stated as, "1--01 combines with 1--00 to produce 1--0-". The rule is that two terms with the same pattern of dashes (meaning that the same variables are present) and with the same pattern of 0s and 1s in all but exactly one position can be united. The result is a term with the same pattern where the two components are the same, and with a dash in the one position in which they differ. A necessary condition then, is that they differ by exactly 1 in the total number of 1s that they contain. With this background the method is introduced below by means of an example.

Example 3.14

A five-variable function is to be processed:

$$Z = \Sigma m(4, 5, 6, 7, 12, 14, 15, 16, 17, 25, 27, 30).$$

We begin by listing, in column a (see Figure 3.26—ignore the check marks for the time being) the vectors corresponding to the 1-points, grouping them according to the number of 1s they contain (in increasing order from top to bottom) with a space separating each adjacent pair of groups. Include don't care points. Column b is generated from column a by comparing each element of column a with each of the elements in the group below it to determine if they can be united. Recall that an element can be united with another element only if the number of 1s in the pair differs by exactly one. Hence, there is no need to compare elements that are not in adjacent groups.

The first group in column a contains only 00100. Compare it, in turn, with each member of the group below, looking for members of the second group that have 1s everywhere that 00100 has a 1. The first, second, and fourth members satisfy this condition and so each of them can be united with it to form the first three members of column b (its top group). Note that a given element can be united with any number of other elements. We next compare 00101, the first member of the second group of column a, with each member of the third group of column a. We are now looking for members with 1s in the third and fifth positions. Matches occur for the first, second, and fourth members, yielding the first three members of the second group in column b. Repeating this process for each member in turn of the second group of column a

a	b	c	d
00100√	0010-√	001--√	0-1--
	001-0√	0-10-√	
00101√	0-100√	0-1-0√	
00110√			
01010√	001-1√	0-1-1√	
01100√	0-101√	-01-1	
11000	-0101√	0-11-√	
	0011-√	011--√	
00111√	0-110√		
01101√	01-10		
01110√	0110-√		
10101√	011-0√		
01111√	0-111√		
10111√	-0111√		
	011-1√		
	0111-√		
	101-1√		

Fig. 3.26. Finding all of the pi's for a function.

generates the remaining five members of the second group of column b. The members of the third and fourth groups of column a produces the third group of column b. Each element that is successfully united with another element is checked. Thus the only unchecked element in column a is 11000.

A similar process is now carried out on the groups of column b to generate column c, the column c groups are used to generate column d, etc. Each element is compared only with elements in the next group of the same column that have the same pattern of dashes. Elements that remain unchecked correspond to the pi's. (Delete any terms covering only don't care points.) Readers should go through the complete process for this example to make sure they understand how it works, and to get an idea of the kind of effort involved. The column b elements correspond to 1-cubes, those in column c correspond to 2-cubes, etc. Since there are several ways to form an n-cube from cubes of smaller dimensionality, there are multiple ways of generating entries in the columns produced after column b. For example, the 0-11- entry in column c, corresponding to the 2-cube $\overline{A}C\overline{D}$, is generated both by 0011- united with 0111- (i.e., $\overline{A}\,\overline{B}CD + \overline{A}BCD$), and by 0-110 united with 0-111 (i.e., $\overline{A}C\overline{D}\overline{E} + \overline{A}CDE$). When a duplicate of a previously generated element is found in the process, we simply omit writing it down again (or we could write each element down as generated and later screen each group to remove duplicates). Inadequate duplication is an error indication.

3.4.2. Finding Minimal Covers

We say that a collection of sets whose union includes the set S is a *cover* of S, or that it *covers* S. The problem of selecting a good subset of pi's can be considered a special case of the following more general problem, called the *covering* problem:

PROBLEM. *Suppose C is a collection of sets, each with a given cost. Assuming C covers a set S, find a minimal-cost subset of C that covers S.*

In our problem, the pi's constitute the sets of C, each having a cost (in gi's) equal to 1 + the number of literals in it (if that number exceeds 1; else its cost is 1). The set S consists of the 1-points of the function. Since this is clearly a finite problem, it is always solvable, given enough time and memory. However, as is the case with many combinatorial problems, computation time increases at an enormous rate with the problem size. The method presented here is relatively efficient and suitable for computer implementation. The most important possibilities for improvement are probably through the use of approximate methods that do not guarantee an exact solution, but which yield reasonably good results. Of course, what has been said about the computation time applies to the K-map approach as well, since it must attack the same problem.

The basic idea is to put the problem in the form of a *covering table* in which the rows represent the sets (pi's) and the columns represent the elements to be covered (1-points). If a set covers an element, then an x is placed in the row of the element column corresponding to that set. Thus, for each set, the associated row of x's indicates the elements covered by that set, and for each element, the associated column of x's indicates the sets covering that element.

Don't care entries do not affect the problem at all. When using the procedure of the preceding subsection to find the pi's, don't cares are treated as 1-points. The covering table columns represent only 1-points, since only these need be covered. The following examples illustrate the method.

Example 3.15

In the covering table shown in Figure 3.27, the parenthesized numbers associated with each row represent costs. The process of choosing the rows parallels the process used to find minimal SOP expressions with K-maps. We look for essential, or good rows by examining columns, one by one. If there is only one x in a column, then the row in which that x appears is clearly the only one covering that column and therefore should be selected.

We begin by selecting row f, which is essential to cover column 6. Our choice of row f is indicated by a $1\sqrt{}$ entry at the right end of the row. We also indicate by 1s and checks at the bottoms of columns 5 and 6 that both of these columns have been covered by selection 1. (A useful technique for hand computation is to cross out the selected rows and covered columns.) Next observe that row b is dominated by row d in that row d covers all of the columns covered by row b, and row d costs no more than row b. Since there is no reason to choose row b, we eliminate it from consideration (step 2) by placing a 2X entry at the right end of row b. This is precisely the same concept of domination introduced earlier (Example 3.3, Subsection 3.3.4) in connection with K-maps. With row

	1	2	3	4	5	6	
a(3)	x		x				
b(3)		x	x				2X
c(3)	x			x			
d(2)		x	x	x			$3\sqrt{}$
e(2)	x		x		x		$4\sqrt{}$
f(3)					x	x	$1\sqrt{}$
	$4\sqrt{}$	$3\sqrt{}$	$3\sqrt{}$	$3\sqrt{}$	$1\sqrt{}$	$1\sqrt{}$	

Fig. 3.27. Covering table with one essential row.

b gone, column 2 is now covered only by row d, and so, in step 3, we choose row d. This also covers columns 3 and 4. Hence we place 3√ symbols at the bottoms of columns 2, 3, and 4, as well as at the right end of row d. Only column 1 remains uncovered, and so we complete the solution in step 4 by choosing the cheapest row that covers column 1, namely row e. Thus, the minimal-cost cover consists of rows d, e, and f, written as {d, e, f}.

In the next example branching is required, following the same general procedure used in connection with K-maps.

Example 3.16

Assume that the costs of all the rows are the same, so they are not specified. There are no essential rows in the covering table of Figure 3.28 to get us started, and so it is necessary to branch. A good choice for the branch is column 6, which is covered by two rows, each of which has a relatively large number of x's. Hence, after the first step, a number of columns will have been covered. The branch is indicated by a 1* entry placed at the foot of column 6 and the right end of row b, the first branch taken. There are also 1s placed at the bottoms of the other columns covered by the first choice. Now, with column 4 gone, row a is dominated by row e, and so in step 2 it is deleted. Similarly, with column 7 gone, row c is dominated by row d, and it too is eliminated (step 3). These deletions make rows d and e good, and so they are chosen in steps 4 and 5. It remains now to cover column 3, which we do with row g to complete the solution for the first branch i.e., {b, d, e, g}.

Next, we return to the last branch point, in this case, the first step, and as shown in Figure 3.29, pursue the next (in this case the only other) branch; i.e., we cover column 6 with row d. In step 2, row e is eliminated, since, with column 2 covered, it is now dominated by row a. Step 3 is the selection of row a, on the basis of column 1. Rows c and f are eliminated in steps 4 and 5, respectively. This leads to the choice of row g in step 6 to cover column 3 and complete the solution, which is {a, d, g}. Since there are no more branches to explore and since this solution is more economical than the one found for the first branch, it is a minimal solution.

Example 3.17

As pointed out in connection with Example 3.7 in Subsection 3.3.4 dealing with K-maps, branching can sometimes be avoided by observing that certain 1-points can be

	1	2	3	4	5	6	7	8	
a	x			x					2X
b				x	x	x	x		1*
c						x		x	3X
d		x				x		x	4√
e	x	x							5√
f			x	x					
g		x			x		x		6√
	5√	4√	6√	1√	1√	1*	1√	4√	

Fig. 3.28. First branch.

	1	2	3	4	5	6	7	8	
a	x				x				3√
b		x	x	x	x				
c							x	x	4X
d		x				x		x	1√
e	x	x							2X
f			x	x					5X
g		x			x		x		6√
	3√	1√	6√	3√	6√	1√	6√	1√	

Fig. 3.29. The second branch.

converted to don't cares because they dominate other 1-points. The same principle can be applied to the covering problem. If all of the x's in column i correspond to x's in the same rows of column j, then any row covering column i will also cover column j. Hence we can delete column j (which is said to *dominate* column i).

This point is illustrated in the table shown in Figure 3.30. Since there are at least two x's in every column, there are no essential rows. But, since column 1 dominates column 3, we can delete column 1, as indicated by the 1X entry at the foot of column 1. Then row a is eliminated by row d (step 2), and this makes row d good, so it is selected in step 3. Row c is eliminated in step 4 due to domination by row b, and then row b is chosen to complete the solution: {b, d}.

A summary of procedure for solving the covering problem:

1. Select any row that uniquely covers a column. Then delete that row and the columns it covers. Repeat this step as long as it is applicable.
2. Delete any row dominated by another row. Note that if row costs are specified, then row domination includes having a cost no greater than the cost of the dominated row. Apply step 1 any time it is applicable. Repeat this step as long as it is applicable.
3. Delete any column dominating any other column. (If several columns are identical, delete all but one of them.) Repeat this step as often as it is applicable. Then go back to step 2. Go to step 4 only after neither step 2 nor step 3 is applicable.
4. If neither step 1 nor step 2 can be applied, and if some columns remain uncovered, then branch, i.e., choose a column and make an arbitrary choice of a covering row. Continue with steps 1, 2, and 3 until a solution has been found. Then return to the branch point, make an alternative choice, etc. (The general

	1	2	3	4	5	
a	x	x				2X
b	x		x		x	5√
c	x		x	x		4X
d		x		x		3√
e				x	x	
	1X	3√	5√	3√	5√	

Fig. 3.30. Table with a dominated column.

procedure for branching is the same as that described in the summary at the end of Subsection 3.3.4 in connection with K-maps.)

3.5. SIMPLIFICATION OF MULTI-OUTPUT LOGIC

Thus far, logic functions have been treated individually. However, in many—perhaps most—cases, logic circuits must be designed with multiple outputs. For example, the inputs might be the bits of a BCD number, and the outputs might be the bits of another BCD numeral twice the size of the first modulo-10. The methods previously described for designing single-output logic circuits are always applicable, since we can design a separate logic circuit for each output signal. However, it is often possible to realize a set of functions more efficiently by sharing logic elements among them. Refinements of both the K-map approach and the Quine-McCluskey method are presented below. As in the previous discussions, only SOP expressions are considered here.

3.5.1. Using K-Maps

Several examples are used to illustrate the concepts involved in sharing logic elements among circuits realizing different functions. In the first case, the way to share is obvious. In the second case, a generalization of the prime implicant concept is required. (Double-rail inputs are assumed for all examples in this section.)

Example 3.18

A pair of functions of three-variables is shown in the maps of Figure 3.31. Solutions are indicated based on the fact that each function has two essential pi's covering all of its 1-points. Thus, if realized separately, two 2-input AND-gates and a 2-input OR-gate would be required to generate each output. But, since both solutions include the *p*-term BC, we need not generate that term twice. A single AND-gate producing a signal corresponding to BC can be used, and its output can be "fanned out" to the OR-gates for Z_1 and Z_2, as shown in Figure 3.32. To indicate such sharing, when we write the expressions for Z_1 and Z_2, we underscore the second appearance of BC, i.e., we write $Z_1 = A\overline{B} + BC$, $Z_2 = \overline{A}\,\overline{B} + \underline{BC}$.

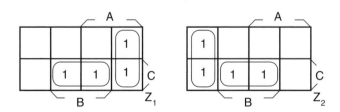

Fig. 3.31. A pair of functions to be realized together.

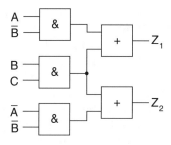

Fig. 3.32. Multi-output logic circuit with fanout from shared gate.

Example 3.19

In Figure 3.33a, a pair of logic functions is shown, along with the subcube selections that would have resulted from the application of the techniques presented earlier for finding minimal SOP expressions for each function separately. Two pi's are used for each function. There is no pi common to the two functions. Hence the solution, requiring a total of six gates and 12 gi's is $Z_1 = A\overline{C} + A\overline{B}$, $Z_2 = AC + BC$.

Now consider Figure 3.33b. Here, after choosing $A\overline{C}$ for Z_1 and BC for Z_2, we find that, for both functions, 101 is the only uncovered 1-point. Hence, if we choose $A\overline{B}C$ for both functions, we have the solution $Z_1 = A\overline{C} + A\overline{B}C$, $Z_2 = BC + A\overline{B}C$. This has a cost of only five gates (one shared) and 11 gi's.

But the $A\overline{B}C$ term is not a pi of either function. How does this fit in with the argument developed earlier to the effect that only pi's should be considered when generating minimal SOP expressions? The answer is that, since we are now dealing with multi-output logic, the concept of a prime implicant must be generalized. A necessary condition for a *p*-term to cover a 1-point in a single function at minimal cost is that it

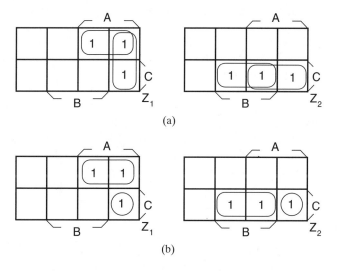

Fig. 3.33. Comparison of separate versus combined logic.

correspond to a maximal-size subcube covering 1-points of that function. A subcube is maximal if it cannot be enlarged and still correspond to a p-term of the function. This is one way to define pi's (retrospectively we rename these as *single-output* prime implicants, or *sopi's*).

The generalization is a p-term that can optimally cover 1-points in a *set* of functions. This corresponds to a subcube that is maximal in the sense that it cannot be enlarged and still be a p-term of *all* of the functions in the set. Such a p-term is called a *multi-output prime implicant* (or *mopi*). The *product* of a set of functions is the function that has 1-points wherever all of the functions in the set have 1-points, and only at these locations. Then a mopi is a sopi of the product of a set of functions. The set may consist of all of the functions being realized together, or of any subset of them. In example 3.19, $A\overline{B}C$ is a mopi by virtue of being a sopi of the product of Z_1 and Z_2 (in this case, it is *equal* to the product).

In general, when considering how to cover a given 1-point of a function, not only must sopi's be considered but, if among the other functions being realized simultaneously, there are some that also have 1-points at the same location, then mopi's corresponding to all subsets of these must also be considered. If a mopi is selected, then it must be used for all of the functions that kept it from being expanded to cover more 1-points. These ideas are further illustrated in the next example.

Example 3.20

A set of three four-variable functions is displayed in Figure 3.34. We begin by looking at 1-points unique to a single function in the hope of finding some essential or good sopi's. The choices labeled 1 and 2 are in this category, each corresponding to a 2-cube. The 1-point at the lower left corner of Z_3 is not helpful, because it is covered by two sopi's. But the 1-point just above it does have a contribution to make. It is covered by the sopi $\overline{A}\,\overline{B}$, and by the mopi $\overline{A}\,\overline{B}D$, which is also a p-term of Z_2. However, both of the points it covers in Z_2, have become don't cares as a result of being covered by the first sopi selected. Hence there is no reason to choose $\overline{A}\,\overline{B}D$ (it is dominated by $\overline{A}\,\overline{B}$). Our third choice is therefore $\overline{A}\,\overline{B}$ for Z_3.

Covering 1001 in Z_2 can be done with the sopi $\overline{B}D$ or with the mopi $A\overline{B}D$. It would be reasonable to choose the latter only if we also use it for Z_1 as well as for Z_2. In a formal sense, this situation calls for branching, since there are no good pi's. ($\overline{B}D$ does not dominate the mopi $A\overline{B}D$ because the latter covers points in Z_1 not covered by the former. Nor, because of its greater cost, is $A\overline{B}D$ dominant.) However, if we look ahead a bit, it becomes clear that we will do better with $A\overline{B}D$, since, at a cost of 1 gi, this will save generating a new p-term to cover 1001 in Z_1. Selections labeled 4 and 5 in the figure reflect this choice. At this stage, 0110 is the last uncovered 1-point, and it is uncovered for all three functions. No cube larger than $\overline{A}BC\overline{D}$ can be used for all of them, so this is a mopi. The solution, completed with the addition of this term for all three expression, is $Z_1 = A\overline{D} + A\overline{B}D + \overline{A}BC\overline{D}$, $Z_2 = \overline{A}D + A\overline{B}D + \overline{A}BC\overline{D}$, $Z_3 = \overline{A}\,\overline{B} + \overline{A}BC\overline{D}$.

Ensuring that the solution produced is minimal requires careful consideration of each choice, taking into account both the 1-point coverage of the proposed term and its cost. If it is indeed important to minimize the result, then the conservative move is to branch whenever in doubt.

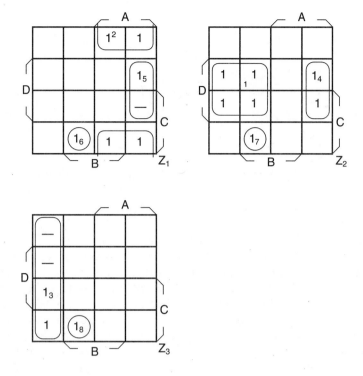

Fig. 3.34. Realizing a set of three four-variable functions.

3.5.2. Using the Quine-McCluskey Method

One could find all of the mopi's by generating all possible products of the functions to be realized, and then finding the pi's for each of them using the method of Subsection 3.4.1. There is, however, a more efficient procedure that does not necessitate generating the product functions. It involves tagging the 1-points and subsequently developing p-terms with vectors indicating which functions they apply to. The following example illustrates the process.

Example 3.21

Consider the set of functions:

$$Z_1(A, B, C) = \Sigma m(1, 2, 3, 4, 6), \qquad Z_2(A, B, C) = \Sigma m(2, 3, 5, 6, 7),$$

$$Z_3(A, B, C) = \Sigma m(0, 1, 4).$$

The process is organized as in Subsection 3.4.2, except that, attached to each 1-point in column a is a parenthesized "tag" that indicates which of the three functions includes that 1-point. (See Figure 3.35.) The tag has one position for each function, starting with the first function at the left end. A 1 in a position indicates the presence of that 1-point, and a 0 indicates its absence. Thus, in the present case, the tag for 010 (represented by octal number 2 in the function description), is 110, since it is present in the first two

a	b	c
000(001)√	00-(001)	-1-(010)
	-00(001)	
001(101)		
010(110)√	0-1(100)	
100(101)	01-(110)	
	-10(110)	
011(110)√	1-0(100)	
101(010)√		
110(110)√	-11(010)√	
	1-1(010)	
111(010)√	11-(010)√	

Fig. 3.35. Generating mopi's.

functions but not in the last. To generate the next column, we begin as in the single-output case, ignoring the tags. When a new term is generated, we then check the tags of the component terms. The new term is assigned a tag with 1s in precisely those positions in which both of the component terms had 1s. (In effect we take the point-by-point Boolean product of the component tags.) If there is *no* position in which both have 1s, then delete the new term altogether. Readers should be able to satisfy themselves that terms so generated are indeed *p*-terms of the functions specified by the tags. Terms are checked off as covered by larger terms if they participate in the formation of a new term that has the *same* tag.

The *un*checked terms in Figure 3.35 specify the mopi's. One must be careful, however, in interpreting the result. Consider the unchecked term 01-(110) corresponding to $\overline{A}B$. It is a mopi of the *pair* of functions Z_1, Z_2. Since it combined with 11-(010) to help form the mopi B, $\overline{A}B$ is *not* a mopi of Z_2 alone. But it *is* a mopi (more specifically in this case, a sopi) of Z_1.

There are various ways of proceeding to formulate the second part of the problem: selecting an optimal subset of the mopi's. The approach illustrated in the next example, which will continue with the same problem initiated in Example 3.21, is relatively straightforward. Each mopi, complete with tag and cost, is listed as a row. The cost is assigned under the assumption that the *p*-term is generated with an AND-gate and used in all of the functions indicated by its tag. Thus, for example, the cost associated with $\overline{A}B$ when it carries the tag (110) is 4 (including two OR-gate inputs), and in the row where its tag is (100), the cost listed is 3. Each 1-point, labeled by its octal abbreviation, is represented by a distinct column for each function in which it appears. The columns are grouped by function.

Example 3.22

The covering table for the set of three functions considered in the last example is shown as Figure 3.36. The first step is easy. There is a unique cover for point 5 under Z_2, namely AC. Hence AC is essential. The symbols 1√ are entered for the AC row, and for the columns it covers, 1-points 5 and 7 for the second function. The next step entails more subtle reasoning. The *p*-term B costs nothing at all to realize, i.e., no AND-gate is needed. While it does not qualify as an essential pi for Z_2, the fact that it costs nothing

	Z_1					Z_2					Z_3			
	{1	2	3	4	6}	{2	3	5	6	7}	{0	1	4}	
$\overline{A}\,BC(101)5$	x											x		10√
$A\overline{B}\,\overline{C}(101)5$			x										x	8X
$\overline{A}\,\overline{B}(001)3$											x	x		
$\overline{B}\,\overline{C}(001)3$											x		x	9√
$\overline{A}C(100)3$	x		x											
$\overline{A}B(110)4$		x	x			x	x							4X
$\overline{A}B(100)3$		x	x											5*
$B\overline{C}(110)4$		x			x	x			x					3X
$B\overline{C}(100)3$		x			x									6X
$A\overline{C}(100)3$				x	x									7√
$AC(010)3$								x		x				1√
$B(010)1$						x	x		x	x				2√
	10√	5*	5√	7√	7√	2√	2√	1√	2√	1√	9√	10√	9√	

Fig. 3.36. Covering table for multi-output function.

to generate gives it special status. What could possibly be gained by *not* using B to cover, say, 011? There might be a mopi that covers 011; in fact, there is, namely $\overline{A}B$. But even if $\overline{A}B$ is generated to cover a 1-point for another function (in this case Z_1), and is hence available for use in Z_2 at no charge at all, it still costs 1 gi to connect it to the Z_2–OR-gate. But for the price of that input we might as well use B, since B covers all of the points covered by $\overline{A}B$ (plus one more as a bonus). Thus we can safely choose B for our second selection, as indicated by the 2√ at the right end of the B-row and at the bottom of the Z_2 columns that it covers.

The next step is based on the fact that, with the Z_2 function 1-points 2 and 6 covered by B, the uncovered points in the $B\overline{C}(110)$ row are all covered by the $B\overline{C}(100)$ row at a lower cost. Hence the $B\overline{C}(110)$ row is deleted at step 3 (indicated by the 3X). This amounts to the simple assertion that since both of the Z_2-points covered by $B\overline{C}$ have already been covered, there is no point in considering using $B\overline{C}$ for *both* Z_1 and Z_2. Similarly, at step 4 we delete $\overline{A}B(110)$ on the grounds that it is dominated by $\overline{A}B(100)$. At this stage there are no row or column dominations, and no good rows. Hence we must branch. This is done by selecting $\overline{A}B(100)$ to cover 1-point 2 for the Z_1 function. This is indicated by the 5* entries for the row and column, and the 5√ entry of the other column covered. In step 6 we eliminate $B\overline{C}(100)$, since it is dominated by $A\overline{C}$. The next step is to select $A\overline{C}$ because of 1-point 6 for Z_1. Row $A\overline{B}\,\overline{C}$ is deleted at step 8 due to dominance by $\overline{B}\,\overline{C}$. This leads to the selection of $\overline{B}\,\overline{C}$ to cover 1-point 4 for the Z_3 function. The last step is to choose $\overline{A}\,BC$ to cover 1-point 1 in both Z_1 and Z_3. This completes the first branch, leading to the result

$$Z_1 = \overline{A}B + A\overline{C} + \overline{A}\,\overline{B}C, \qquad Z_2 = AC + B, \qquad Z_3 = \overline{B}\,\overline{C} + \overline{A}\,\overline{B}C.$$

The next step is to go back to the branch point at step 5 and make the alternative choice to cover 1-point 2 for the Z_1 function, i.e., to choose $B\overline{C}(100)$. Readers who follow through on this will find that the alternative solution has the same cost as that for the solution just found.

3.6. Unate Functions—An Interesting Special Case

Consider a symmetric function f of four variables that is defined to be 1 if at least two of its inputs, any two, are equal to 1. It follows then that for any input state, if some input variable is turned on (i.e., changes from 0 to 1), then the value of f cannot change from 1 to 0. If we were talking about conventional algebraic functions, then we would say that such a function is *monotone increasing*, or that it is a *positive* function. It is useful to depart here from our usual assertion that the 0s and 1s of Boolean algebra are *not* to be considered as numbers. If we do so, then the function f is indeed positive.

Here is another example of a positive Boolean function: $Z = ABC + AD + BD$. It must be positive, because it has the value 1 only when at least one of its p-terms is on, and a p-term containing only uncomplemented literals cannot be turned *off* by turning any variable *on*. It is easy to generalize this case to establish that:

Any logic expression with no complemented terms (or complemented subexpressions) must be positive.

It follows that logic circuits composed only of AND- and OR-gates can only realize positive functions. What can be said about the converse? Is it true that no expression with complemented terms can describe a positive function?

The refutation of that assertion is implied by the expression $ABC + AD + ABD + \overline{A}BD$. Applying the uniting theorem to the last two p-terms reduces this to the expression of the preceding paragraph. Thus, a positive function may indeed be expressed by an algebraic expression that includes complemented terms. However, by modifying the converse statement, an important valid statement can be made, namely that:

No complemented literal can appear in any pi of a positive function.

This assertion is proved below.

Suppose that Q is a pi of a function f, and that $\overline{A}$ is a literal of Q. (For example, suppose $Q = \overline{A}BD$ is a pi of $f(A, B, C, D)$.) Consider the p-term Q' obtained by replacing the $\overline{A}$ in Q with an A. (In our example, $Q' = ABD$.) Can Q' be a p-term of f? If it were, then we could apply the uniting theorem to Q and Q' to obtain a more powerful p-term (consisting of Q with the $\overline{A}$-literal deleted) that would be implied by Q. (In our example $Q + Q' = \overline{A}BD + ABD = BD$.) But this contradicts the initial assumption that Q is a pi. Therefore, it follows that Q' is *not* a p-term of f. Hence there must be at least one 0-point of f in Q'. Let one such 0-point be labeled z'. (Suppose that $z' = 1101$ is such a point of $Q' = ABD$ in our example.) Note now that for all points in Q', changing the value of A from 1 to 0 takes us to a point in Q (since Q and Q' differ only in A). Therefore, the point z, differing from z' only in that $A = 0$ is a point in Q. (In the example, $z = 0101$.) Since all points in Q are 1-points of f, we now have an instance where changing A from 1 to 0 causes f to change from 0 to 1. (In our example, this is the transition 1101 to 0101.) Thus we have proved that f is not a positive function, establishing that pi's of positive functions cannot contain complemented literals.

We can summarize the preceding discussion by stating the theorem:

A function is positive if and only if none of its pi's contains a complemented literal.

The fact that all pi's of positive functions are composed of only uncomplemented literals leads immediately to the observation that:

Every pi of a positive function covers the all-1-point, the point for which all variables are equal to 1.

An important property of positive functions is that:

All prime implicants of a positive function are essential.

Thus positive functions are exceptionally amenable to logic minimization. We need only generate the set of all pi's and we have the unique minimal SOPs expression. The following proof of this property depends on identifying a 1-point for each pi that is uniquely covered by that pi. (The positive function mapped in Figure 3.37, with the pi's ovalled will be used to illustrate the argument. (Note incidentally, that each pi covers the point ABCD = 1111.)

Recall first from the discussion of the Quine-McCluskey method (Section 3.4) that p-terms (or cubes) can be represented by ordered sequences of 0s, 1s, and dashes. An uncomplemented literal is replaced by a 1, a complemented literal by a 0, and each missing literal is represented by a dash. (For example, for the variables $\{A, B, C, D\}$, the p term $\overline{A}CD$ can be specified as 1-11.) Since every pi of a positive function is a product of uncomplemented literals, they can be described with only 1s and dashes. (Thus BD, the pi labeled 2 in Figure 3.37, can be specified as -1-1.) Let Q be a pi of a positive function f. Then $m(Q)$, the 1-point of Q furthest from the all-1-point, can be found by setting to 0 all of the dashes in the description of Q. (In the example, $m(BD) = m(-1-1) = 0101$.) Suppose Q', some other pi of a positive function f, also covered $m(Q)$. Then the description of Q' cannot have 1s in any of the positions where $m(Q)$ has 0s. That is, the 1-positions in the description of Q' are a subset of the 1-positions of Q and all the other positions must have dashes. Then the literals in the product constituting Q' form a subset of the literals of the Q-product. Hence Q' is implied by Q, which

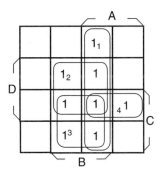

Fig. 3.37. An example of a positive function.

contradicts the assertion that Q is a pi. (Recall that a pi of a function cannot imply any other product term of that function.) So every pi, Q, of f uniquely covers the 1-point m(Q), and is, therefore, essential.

Monotone *decreasing* or *negative* functions are defined analogously to positive functions. They have pi's consisting entirely of *complemented* literals, and all cover the all-0-point. More generally, we can define the class of functions that are monotone increasing *or* decreasing in each variable as *unate*. If we take a positive function and complement any subset of its inputs, we obtain a unate function. It is negative in each of the complemented variables and positive in each of the others. For each variable in which it is negative, the function cannot increase when that variable increases, and for each variable in which it is positive, the function cannot decrease when that variable increases.

A unate function that is neither positive nor negative is mapped in Figure 3.38a. All of its pi's cover the point 0101. Readers should verify that only the literals $\overline{A}$, B, $\overline{C}$, and D appear in those pi's, all of which are obviously essential.

Examine the function mapped in Figure 3.38b. In particular note the transitions marked p and q. Each of these crosses the boundary marking the A = 1 region of the map from outside to inside, i.e., each marks a change of A from 0 to 1. In the case of p, the value of f changes from 0 to 1, so that we know the function is not *negative* in A. For the q-transition, the function value changes from 1 to 0, so that the function is not *positive* in A. Hence we know that the function is not unate. It is also not monotone in B as indicated by the transitions r and s into the B-region. It is, however, negative in C and positive in D. This means that its pi's will be composed of the literals, A, $\overline{A}$, B, $\overline{B}$, $\overline{C}$, and D. Readers may verify this by finding the pi's.

3.7. MULTISTAGE LOGIC, FANIN, FANOUT

There are two principal reasons for continuing our design efforts beyond 2-stage logic. One is to reduce circuit cost in terms of gi (gate-inputs) and, perhaps, g (gates). The other is to reduce fanin.

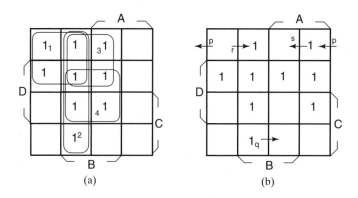

Fig. 3.38. (a) Unate and (b) non-unate functions.

A principal technique for multilevel logic design is factoring. This can reduce both the number of gates and the number of gi's. For example, the cost of the circuit corresponding to Z = ABC + AD + AE is 10 gi (gate inputs), and 4 g (gates). Factoring out the A to obtain Z = A(BC + D + E) reduces the cost to 7 gi and 3 gi. There are cases, however, in which the number of gates can be *increased* by factoring. For example, Z = ABC + ADE has a gate count of 3. Factoring to obtain Z = A(BC + DE) increases the gate count to 4. (The gi count remains at 8.) In the first example, two terms with two literals each were reduced to terms with one literal, thereby eliminating the need for gates to generate them. Let us see how we can quantify the effects of a particular factoring operation.

Suppose we factor f literals out of t terms. For each literal factored out, we eliminate an AND-gate input corresponding to each term, for a savings of ft gi. But we need a new AND-gate with $f + 1$ inputs, thus leaving an apparent net saving of

$$S_{gi} = ft - (f + 1) = f(t - 1) - 1 \text{ gi.}$$

This is indeed correct in the case where we factor AB out of the expression ABCD + ABEF + ABGH to obtain AB(CD + EF + GH), where $f = 2$, $t = 3$, and gi is reduced from 15 to 12. But consider the expression ABCD + ABEF + ABGH + JK. Two literals (AB) can be factored out of three terms, so $f = 2$ and $t = 3$. The total number of terms is $t' = 4$. The result is the expression AB(CD + EF + GH) + JK. The unfactored term must be added to the factored terms, which requires another OR-gate input. So the net gain is reduced by one. A convenient way to express this in general is to introduce a variable r that is 0 if all of the terms are factored and is otherwise 1. This transforms our formula to

$$S_{gi} = f(t - 1) - 1 - r.$$

Suppose that the factoring reduces i of the terms to a single literal. Then for each of these terms, no AND-gate is required (a 1-input AND-gate is just a wire), so that an additional gi is saved. The formula then becomes

$$S_{gi} = f(t - 1) - 1 + i - r.$$

Applied to the case where ABC + ABD + ABEF + GH is factored to obtain AB(C + D + EF) + GH, where $f = 2$, $t = 3$, $i = 2$, and $r = 1$, we have a gi reduction from 16 to 12 corresponding to $S_{gi} = 4$. In the worst case (see Problem 3.46), factoring may actually *increase* gi, though not by much.

The total number of *gates* (g) is often *increased* as a result of factoring. The only way factoring can decrease g is when, as just mentioned, factoring reduces the number of literals in a term to one, thereby eliminating an AND-gate. This corresponds to the variable i in the formula for gate input savings. If not all terms are factored, another OR-gate is necessary. Using the same variables just introduced, the formula for the reduction in g is

$$S_g = i - 1 - r.$$

Applying this to the situation where ABC + ABD + ABE + ABFG + H is factored to produce AB(C + D + E + FG) + H, we see that $i = 3$, $r = 1$, and the gate count is reduced from 5 to 4. In the case where ABCD + ABEF + ABGH + J is factored to produce AB(CD + EF + GH) + J, our formula indicates that S_g is -2, reflecting the fact that g increases from 4 to 6. (Note that $S_{gi} = 2$ for this case.)

Although both of the preceding formulas were discussed in terms of SOP expressions, they can be applied just as well if the literals are replaced by subexpressions. For example, if (A + B)(C + D) is factored out to convert (A + B)(C + D)EF + (A + B) (C + D)G + H to (A + B)(C + D)(EF + G) + H, we have $f = 2$, $t = 2$, $i = 1$, $r = 1$, $S_{gi} = 1$, and $S_g = -1$. (It is assumed here that in implementing the original expression, the subexpressions would each be generated once and the results fanned out to two places each.)

There are no known good, generally applicable, systematic techniques for generating economical multistage logic circuits. In subsequent chapters, techniques useful for various special cases are introduced. Here, a few examples are presented illustrating an approach based on factoring and on the exploitation of repeated subexpressions. The discussion is in terms of AND–OR–NOT logic. Conversions can be made afterward to NOR- or NAND-circuits as discussed in Section 2.6. Techniques for generating multistage NOR- or NAND-circuits directly are not presented here.

Example 3.23

A minimal SOP expression for the five-variable function mapped in Figure 3.39 is

$$Z = A\overline{C}D + AC\overline{D} + B\overline{C}D + BC\overline{D} + C\overline{D}E. \qquad (5)$$

The cost of this expression (of course, this means the cost of the *circuit* corresponding to it) is 20 gi, 6 g. Maximum fanin (mfi) is 5, to the OR-gate. Maximum path length (mpl) is 2, excluding inverters at the inputs; we assume double-rail inputs.

An examination of the map indicates that the selected cubes can be partitioned into two sets, one in the upper half and one in the lower half. These represent the $\overline{D}$ and D regions, respectively, and so suggest a factoring out of D and $\overline{D}$ from the two sets. Similarly, a partitioning can be made with respect to the C and $\overline{C}$ regions. This suggests

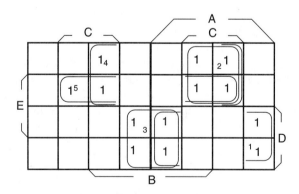

Fig. 3.39. Function to be realized with multistage logic.

factoring out literals involving C and D. (The same observations might have been made directly from (5)). Such factoring yields

$$Z = (A + B)\overline{C}D + (A + B + E)C\overline{D}. \tag{6}$$

The cost of this expression is 13 gi, 5 g. The mfi is 3, and the mpl is 3. This is a substantial improvement, but we can do a little better.

The subexpression $A + B$ appears twice in (6). A signal corresponding to it is generated by an OR-gate corresponding to the $(A + B)$ at the left end of the expression in (6). This signal might be used a second time to feed an OR-gate with E as the other input to produce the $A + B + E$ subexpression. This can be described by the expression:

$$Z = (A + B)\overline{C}D + \left(\left[A + B\right] + E\right)C\overline{D}. \tag{7}$$

The corresponding circuit is shown as Figure 3.40. The cost is 12 gi, 5 g. The mfi is 3, and the mpl is increased to 4. As shown in the next example, subexpressions can sometimes be a bit harder to find.

Example 3.24

The following is a minimal SOP expression for another five-variable function

$$Z = B\overline{C}D + A\overline{C}D + \overline{A}\,\overline{B}C\overline{D} + C\overline{D}E. \tag{8}$$

The cost is 17 gi and 5 g, the mfi is 4, and the mpl is 2. Factoring yields

$$Z = (A + B)\overline{C}D + (\overline{A}\,\overline{B} + E)C\overline{D}. \tag{9}$$

This makes the cost 14 gi and 6, with an mfi of 2 and an mpl of 4. Observe that $\overline{A}\,\overline{B} = \overline{A + B}$, so that the output of the gate generating $A + B$ can be inverted and used a second time. This corresponds to

$$Z = (A + B)\overline{C}D + \left(\left[\overline{A + B}\right] + E\right)C\overline{D}. \tag{10}$$

This reduces the cost in gi to 13 while increasing the mpl to 5.

Other algebraic manipulations are sometimes useful. For example, conversions between product and sum forms, such as the identity $(P + Q)(\overline{P} + R) = PR + \overline{P}Q$, can sometimes be exploited. In general, both SOP and POS forms should be considered as starting points.

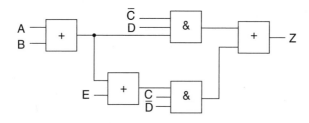

Fig. 3.40. Logic circuit corresponding to expression (7).

All other things being equal, increasing the number of stages increases delay, as signals have longer paths to traverse. However, in some technologies, reducing fanin can reduce delays to the extent that the effect of longer paths is more than canceled out. Increased fanout is a by-product of some methods generating multistage logic, and this may increase delays. Thus it is not always obvious how certain changes in logic design will affect delays, and careful analysis of the detailed circuits is necessary.

3.8. Designing Pass-Transistor Circuits

It is often possible to substantially reduce the number of transistors needed to realize a logic function by using pass-transistor networks. These were introduced in Section 1.2.1, and the technology for n-channel MOS (nMOS) and complementary MOS (CMOS) pass-transistor circuits is discussed in Appendix A.3.

In a pass-transistor network, one or more of the places where V_{dd} (high voltage), or ground (0-volt terminal) would normally be connected, are instead connected to an input variable. (The input terminal may be thought of as the output of some other gate.) Two rules must be followed:

(1) For every output terminal, except during brief transient intervals, there must always be a path with transmission 1 between that terminal and either V_{dd} or ground. (This rule might be stretched in cases where it is reliably known that the interval during which the node may "float" is small compared to certain time constants.)
(2) No output terminal may ever be connected simultaneously to two opposite signal values.

The design of an XOR-gate (Section 2.9) where only single-rail inputs are available is an important problem that allows us to compare a number of different types of logic circuits. First, the most efficient design in terms of NAND-gates is shown in Figure 2.18c. Four 2-input NAND-gates are used, which would require a total of 12 transistors in nMOS and 16 in CMOS. But we can find much more economical designs. (In the subsequent discussion, only CMOS technology will be considered as this is now dominant.)

We begin with a circuit employing a complex gate. Based on the Boolean manipulations $\overline{A \oplus B} = AB + \overline{A}\overline{B} = AB + (\overline{A + B})$, we can derive the efficient CMOS design, shown in Figure 3.41, using a 2-input NOR-gate and a complex gate. This design requires 10 transistors.

Now let us see what can be done with pass transistors. A straightforward use of two transmission gates (Figure A.17 in Appendix A.3.2) leads to the XOR-gate of Figure 3.42. The upper transmission gate transmits $\overline{B}$ when $A = 1$, thereby realizing the XOR function when $A = 1$. The lower gate transmits B when $A = 0$, taking care of the situation when $A = 0$. Since the complements of both A and B are required as inputs to this circuit, two inverters must be used with it, and therefore, the total transistor count is 8. This is a nice improvement over the previous circuit, but we can do still better.

Consider the circuit of Figure 3.43. The lower two transistors constitute a transmission gate with the same inputs as the lower transmission gate in Figure 3.42. When $A = 1$ this gate acts as an open circuit, blocking all current flow. The upper two transistors are connected as a CMOS inverter (see Figure A.14 in Appendix A.3.2) with A acting as V_{dd}

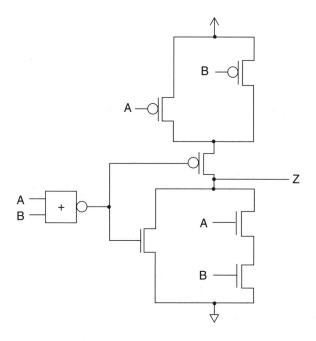

Fig. 3.41. Complex CMOS XOR-gate.

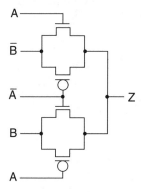

Fig. 3.42. XOR using two transmission gates.

and $\overline{\text{A}}$ acting as ground, with B as the input. Thus, when A = 1, this part of the circuit sends $\overline{\text{B}}$ to the output. So the circuit correctly generates A $\oplus$ B when A = 1. When A = 0, the lower two transistors, acting as a transmission gate, pass B to the output—again the correct response for A $\oplus$ B. The complement of B is not required, only one inverter is necessary, and so the total transistor count is 6.

A drawback of pass-transistor circuits, shared by the last two XOR realizations, is that if there are chains of pass-transistor circuits, there will be delays rising approximately as the square of the number of units in the chain. We are concerned here with the maximum number of transistors connected in series between the output terminal and

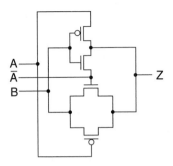

Fig. 3.43. Improved XOR-gate.

either ground or V_{dd}. If the input to a chain of k XOR-gates comes from an inverter, then there will be a chain of $k + 1$ transistors, including one in the inverter circuit. A rough rule is to keep the lengths of such chains below 5.

An instance in which many XOR-gates must be cascaded is where 2-input XOR-gates are used to construct n-input XOR-gates, for n much greater than 2. The solution is to insert inverters in the chains at appropriate intervals.

3.9. Fault Detection in Combinational Logic

Modern digital logic circuits are remarkably reliable if we measure reliability in terms of probability of a particular gate output being incorrect on any specific occasion. But the likelihood of an incorrect response from a complete system over a period of, say, 24 hours, is far from being negligible. Some failures result from transient effects, such as electrical noise entering through the power supply, or cosmic rays creating a burst of ions on a chip at just the wrong time and place. These are called *soft* faults, to distinguish them from *hard* faults that are of a permanent nature. Soft faults are difficult to detect, and defenses against them tend to be in the realm of software and overall system design (see Appendix A.1.12 on error-correcting codes). Hard faults can result from vibration or other mechanical effects, perhaps initiated by temperature swings, or chemical effects, perhaps long-term corrosion, or such subtle effects as metal migration, which is a type of electrolytic process whereby current flow through a thin conductor on a chip causes the gradual displacement of molecules. Given the minuscule sizes of elements on integrated circuit (IC) chips, it does not take much to cause a previously functioning element to fail. Thus, a necessary consideration in digital system design is the question of how the resulting devices can be tested to ensure that they are free of hard faults.

The issue goes beyond that of detecting failures of previously functioning circuits. The process by which IC chips are reproduced is a very delicate one. Tiny flaws on a silicon chip can render the circuit implanted on it inoperative. A process with a *yield* of 90 percent i.e., where 90 percent of the chips produced work properly, would be considered quite acceptable. It is thus necessary to test all of the chips produced, in order to reject those that are flawed. Such testing is an integral part of IC manufacturing. Since we are discussing enormously complex systems, with perhaps millions of gates embedded on a single thumbnail size chip, the problem of devising thorough, but economically feasible testing procedures is clearly a very difficult one. Our goal, incidentally, is to be able to detect the *existence* of faults in a circuit. It might be nice to be able to pinpoint the exact *nature* and *location* of a fault within a circuit, but this is not necessary for most purposes. A brief introduction to the requirements and to some basic techniques for meeting them with respect to combinational logic follows in this section. Further discussion of testing with respect to sequential logic is in Chapter 7. A technique for *correcting* faulty signals is presented in Appendix A.1.12.

Since we would like to deal with the testing problem in terms of Boolean variables, it is necessary to devise a digital model for characterizing the faults to be detected. This is not a simple task, since the faults themselves are physical in nature, and are not constrained to manifest themselves in digital terms. The most widely used model characterizes all faults as causing signals at gate inputs or outputs to be *stuck-at-0* (@0) or *stuck-at-1* (@1). A @0 fault exists at a node in a circuit if the signal at that point becomes fixed at 0, regardless of the values of all other signals. For example, if the pullup resistor in an inverter gate, or a wire in series with it, is broken, resulting in an open circuit, the output of that gate will be @0. On the other hand, if an open circuit develops in series with the pulldown transistor, in the same gate, the gate output will be @1. (We write X@0 to signify that the signal X, which may be an input or an internal signal, is stuck-at-0. Stuck-at-1 faults are expressed as X@1.) Experience indicates that tests for stuck faults will detect the great majority of all actual faults. A fault type not discussed in this work, but treated to some extent in the literature, is the *bridging* fault. This refers to an unintended connection between two signal nodes, forcing the signals to be equal. The difficulty is that, depending on the details of the technology, the combined signal may or may not be digital in nature. The discussion of wired logic in Subsection 1.3.1 is relevant here.

Another assumption made in devising test procedures is that there is at most one fault in the circuit. Multiple faults are indeed extremely unlikely events in situations where the fault mechanism is such that each possible fault is statistically independent of all other faults. Unfortunately, it is not hard to think of situations where the same condition is likely to produce a multiplicity of logic faults. For example, a flawed region on a chip may contain several gates. Or a power supply surge might damage several wires or elements on a chip. Nevertheless, as in the case of the stuck fault assumption, a set of tests that detects all single faults also detects almost all multiple faults. The problem addressed in the subsequent discussion is that of testing for single stuck faults in combinational logic circuits.

A *test* of a combinational logic circuit consists of the application of an input state and a check of the resulting output against the correct output. A *complete test set* is a set of tests that will detect all faults of the type being investigated. For a circuit with n input variables, the set of all 2^n possible input states is obviously complete. While this may be a reasonable solution if n is very small, it rapidly becomes excessively clumsy as n increases. Fortunately, much smaller complete test sets usually exist, although *minimal* sets are not easy to find. We begin by considering the inverse problem: Given an input state, what single stuck faults can it find?

Example 3.25

Suppose that the input ABCDE = 11110 is applied to the circuit shown in Figure 3.44. What single stuck faults would be revealed by this test? Since B = 1, input h to AND-gate 2 is turned on. Then, assuming no fault, since A = 1, the signal a at the other input to AND-gate 2 is also on. This means that the AND-gate output is on, which turns

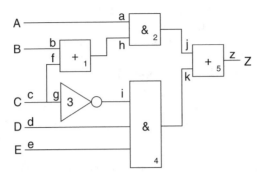

Fig. 3.44. Logic circuit being tested.

on OR-gate 5 and makes Z = 1. Therefore, if the result of the test is a 0-output, there must be a fault in the circuit. In order to determine what faults might force the output to 0, let us trace our way back from the output. Certainly a @0 fault at z (the node at the output of OR-gate 5) would do it. Observing that input k to OR-gate 5 is 0 (due to the E = 0 signal that turns off AND-gate 4), we see that j@0 would also lead to this result. (Note that j@0 can result from a fault at the output of AND-gate 2 or at the input to OR-gate 5, or in the wiring in between. We don't care which.) Working back to the source of signal j, we see that @0 faults at either input to AND-gate 2, i.e., at a or h, would account for the Z = 0 result. Signal h is produced by OR-gate 1. Since B = C = 1, *both* of the inputs to that gate are on. Therefore, under our one-fault assumption, neither b@0 nor f@0 would affect h. This illustrates a key point. If an input to an OR-gate is 1, then the output of that gate will be 1, *regardless* of the value of any other input to that gate. That is, a 1-input to an OR-gate makes the output of that gate insensitive to all other inputs to that gate. In this instance, two of the inputs are 1s, so that the gate output is insensitive to *both* of them. Applying this observation to OR-gate 5, we can see that, since j = 1, the gate output is insensitive to any change in k (which is 0 for the no fault case, so that it does *not* desensitize the gate to j). This means that there is no need to trace back from k. Thus we see that single @0 faults at a, h, j, or z constitute the complete set of faults detected by this test.

Next, the problem of determining which tests detect a particular fault is considered. Referring again to the Figure 3.44 circuit, suppose it is necessary to test for i@1. The set of all such tests can be found in a systematic fashion. Each test must satisfy the following two requirements:

1. If there is no fault, the signal at i must be a 0 (otherwise, i@1 would have no effect at all); and
2. the output at Z must be *sensitive* to the signal at i, in the sense that the input variables must have values such that a change in the signal at i must change the signal seen at Z (otherwise there would be no way of determining if the signal at i is incorrect; it is assumed that the only information available about the circuit state comes from a circuit output).

A more general statement of condition (1) is that, in order to be able to test for a fault at a node i, the signal at i must be *controllable*, in the sense that, it must be possible to set i at a specified value. In other words, at "test time," we must be able to inject inputs to the circuit that control the value of i. Furthermore, condition (2) states that the value of the signal actually present at i must be *observable*, in the sense that the value of some output signal must indicate what the value of i is. That is, at "test time," it must be possible to set the values of certain inputs so as to make at least one accessible output signal a function of i.

An examination of the circuit reveals that setting C = 1 is necessary and sufficient to make i = 0 and thereby to satisfy condition (1), the controllability requirement. In order to satisfy condition (2), the observability requirement, we must assign the other input values in such a way as to *sensitize* a path from i to Z. Starting at i, the signal must pass through AND-gate 4. If any other input to that gate is a 0, then clearly the output will be 0, regardless of the value of i. (This is the *dual* of the situation discussed earlier in connection with OR-gate 1.) Hence, D = E = 1 is necessary to sensitize the path from i through AND-gate 4 to k. Now the signal must pass through OR-gate 5 to reach Z. The output of this gate is sensitized to k only if its other input (signal j) is 0. Therefore, we must make sure that the other circuit inputs are such as to make j = 0. Working back from j, we come to AND-gate 2. The requirement that j = 0 is met if A = 0, or if h = 0. But, since C has already been specified at 1, there is no way to make h = 0. Hence, A *must* be set to 0. This completes the set of conditions for sensitizing the path from i to Z. We require A = 0 and D = E = 1. There is no constraint on B, because the only gate it feeds, OR-gate 1, is desensitized by the C = 1 input. Therefore the set of tests that detect i@1 is characterized by ABCDE = 0-111. For either of these input states, the Z output is 0 for no fault and 1 for i@1. (As suggested in Problem 3.48, these tests also detect other faults.)

The path sensitizing technique applied earlier is the basis for many detailed procedures for test generation, and so is worth restating in more general terms. A path from a node p to a node q is sensitized for a given input state if a change in the value of p causes a change in the values of all signals on that path, including q. If the path passes through a gate G, then all other inputs to G must be such that the output of G is sensitive to the path signal, in that it changes if the path signal changes. Thus, if G is an AND- or NAND-gate, all other inputs to that gate must be 1s. The dual situation applies to OR- or NOR-gates. Note that inverters are always sensitive to their inputs. Thus, to sensitize a path, all gates along the path must be examined, and their inputs constrained as described.

If a particular stuck fault has an effect on an output for some input state, then there must be a path from the fault node to that output that is sensitized for that input state. There may be more than one such path. Conversely, if no path can be sensitized for a particular fault, then that fault can never affect an output. Such a circuit is redundant in the sense that it could be redesigned with fewer gates. However, there are situations where redundancy serves a useful function, for example, in combating combinational logic hazards (Subsection 6.2.6). (In such cases the circuit can be made testable by adding one or more extra inputs to disable redundant elements during tests.) The application of path sensitizing to overall circuit testing is treated next.

For circuits with moderate numbers of inputs, exhaustive testing may be a reasonable approach. It has the great advantage of ensuring that *all* hard faults will be detected. But if a circuit has, say, 20 or more inputs, the time involved might be significant and the results of all of the tests would have to be stored somewhere, implying a cost in memory. It is therefore worth considering how smaller complete test sets might be generated. Testing the circuit of Figure 3.44 by applying all inputs would require $2^5 = 32$ tests. We shall do a great deal better with the following approach that is very straightforward and produces good, though generally not optimum, solutions, with reasonable amounts of computation (carried out on computers for large problems).

1. Form a set L of all faults to be checked.
2. Select any member of L and find a test for it by sensitizing an appropriate path.
3. Find the set of all faults in L that this test detects and delete them from L. If L is not empty, go back to step (2), else stop.

Let us see how this applies to the Figure 3.44 circuit. The points at which stuck faults might occur are indicated in the diagram with lowercase letters. Note that, corresponding to the point at which input C enters the circuit, there are three labeled nodes (c, f, g). How could one of these nodes be stuck at some value without having all of them stuck at the same value, since they are all connected together? The answer is that stuck faults can be caused by conditions that do not propagate along wires. If, within OR-gate 1, the f input wire were shorted to the V_{dd} terminal, putting a voltage corresponding to a logic 1 on that line, this *would* make the signals at c, f, and g all @1. But if this is an nMOS gate and the transistor to which f is connected became open circuited, the effect would be f@0, but this would have no effect on g. Of course, c@1 *would* mean both f@1 and g@1. The total number of possible single stuck faults is 24. Now observe that any test for b@0 must involve sensitizing the path that goes through h, j, and z, since this is the *only* path to the output. When this path is sensitized, paths from h and j to z are obviously also sensitized, as they are subpaths of the longer path. Thus a test for b@0 is also a test for h@0, j@0, and z@0. In such cases, we say that the last three faults are *dominated* by b@0, so they need not be placed on the fault list. Wherever there is a unique path from a node to the output, there will be such dominated faults. The reduced fault list for this problem is thus {@0: a, b, c, d, e, f, g; @1: a, b, c, d, e, f, g.}

A test for a@0 is ABCDE = 11000. This is also a test for b@0, so both may be deleted from the list. A test for a@1 is 01001, which also sensitizes a path from d, thereby testing for d@1 and removing that fault from the list. Choosing b@1 leads to test 10010, which also tests for c@1, f@1, and e@1. A test for c@0 is 10100, which also checks for f@0. Testing for d@0 is accomplished by 00011, which also tests for g@1 and e@0. This leaves only g@0 to be tested, and 00111 accomplished this. Thus a complete test set for this circuit consists of the following six tests (the symbol after the slash indicating the test result for no fault): {11000/1, 01001/0, 10010/0, 10100/1, 00011/1, 00111/0}. This happens to be a minimal complete test set, comparing very favorably with the 32 tests corresponding to the application of *all* input states.

OVERVIEW AND SUMMARY

Minimizing logic is important because it can lead to savings in chip area and in power consumption (important in high-performance systems because it eases the heat disposal burden). It also increases reliability by reducing the number of points at which failures can occur, and often makes the resulting design easier to understand. There are situations, however, in which these benefits are less important. If, for example, the circuit to be implemented on a chip takes up only a fraction of the space available for it, then area is not a factor. There are several possible reasons why the circuit outputs for certain input states may be optional (a don't care condition), e.g., if an input state can never occur. Don't cares can be useful to simplify designs. Prime implicants (pi's) of a function are *p*-terms with no redundant literals. They are the building blocks from which minimal SOP expressions are constructed. Finding such expressions consists of selecting appropriate sets of pi's, since various irredundant sums of pi's equivalent to the same function may have different costs. A powerful tool for finding minimal SOP expressions is the Karnaugh map (K-map), which has many other uses in helping us visualize problems involving Boolean variables. K-maps can be thought of as representations of n-dimensional cubes in which the vertices, corresponding to input states, are represented by cells. Or they can be thought of as Venn diagrams, with the cells representing intersections of regions, each of which corresponds to a particular variable being on. Entries in variable-entered maps are not limited to 1s, 0s, or $-$'s (don't cares); they may also be variables or even functions of variables. This extension often makes it possible to deal with functions of more variables than would otherwise be feasible.

In terms of K-maps, pi's correspond to maximal filled subcubes. The key to using K-maps to find minimal SOP expressions is to find 1-points of the functions for which arguments can be made that particular pi's cover those points at least as economically as any other pi's. A pi Q covering a particular 1-point can safely be selected as part of the minimal-sum expression if no other pi covering that 1-point is cheaper, *and* if no other pi covering that 1-point also covers another 1-point not covered by Q. The selection of a pi may facilitate subsequent choices by altering the relative covering power of competing pi's. When no choices can be made on this basis, an enumeration process called branching is usually necessary.

An alternative approach to finding minimal logic expressions is the Quine-McCluskey method. The data structure and the individual steps of this process are tailored for implementation on a digital computer. It consists of two distinct phases. In the first, the set of *all* pi's is generated by the repeated, systematic application of the uniting theorem.

In the second phase, a covering table is constructed that shows exactly which pi's cover each 1-point of the function. A systematic method, incorporating branching as a last resort, is then applied to find an optimal selection of pi's that covers all the 1-points. For hand computation, K-maps are much easier to use. For the multi-output case, the situation is rather more complicated. The concept of the essential prime implicant is generalized to the multi-output prime implicant, or mopi. These are pi's of the various products of subsets of the functions to be realized.

If we consider the 0 and 1 signals of logic circuits to be numbers, then we can talk about logic functions as being positive or negative (monotone increasing or decreasing) in the same sense that we apply the same description to ordinary algebraic expressions. Unate functions are those that, with respect to each variable, are either positive or negative. Unate functions have interesting and useful properties. For example, all their pi's are essential.

The potential advantages of multistage logic are reductions in gate inputs, number of gates, and fan in. Although there are no really good general algorithms for the design of multilevel logic, the careful use of factoring and the exploitation of common subfunctions is generally effective. XOR-gates can be efficiently realized with complex CMOS gates, or, perhaps most efficiently, with a pass-transistor circuit using a CMOS transmission gate.

The testing of logic circuits is an integral part of the manufacturing process, as well as a necessity for maintaining digital equipment. *Stuck* faults (@0 and @1) are the basis for most fault-detection procedures. For combinational logic, path sensitization is a basic technique for finding tests for particular faults, and can be used systematically to find near-minimal complete test sets for a circuit.

SOURCES

The four-dimensional hypercube of Figure 3.3 was introduced in [Unger, 1953] along with similar diagrams for higher dimensions and another way to represent hypercubes more simply. K-maps were introduced in [Karnaugh, 1953], and the Quine-McCluskey method was introduced in [McCluskey, 1956] (see [McCluskey, 1986] for the latest refinements). Additional material on testing can be found in [McCluskey, 1986]. The XOR-gate of Figure 3.43 is in [Mukherjee, 1986]. More on testing can be found in [Breuer, 1976], [Abramovici 1994], in [Fujiwara, 1985], in [Miczo, 1988], and in [McCluskey, 1986].

PROBLEMS

3.1. For each of the following expressions, calculate the costs in gates and in gi's of the logic circuits to which they correspond directly, assuming only uncomplemented inputs, and determine the mpl, the maximum number of stages through which any signal must pass. (One approach, of course, is to draw the circuits; but this is an unnecessarily lengthy process. Readers should learn to make the calculations directly from the expressions.)

   *(a) $A\overline{B}C + D + B\overline{C}$

   (b) $A(B\overline{C} + D)(E + F\overline{G})$

   (c) $AB + C(\overline{DE} + \overline{A}(\overline{B} + CD))$

3.2. Show that a *different* literal might have been deleted from $\overline{A}BC$ in Section 3.2. (p. 47) Could we delete both this other literal and C without changing the value of f?

3.3. Which, if any, of the following expressions are implied by $(A + B)C\overline{D}$?

   (a) $B\overline{D} + AC$

   (b) $\overline{A}\,\overline{B} + C\overline{D}$

   (c) $\overline{B}D + (\overline{A}B + A\overline{B})C$

   (d) $\overline{B}C + AB + \overline{A}\,\overline{D}$

3.4. Using algebraic theorems, then truth tables, then K-maps, prove that expressions (3) and (4) of Section 3.2 (p. 48) are equivalent.

3.5. If $AB\overline{C}$ is a pi of the function $f(A, B, C, D)$, then which of the following p-terms can*not* possibly be pi's of f? $AB, BCD, AB\overline{C}D, A\overline{B}, ABC, \overline{A}C, \overline{B}.$

3.6. (a) How many different 3-cubes are there in a 4-cube?

   (b) If $j > i$, then how many different i-cubes are there in a j-cube?

3.7. *(a) Within a 5-cube, how many different 3-cubes contain each 1-cube?

   (b) Within a 5-cube, how many different 4-cubes contain each 2-cube?

   (c) Within a k-cube, how many different j-cubes contain each i-cube, where $k > j > i$?

3.8. Plot each of the following expressions on K-maps; do this directly, without expanding to sums of minterms.

   *(a) $A\overline{B}CD + \overline{A}\,\overline{D}E + BDE + \overline{B}CD + BC\overline{E}$

   (b) $ACD + A\overline{B}\,\overline{C}E + \overline{B}\,\overline{C}EF + \overline{A}CF$

   (c) $(A + \overline{B})C + \overline{A}(\overline{C} + D)$

3.9. Plot the following function on a K-map and then find a minimal-sum expression for it:

$$Z = \overline{B}\,\overline{C} + \overline{A}BD + ABC\overline{D} + \overline{B}C.$$

3.10. (a) How far apart are the points 11011 and 00001?

(b) How many different minimal-length paths are there between the points 11011 and 00001?

3.11. (a) Using the same principles governing the construction of K-maps for two through six variables, sketch a K-map for seven variables. Indicate the various types of neighboring pairs of points.

(b) Repeat for eight variables.

3.12. Use a K-map to find two different Gray code sequences of nine 4-bit words that start at 1100 and end at 1010.

3.13. Use K-maps to find Gray code sequences of 4-bit words for each in the following cases (otherwise explain why it can't be done):

(a) A closed path containing 13 states including 1111.

(b) A path containing 10 states starting at 0000, ending at 0001, and including 1010.

3.14. Find minimal-SOP and minimal-POS expressions for each of the following functions:

*(a) $f(A, B, C, D) = \Sigma m(2, 7, 11, 12, 13, 14, 16, 17)$

(b) $f(A, B, C, D) = \Sigma m(2, 4, 7, 10, 11, 12, 13, 14, 15, 16, 17)$

3.15. Find a minimal two-stage NAND-gate realization for the function plotted on the K-map of Figure 3.45. Then find a minimal two-stage NOR-gate realization for the same function.

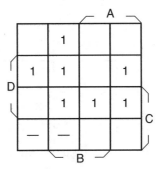

Fig. 3.45.

3.16. Find minimal-SOP expressions for each of the two functions plotted in Figure 3.46.

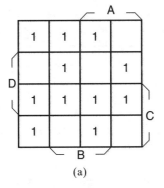

(a)

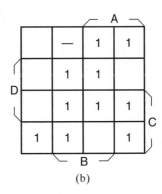

(b)

Fig. 3.46.

3.17. Find a minimal-SOP expression for the function plotted in Figure 3.47.

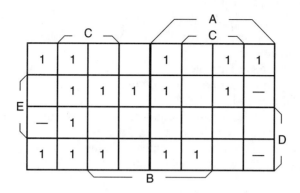

Fig. 3.47.

3.18. Find a minimal-SOP expression for the function plotted in Figure 3.48.

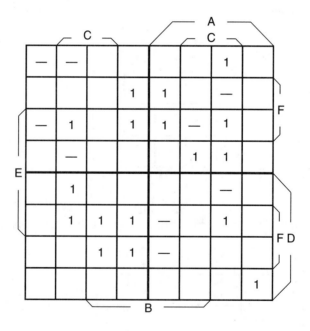

Fig. 3.48.

*3.19. The octal numbers grouped after the Σd symbol represent don't care points. Find a minimal-SOP expression for the function:

$$f(A, B, C, D, E) = \Sigma m(1, 5, 7, 13, 17, 21, 31, 33) + \Sigma d(0, 4, 12, 20, 32).$$

3.20. Find a minimal-SOP expression for the function:

$$f(A, B, C, D, E, F) = \Sigma m(0, 1, 10, 14, 16, 23, 32, 33, 36, 44, 50, 63, 74, 75, 77)$$

$$+ \Sigma d(3, 4, 11, 21, 40, 54, 64, 67, 70, 73).$$

3.21. As mentioned at the end of Subsection 3.3.3, it is sometimes appropriate to represent a K-map in a nonstandard form. Construct a 4-map with two rows (representing one variable) and eight columns (representing three variables). Plot the function of Problem 3.9 on this map and find a minimal SOP expression for it.

3.22. Use a K-map to find *all* of the pi's of the function described by:

$$Z = (\overline{A} \oplus B)(\overline{C} + D) + (A \oplus B)(C + \overline{D}).$$

3.23. For Figures 3.22 through 3.24 (Subsection 3.3.5, on p. 69, dealing with variable-entered maps), plot the functions on K-maps in the conventional way and find minimal SOP expressions to check against the given solutions.

3.24. Find minimal-SOP expressions for each function mapped in Figure 3.49.

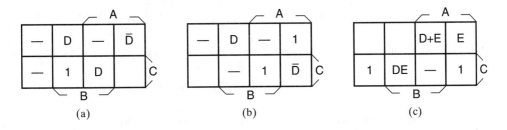

(a)   (b)   (c)

Fig. 3.49.

3.25. Find a minimal-SOP expression for the function mapped in Figure 3.50.

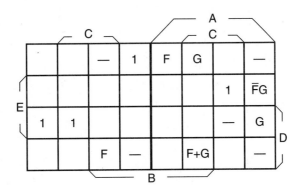

Fig. 3.50.

3.26. Find a minimal-SOP expression for the function mapped in Figure 3.51, in which the circle around the D means that there is a don't care in that position for D = 1, and a 0 there for D = 0.

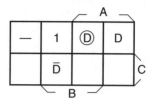

Fig. 3.51.

3.27. Use the Quine-McCluskey method to find all the pi's of the function:

$$f(A, B, C, D) = \Sigma m(0, 3, 4, 6, 7, 10, 13, 15) + \Sigma d(2, 11, 17).$$

3.28. Use the Quine-McCluskey method to find all the pi's of the function:

$$f(A, B, C, D, E) = \Sigma m(1, 3, 6, 10, 16, 22, 23, 27, 36, 37) + \Sigma d(7, 17, 26, 31).$$

3.29. Find minimal-cost coverings for each of the tables in Figure 3.52.

	1	2	3	4	5	6
a(3)	x			x		
b(3)		x		x	x	
c(3)	x				x	x
d(2)	x		x	x		
e(2)					x	x

(a)

	1	2	3	4	5	6
a(2)	x	x		x		
b(3)		x		x		x
c(3)			x	x	x	
d(3)		x			x	x
e(4)	x		x		x	

(b)

Fig. 3.52.

3.30. Find a minimal-cost covering of the table in Figure 3.53.

	1	2	3	4	5	6	7
a(1)	x	x		x		x	
b(1)		x	x	x			x
c(2)	x		x	x			
d(2)	x		x		x		
e(2)			x			x	x
f(1)		x			x	x	x

Fig. 3.53.

*3.31. Using the Quine-McCluskey method, find minimal-SOP expressions for each of the following octally specified functions:
(a) $f(A, B, C) = \Sigma m(2, 5, 7) + \Sigma d(1, 6)$
(b) $f(A, B, C, D) = \Sigma m(1, 3, 11, 13, 16) + \Sigma d(0, 7)$

3.32. Using the Quine-McCluskey method, find minimal-SOP expressions for the function:

$$f(A, B, C, D, E) = \Sigma m(0, 1, 3, 10, 12, 13, 17, 20, 22, 23, 33, 35, 37) + \Sigma d(15, 31).$$

Then solve the same problem with K-maps.

3.33. In the Quine-McCluskey process for generating pi's, how many times should a j-cube be generated? Is the answer dependent on the total number of variables?

3.34. Assuming double-rail inputs, use K-maps to find minimal two-stage NAND-gate circuits realizing each of the following multi-output three-variable functions.
*(a) $Z_1 = \Sigma m(2, 3, 6) + \Sigma d(4)$; $Z_2 = \Sigma m(1, 4, 5) + \Sigma d(6)$
(b) $Z_1 = \Sigma m(0, 1, 2) + \Sigma d(4)$; $Z_2 = \Sigma m(2, 3, 7) + \Sigma d(4)$

3.35. Assuming double-rail inputs, use K-maps to find a minimal AND-to-OR realization for the following multi-output function. Specify the logic expressions, appropriately indicating shared gates; you need not draw the circuits.

$$Z_1 = \Sigma m(0, 2, 5, 6, 7) + \Sigma d(1) \quad Z_2 = \Sigma m(5, 6) + \Sigma d(7) \quad Z_3 = \Sigma m(0, 1, 3, 5).$$

3.36. As in the previous problem, find the logic expressions for a minimal two-stage AND-to-OR realization of the pair of functions mapped in Figure 3.54.

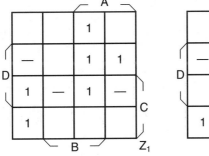
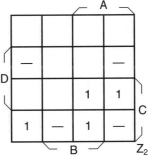

Fig. 3.54.

3.37. Let inputs $X_1X_2X_3X_4$ be the bits of a BCD digit Q (X_1 most significant), and let $Z_1Z_2Z_3Z_4$ be the BCD representation of the value, modulo-10, of 3Q. Assuming double-rail inputs, use K-maps to generate logic expressions for the Z_i corresponding to a minimal, multi-output, SOP logic circuit.

3.38. Solve Problem 3.34(b) using the Quine-McCluskey method.

3.39. For each function defined below, determine if it is monotone (and if so, in what direction) with respect to each of the variables. For each of the functions that are unate (if any), specify a 1-point common to all of its pi's.

    (a) $Z = (\overline{A}\overline{B} + AB)\overline{C}$

  *(b) $Z = \Sigma m(1, 4, 5, 7, 15, 17)$

    (c) $Z = \Sigma m(4, 6, 7, 13, 14, 15, 16, 17)$

    (d) $Z = \Sigma m(0, 1, 2, 10, 11, 12, 13, 14, 15, 16)$

3.40. Identify $m(Q)$ for each pi Q of the functions mapped in Figure 3.37 (p. 83).

3.41. (a) Construct a unate function $f(A, B, C)$ that is positive in B and negative in both A and C.

    (b) Construct a unate function $f(A, B, C, D)$ that is positive in both A and B, and negative in both C and D.

3.42. Every pi of a unate function is essential. The converse of this theorem might be stated as "every function whose pi's are all essential is unate." Either prove the validity of this converse statement, or else show, by means of a counterexample, that it is false.

3.43. Specify expressions corresponding to multilevel logic circuits for each of the functions specified below. Try to minimize the cost in (gi's). Is it possible in every case to reduce the gi count by increasing the number of stages? In each case, determine the gi count, the g count, the maximum fanin (mfi), and the maximum path length (mpl) for the original and final expressions. Assume single-rail inputs.

  *(a) $Z = AB + AC + D$

    (b) $Z = ABCD + AB\overline{C}\overline{D} + E$

    (c) $Z = ABC + ADE + F$

    (d) $Z = ABC + ABD + CEF + DEF$

3.44. Find an equivalent to the following expression that minimizes gi. Show what happens to g, the mfi, and the mpl:

$$Z = AD + BCD + AEF + BCEF$$

3.45. Show how to realize the following function with mfi of 2 and with gi less than 13. Then find another realization with a smaller value of gi, but with mfi = 3:

$$Z = ABC + ADE + BCDE$$

3.46. Construct a three-term SOP expression such that a nontrivial factoring *increases* gi by 1. Is it possible, by a single factoring operation, to do worse than this for such an expression?

3.47. Construct a three-term SOP expression such that a nontrivial factoring *increases* g by 1. Is is possible, by a single factoring operation, to do worse that this for such an expression?

3.48. For the circuit of Figure 3.44, what is the complete set of faults detected by each of the tests in the set ABCDE = 0–111?

3.49. Refer to the circuit of Figure 3.55.
(a) Find the set of all tests for a@0.
(b) Find the set of all tests for b@1.
(c) Find all the single stuck faults detected by ABCD = 1100.
(d) Repeat (a), (b), and (c) assuming that the NOR-gates have been replaced by NAND-gates.

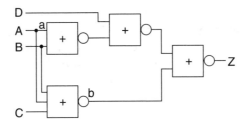

Fig. 3.55.

3.50. (a) Find the smallest complete test set (or at least a reasonably small set) for the circuit of Figure 3.55.
(b) Repeat (a) for the NAND-gate version of the same circuit.

3.51. Refer to the logic circuit of Figure 2.5 (p. 25).
(a) Find a test for a @1 fault at the C + D input to the lower left AND-gate.
(b) Find a test for a @1 fault at the C + D input to the upper left AND-gate. Is there any solution for this part that also satisfies (a)?
(c) Find a test for a @1 fault at the output of the OR-gate generating C + D.
(d) Are there any solutions to (a) or (b) that do *not* also satisfy (c)?

3.52. Find a minimal complete test set for the tree of XOR-gates shown in Figure 5.15 (see p. 142).

3.53. For the circuit of Figure 3.56, show that 11110 is a test for g@0, and that 11111 tests for h@0. Now note what happens if *both* g and h are @0. Find a pair of tests that detects both of these single stuck faults and *also* detects the double fault.

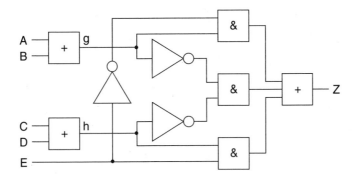

Fig. 3.56.

3.54. Modify the circuit of Figure 3.44 by deleting the inverter. Verify that 11011 is a test for a@0, and that 10000 is a test for c@1. For the *double* fault a@0 and c@1, which, if either, of the preceding tests indicates a failure?

3.55. Show how to realize the complement of a 2-input XOR-gate in CMOS technology using six transistors. Inputs are single-rail and outputs are never degraded (i.e., they swing from 0 to V_{DD}). (*Hint:* See Figure 3.43.)

3.56. Show how to realize a 2-input AND-gate with a CMOS pass-transistor circuit using five transistors. Assume single-rail inputs. No output is to be degraded.

3.57. Suppose a 32-bit parity-check circuit is to be realized in CMOS. Assume the inputs (single-rail) come from circuits having paths to ground or Vdd through one transistor. A constraint, related to timing, is that no paths to the output should go through more than four transistors. Specify an economical design, base on XOR-gate designs presented in Section 3.8.

4
Combinational Logic Circuits in Regular Forms

Regularity is a highly desirable feature in the design of almost anything. This is certainly true for the design of highly complex systems such as logic circuits. The most time-consuming and error-prone part of the manufacturing process for integrated-circuit (IC) chips is the production of the masks. These are detailed pictures of the various layers of wiring and types of regions that constitute the device. If substantial areas on the chip are composed of repetitive patterns, this eases the problem of producing correct masks. This is true even if the repetition involves several, but not all the masks, e.g., if it involves the layout of the transistors and one or two layers of wiring. The subject of Section 4.2 is methods for realizing arbitrary combinational logic functions with arrays that are largely regular.

Another kind of regularity is treated in Section 4.1. This consists of the use of standard subcircuits that have been designed previously and perhaps used in other chips. Any time that previously developed designs can simply be copied, substantial savings in time and effort result. In the next chapter the implementation of regular *functions* is treated.

4.1. IMPORTANT LOGIC MACROS

There are a number of classes of logic functions that perform operations important in a wide variety of situations. Hence, their realizations and uses are worth special attention. They may be thought of as building blocks helpful in constructing more elaborate systems. Those familiar with software will recognize the analogy with the macros and subroutines used by programmers.

In modern digital systems, logic macros are used in two ways. In one form, they are used as off-the-shelf components in the form of small- or medium-scale integrated-circuit chips. Although they are still used in this form, the advent of large-scale integration (LSI) and very-large-scale integration (VLSI) technology has made them more important as on-chip modules. That is, decoders, multiplexers, and other logic macros are incorporated in complex systems appearing on large chips. They are often injected onto the chips by means of standard designs generated by computer programs. These programs usually allow key parameters to be specified by the logic designer. Therefore, designers are not limited to a limited number of available module sizes, as is the case when the aforementioned off-the-shelf units must be used. Less waste is involved, since, for example, it is not necessary to use a 4–16 decoder when a 4–12 decoder is needed.

4.1.1. Decoders

A common need in digital systems is to indicate a choice of one out of n objects. For example, it might be necessary to indicate which of n registers is to receive a data word, or to indicate which of n instructions is to be executed. This information could be transmitted as an n-bit word, in which a 1-signal appears in the position corresponding to the selected object and 0s appear in all other positions. But where a price must be paid for each bit used, it is desirable to convey the information by means of a shorter word. This can be done by encoding the choice—perhaps by using a binary number. Since a k-bit binary number can take on any of 2^k values, it is clear that, instead of using an n-bit word to indicate our choice, a word with about $\log_2 n$ bits is sufficient.

When data is encoded in such a manner, it is necessary to employ a logic circuit to *decode* the data. Such a circuit, called a *decoder*, accepts as its input a k-bit binary word. It generates a 1-signal at exactly one of its n output terminals, corresponding of course to the selection implied by the input. An example of such a circuit is a 2–4 decoder (pronounced "two to four decoder"), which has two inputs and four outputs. It behaves according to the truth table of Figure 4.1a, and is represented by the symbol shown in Figure 4.1b.

Realizing such a device is a straightforward matter. Complements of the inputs must be generated, and then a single AND-gate is used for each output, as illustrated in Figure 4.2. The use of NOR-gates instead of AND-gates does not add to the complexity of the circuit if we consider the NOR-gate to be an AND-gate with complemented inputs as shown in Figure 2.10b.

Where the number of inputs is large enough to cause a fanin problem, two or more stages of smaller decoders can be used. This technique is illustrated in Figure 4.3 where the outputs of two 2–4 decoders are fed in pairs (one from each) to 16 2-input AND-gates. In the illustration, one of the 16 final-stage AND gates is shown. It generates the output Z_{10} from the two signals at the intersection—one from the decoding of $X_0 X_1$ and the other from the decoding of $X_2 X_3$. In this case, a total of four inverters (for the inputs), and 24 2-input AND gates are used (four AND-gates for each of the first-stage decoders), for a total of 52 gi. A one-stage realization of the type depicted in Figure 4.2 would require four inverters and 16 4-input AND-gates, for a total

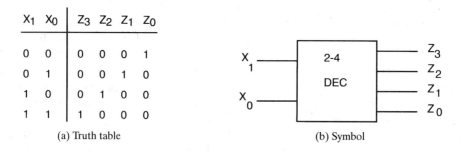

X_1 X_0	Z_3 Z_2 Z_1 Z_0
0 0	0 0 0 1
0 1	0 0 1 0
1 0	0 1 0 0
1 1	1 0 0 0

(a) Truth table   (b) Symbol

Fig. 4.1. 2–4 decoder.

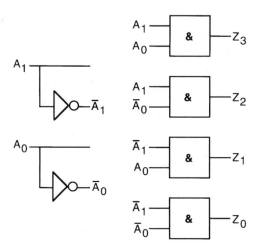

Fig. 4.2. Realization of a 2–4 decoder.

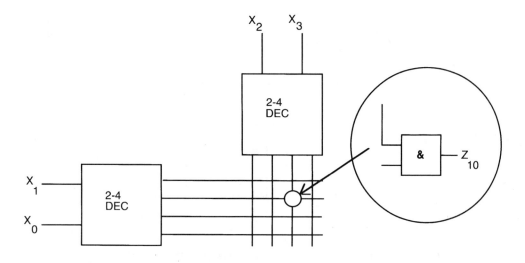

Fig. 4.3. Realization of a 4–16 decoder using two 2–4 decoders.

of 68 gi. This technique is commonly used in the address circuits of random-access memories (RAMs).

A feature found in many decoders is an *enable* input. The enable signal, En, is ANDed to all the decoder outputs. That is, En would be an additional input to each of the AND-gates shown in Figure 4.2 or to each of the AND-gates at the intersections in Figure 4.3. Thus, when En = 1, the decoder is said to be *enabled* and it behaves as just discussed. When En = 0, the decoder is said to be *dis*abled, and all of its outputs are forced to 0.

One application of the enable input is a method of combining decoders to produce larger decoders. This is illustrated in Figure 4.4, where two 2–4 decoders and a 1–2 decoder (this consists of just a single inverter) are interconnected to form a 3–8 decoder. Readers should have no difficulty in verifying the validity of the method and in generalizing it to more complex cases.

A decoder with an enable input can also be used as a *demultiplexer* or *signal distributor*. That is, we can use it to place a given input signal on any of n output leads. This emulates the behavior of the mechanical switch shown in Figure 4.5a, where the A-signals serve to control the position of the switch arm. (Except that, unlike the mechanical switch, the signal flow through the decoder is only from the X-input to one of the Z-outputs, never in the reverse direction.) A 2–4 decoder configured as a demultiplexer is shown in Figure 4.5b.

Another interesting application of a decoder that is occasionally useful is to implement arbitrary logic functions. Assume, for example, that the pair of signals A, B must be decoded and so is being fed to the input of a 2–4 decoder. Now suppose that we also need to implement some other function of A and B, say A $\oplus$ B. Observe that each

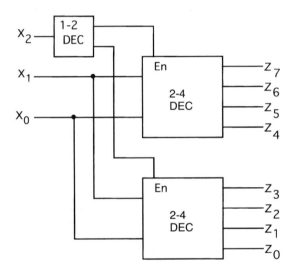

Fig. 4.4. Using decoders as building blocks to produce a larger decoder.

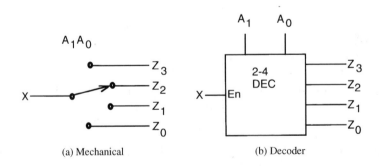

Fig. 4.5. A signal distributor implemented mechanically, and with a decoder.

output of a decoder realizes a minterm of the set of input variables. We can thus implement any function of the input variables, by feeding the outputs corresponding to the minterms of that function to an OR-gate. In this case, to implement $A \oplus B$ we connect the Z_1 and Z_3 signals to the OR-gate, as shown in Figure 4.6. It is generally not economical to use a decoder in this manner to realize a single function. It amounts to generating *all* the complete product terms of the input variables and then using a subset of them to realize the desired function. This technique should be considered only if (1) the decoder is needed for other purposes (as was assumed in our example), or (2) in some special cases where a sufficiently large number of functions of the same set of variables must be realized and the time of the designer is a factor.

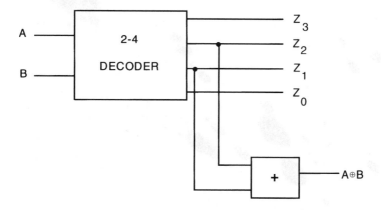

Fig. 4.6. Implementing $A \oplus B$ with a 2–4 decoder and an OR-gate.

The term *decoder* is also applied to circuits that perform such tasks as converting binary-coded decimal (BCD) data to drive a 7-segment display element.

4.1.2. Multiplexers

The Figure 4.5 circuit gates a signal to one of n outputs ($n = 4$ in the example shown). The inverse is a circuit that selects one out of n inputs and gates it to a single output terminal. Such a circuit, called a *multiplexer* or *selector*, is shown in Figure 4.7, with

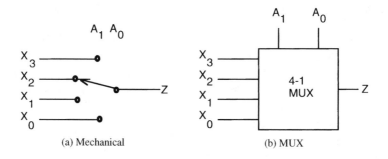

Fig. 4.7. A mechanical switch that performs the MUX function, and the MUX circuit symbol.

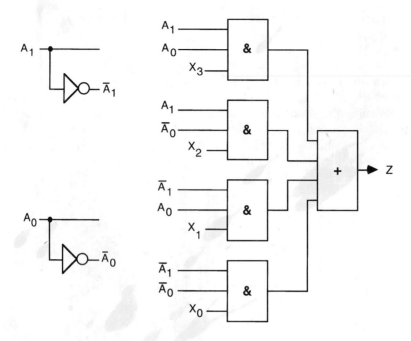

Fig. 4.8. Realization of a 4–1 MUX.

$n = 4$. It is often referred to as a *MUX*. Note that in both parts of the figure the signal flow is from left to right.

MUXs are only slightly more difficult to implement than decoders, the principal difference being the need for a second-stage OR-gate. A realization of a 4–1 MUX is shown in Figure 4.8.

Because of the importance of the MUX, it is worth spending a bit more time on how it can be designed. In particular, we shall show how it can be realized with a complementary MOS (CMOS) pass-transistor circuit. We start with the 2–1 MUX. It is easy to see how this can be realized with three 2-input NAND-gates (see Figure 4.9a). In CMOS this requires 14 transistors, including the inverter needed for A. Using a pair of transmission gates (see Figure A.17, on p. 313), a very satisfactory version can be constructed with just six transistors (again including an inverter—not in the figure), as shown in Figure 4.9b.

The 2–1 MUX can be used as a building block to construct larger MUXs in tree form. For example a 4–1 MUX made this way is shown in Figure 4.10. In general, it takes $k-1$ 2–1 MUXs to build a k–1 MUX. Only one inverter is needed for each of the A_i-variables, so for a 2^k–1 MUX, the number of inverters needed is k, and the total number of transistors needed, if we use transmission gates, is $2^{k+2} + 2k - 4$.

It is often necessary to switch, not just single bits, but multibit words. A MUX designed to select any one of n k-bit words is referred to as a *j-wide n–1 MUX*. It can be implemented as j simple n–1 MUXs. But, especially if j is large, say 8 or greater, logic can be saved by using a decoder to generate selection signals to control the MUX AND-gates, each decoder output feeding j AND-gates

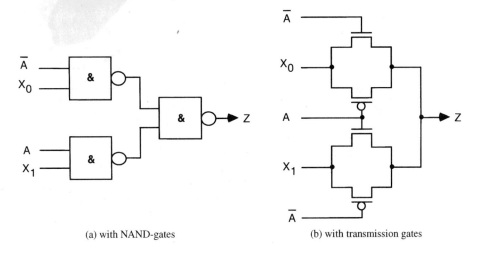

(a) with NAND-gates   (b) with transmission gates

Fig. 4.9. 2–1 MUX.

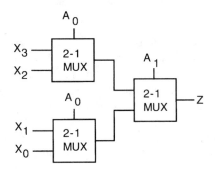

Fig. 4.10. Realization of a 4–1 MUX.

Another application of MUXs is to implement *scalers* or *shifters*, combinational circuits used to shift strings of 0s and 1s one or more places to the left or right. These are used, among other places in arithmetic logic units (ALUs) of digital computers.

A typical 8-bit scaler might work as follows: If the input signal is the binary vector $\mathbf{X} = 10010101$, and if the 3-bit control signal vector, $\mathbf{S}$, to the scaler calls for a right shift of two places, then the output will be the vector $\mathbf{Z} = 00100101$. A realization of a simple 4-bit scaler is shown in Figure 4.11. When control signals S_1 and S_0 are set to 00, there is no shift, i.e., $Z_i = X_i$. Control states 01, 10, and 11, respectively, call for right shifts of one, two, and three positions. Note that 0s are shifted in from the left side. One variation of the scaler is to make the shift circular in the sense that, for a right shift, the rightmost bits, rather than 0s, would be shifted into the left end.

In the previous subsection, it was shown how a decoder can be used to realize arbitrary logic functions, one OR-gate being required for each function. It is possible to implement any one arbitrary k-variable logic function with a 2^k-1 MUX without the use of any other components. As typified by the logic circuit of Figure 4.8, an $n-1$ MUX can be considered as consisting of n AND-gates feeding an OR-gate, with each AND-gate

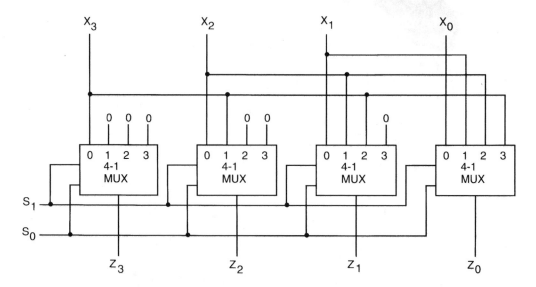

Fig. 4.11. Realization of a 4-bit scaler using MUXs.

corresponding to one of the complete product terms of the A-variables multiplied by one of the n X_i-inputs. A function can be realized simply by using the X-inputs to "enable" precisely those AND-gates that correspond to its minterms.

Suppose that we wish to realize the function $f(A_0, A_1) = A_1\overline{A}_0 + \overline{A}_1 A_0$. Referring to the Figure 4.8 circuit, this can be accomplished by setting $X_1 = X_2 = 1$ to select the two minterms of this function (which, incidentally, is $A_0 \oplus A_1$), and setting $X_0 = X_3 = 0$ to reject the other complete product terms. In other words, fixed 0- and 1-signals are connected to the X-inputs of the MUX, and the function variables are connected to the A-inputs.

Actually, we can do a little better. If, in addition to fixed 0s and 1s, signals corresponding to another input variable (or its complement) can be connected to the X-terminals, then any function of $k + 1$ variables can be realized with a 2^k MUX (see Problem 4.11). For circuits implemented as part of larger systems on LSI or VLSI chips, there is little reason to consider this use of a MUX, since it almost always wasteful in its use of chip area; functions are, in effect, implemented by generating many minterms rather than a smaller number of cheaper p-terms. On the other hand, where small-scale integrated circuits are being used, it may sometimes make sense to use a standard MUX chip to implement a complete function, rather than using one or more chips, each with a small number of gates on it.

4.1.3. Encoders

The inverse of a decoder is, as we might expect, called an *en*coder. A full encoder has 2^n input terminals and n output terminals. Assuming that exactly one of the input signals is on (i.e., has the value 1), the output state corresponds to a binary number indicating which input is on. Such a device is easily implemented with an OR-gate for each output. Each input signal is connected to the output gates corresponding to the positions where its binary number representation has a 1. Thus, if $n = 3$, the X_6-input would be connected to the OR-gates in positions Z_2 and Z_1, but not to the Z_0 output (since 6 is written as 110 in binary). Note that the X_0-input is never used. A 4–2 encoder is shown in Figure 4.12. Encoders of this type are not very useful, perhaps because of the constraint that only one input at a time may be on. A more important device is a variation known as a *priority encoder*. This has the same general form as the ordinary encoder, but there is no restriction on the number of inputs that may be on at any time. The inputs X_i are assumed to be ordered, with the higher priority inputs having the highest values of i. The output state is a binary encoding of the highest priority i among the inputs that are on. Thus, if X_2, X_5, and X_7 are all on, and the other inputs are all off, the output would be an encoding of 7. (According to this specification, if *none* of the inputs are on, the output would be all 0s, the same output as if X_0 were the only input on. This would be all right if there were some other indication available of the no-input-on condition. Otherwise some variation of the specifications is necessary.) A design for a 4–2 priority encoder is shown in Figure 4.13; note that X_0 is not shown among the inputs.

Encoders are far less widely used than are decoders or MUXs. Probably the most common application of priority encoders is in digital computers of all sizes, where, when multiple interrupts occur, the most urgent one must be identified.

4.2. Implementing Irregular Logic with Regular Arrays

A large proportion of the logic in most digital systems is there to perform operations that are regular in the sense that they are easily described and are repetitive in some sense. One example is the gating employed to transfer data between various registers. Although the number of gates needed to control the transfer of a 32-bit word is about double that required for a 16-bit word, the logic has the same simple pattern in both

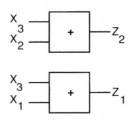

Fig. 4.12. A 4–2 encoder.

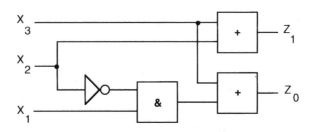

Fig. 4.13. A 4–2 priority encoder.

cases. The same is true for the portion of logic that operates on the data to execute arithmetic operations. An important class of regular functions is treated in the next chapter.

There are also substantial amounts of *ir*regular logic in most systems. Frequently referred to as *random* logic, this part of a system does not fall into any simple pattern. The decision-making logic that generates the signals controlling the transfer of data within a machine is an important instance of random logic. From the point of view of the logic designer, regular logic is easier to implement, in that the amount of thinking required per gate in the eventual circuit is much smaller. Furthermore, when it comes to the work of planning the physical layout of the circuit on a chip and the generation of the masks required for chip production, regularity has great advantages.

Various techniques have been developed for realizing random logic in regular forms. The aforementioned methods for realizing arbitrary logic functions with decoders or MUXs might be thought of as being in this category, although, as was pointed out, they are of very limited value. More powerful methods are the subject of this section.

4.2.1. ROM

A *read-only-memory* (ROM) is a device with k address inputs, designated as **A**, and n outputs. For each state of **A**, the output state may be arbitrarily specified by the designer. Thus, a ROM behaves very much like a random-access memory (RAM), except that, during operation, the contents of memory cannot be changed; i.e., we can only *read* the memory—hence the name. (There are variations that allow off-line changes of ROM contents or even some form of slow writing to take place on-line.) There are a number of ways to implement ROMs, all with the characteristic that the implementation is in a very regular form, allowing for careful attention to be paid to the design of the basic pattern to be iterated.

As previously introduced, the ROM may be used as a fixed memory unit. Indeed that is perhaps its most common application. Many, if not most, computers use ROMs to store programs, usually systems programs, in the same way that programs are stored in RAMs. They are also commonly used to store microprograms that control the execution of instructions, and to store certain types of fixed data. But, since a ROM maps each input state into an arbitrary output state, it can also be thought of as implementing a k-input, n-output combinational logic function. We can think of each input state as corresponding to a complete product term of the input variables. Then, for each output variable Z_i, we can specify the ROM contents so that Z_i has the value 1 for all of the input states that are minterms of the function corresponding to Z_i. Thus, a ROM with 8 input variables and 10 output variables can be specified to realize any 10 functions of the 8 input variables. The designer need only specify the truth table for the set of functions—no minimization procedures are applicable.

From the logic-circuit point of view, ROMs are typified by the simple 2-input, 3-output example shown in Figure 4.14. It consists of a two-stage (not counting the inverters at the inputs) AND-OR circuit, in which the AND-gates realize all the complete product terms of the input variables. In its initial form, each OR-gate receives an input from each of the AND-gates. The contents of the ROM, or the functions that it realizes, are specified by indicating which of the input lines to the OR-gates are to be cut. In Figure 4.14, this is done by marking the lines to be cut with x's. Thus, the first OR-gate, generating Z_1, is permitted to receive the inputs from the second and third AND-gates,

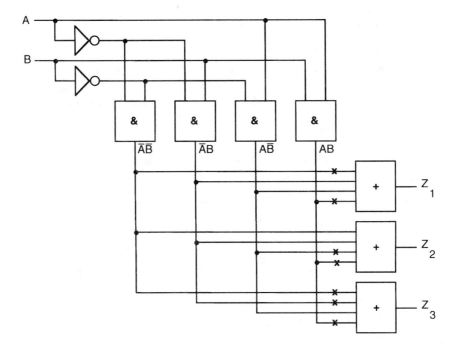

Fig. 4.14. Logic circuit for a 2–3 ROM.

i.e., the p-terms $\overline{A}B$ and $A\overline{B}$, so that the function generated is $A \oplus B$. The other two functions realized are $Z_2 = \overline{A}$, and $Z_3 = A\overline{B}$. Note that for a given number of inputs and outputs, the basic ROM configuration is the same. The process of specifying which connections are to be cut, i.e., of defining the functions to be realized, is called *programming* or *personalizing* the ROM.

Rather than using AND- and OR-gates, ROMs are actually implemented with either NOR- or NAND-gates, depending on the technology. Since, as was shown in Section 2.6, two-stage NOR- and two-stage NAND-logic are both essentially equivalent to two-stage OR-AND or AND-or logic, it makes no difference from the point of view of logic design. ROMs are available as standard IC chips in various sizes and configurations. They are sometimes implemented on a portion of a VLSI chip along with the rest of a system in which they are incorporated.

4.2.2. PAL

As mentioned earlier, ROMs are usually rather wasteful with respect to chip area when used to realize logic functions. For functions of n variables, a ROM must include 2^n AND-gates. Since, the methods discussed in the previous chapter make it possible to realize most functions with far fewer AND-gates, it would be desirable to have a standard unit flexible enough to allow for the use of product terms that cover more than one 1-point of a function. One answer is the *programmable array logic* unit (*PAL*).

A PAL is a device incorporating a relatively small number of AND-gates, but allowing the inputs to each gate to be independently specified by the designer. For an n-input PAL, any input variable, complemented or not, may be an input to any gate. Thus, each gate can be programmed to generate any product term. The output of each AND-gate is fed to exactly one of the second-stage OR-gates that in turn produce the PAL outputs. Thus, for example, a particular unit may have 12 inputs and 6 outputs with 16 AND-gates feeding the 6 OR-gates (perhaps 2 AND-gates feeding each of 4 OR-gates and the other 2 OR-gates receiving inputs from 4 AND-gates each). A simple example of a PAL with 3 inputs, 4 AND-gates, and 2 outputs is shown in Figure 4.15.

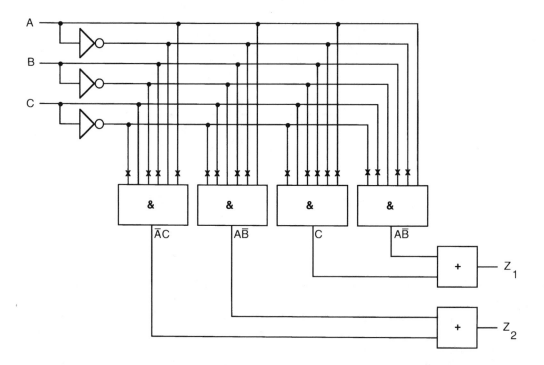

Fig. 4.15. A simple PAL with three inputs and two outputs.

In this example, the two rightmost AND-gates are connected to the Z_1 OR-gate, and the two leftmost AND-gates generate the inputs to the Z_2 OR-gate. As in the case of Figure 4.14 depicting a ROM, cut connections are designated by x's. Here the cuts are made at the inputs to the AND-gates to specify the products each is to produce. The functions realized are $Z_1 = A\overline{B} + C$ and $Z_2 = A\overline{B} + \overline{A}C$. Note that there is no fanout from the AND-gates, each of which is committed to a particular OR-gate. Thus, although both functions have $A\overline{B}$ as pi's, this term must be generated separately for each by two different AND-gates.

In order to exploit the capabilities of PALs, it is necessary to generate minimal sum-of-product (SOP) or product-of-sum (POS) expressions. This must be done independently for each output function. In practice, each output is usually made available in both polarities (i.e., in complemented as well as uncomplemented form). Hence, for each output, we can choose either the SOP or the POS form, whichever is more economical. Limitations on the number of AND-gates available for any output can sometimes be overcome by generating some of the *p*-terms needed at one output terminal, feeding this subexpression back into an otherwise unused input terminal, and gating it through an

AND-gate that feeds the same OR-gate served by other AND-gates generating the other p-terms of the desired function.

PALs are used almost exclusively as medium-scale IC chips available as standard parts. In some cases the customers send in specifications for how the chips they order are to be programmed, and this is implemented by the manufacturers as the last step of the fabrication process. *Field programmable* PALs are also available. These PALs can be programmed after delivery, for example, by applying suitably high voltages at certain terminals to "blow" key links on the chips. As was mentioned in connection with ROMs, PALs are generally realized with NAND- or NOR-gates. Next we consider an elaboration of the PAL that increases its efficiency.

4.2.3. PLA

The *programmable logic array* (*PLA*) has the same purpose as the PAL, namely to implement random logic with circuitry having a relatively regular structure. The PLA differs from the PAL only in that each AND-gate output is connected to an input of each OR-gate, and these connections, in addition to the connections to the AND-gate inputs, may be specified by the logic designer. A logic diagram for a simple PLA with three inputs, four AND-gates, and three outputs is shown in Figure 4.16. Compare this

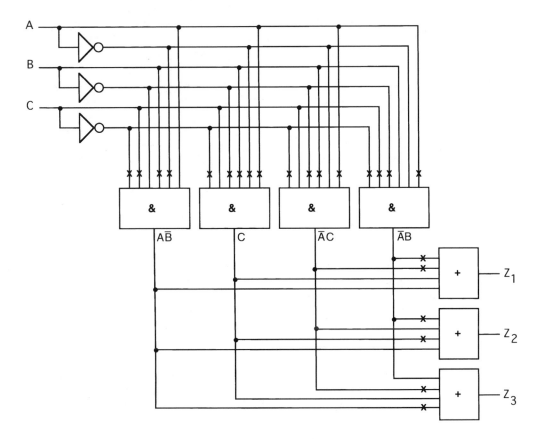

Fig. 4.16. Logic diagram for a simple PLA.

with the corresponding diagram for a PAL shown in Figure 4.15. The basic difference is in the extra connections from AND-gates to OR-gates, all of which are subject to being severed as part of the programming process. As before, x's indicate connections to be cut. (As in the case of the PAL, PLAs are usually realized with NOR- or NAND-gates. This is illustrated by examples later in this subsection.) The functions realized by this PLA are $Z_1 = A\overline{B} + C$, $Z_2 = A\overline{B} + \overline{A}C$, $Z_3 = \overline{A}B + C$. Note how the outputs of two of the AND-gates are each shared by two outputs.

An electronic circuit diagram of another simple PLA is shown in Figure 4.17. This has two inputs, A and B, with their complements generated with inverters. Outputs are Z_1 and Z_2. The technology here is n-channel MOS (nMOS), with the transistors organized in a rectangular array to implement NOR-NOR (or POS) logic. Each of the three vertical lines corresponds to a NOR-gate, with a pullup resistor (in practice the resistors would be implemented with what are called *depletion*-mode transistors) and the four transistors connected in parallel from each vertical line to ground form a pulldown network. The result is that each vertical line constitutes a NOR-gate with four inputs, A, $\overline{A}$, B, and $\overline{B}$. By cutting connections to any subset of these (cuts are again indicated by x's), a particular sum term may be realized. The leftmost gate realizes $\overline{A + B}$. The bottom two horizontal lines, with pullup resistors on the left, similarly constitute NOR-gates, with inputs coming from the three NOR-gates just described. The Z_1-output gate receives inputs from the two leftmost NOR-gates, realizing the function $(A + B)(\overline{A} + \overline{B})$. The other function realized is $Z_2 = A(\overline{A} + \overline{B}) = A\overline{B}$.

The key point to observe is the orderly nature of this circuit. This makes it easy to place arrays of this type on an IC chip, with the number of columns, inputs rows, and output rows a parameter of the design that can be specified by the logic designer. Thus, in addition to being available as individual components, PLAs are commonly incorporated on LSI or VLSI chips as parts of larger systems. In this context, they are often

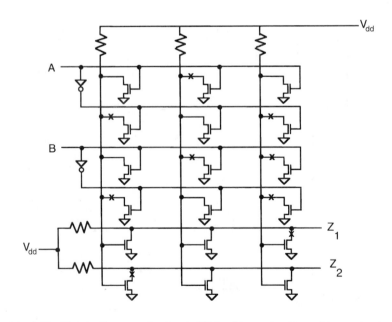

Fig. 4.17. Circuit diagram of a simple PLA with two inputs and two outputs.

referred to as *custom* PLAs. It is possible to specify the programming of the arrays (i.e., the connections to be cut) as part of the array design, so that, instead of breaking connections to unwanted transistors, they are simply omitted from the arrays. Where they are not needed for particular inputs, the input inverters can also be dropped, along with the rows that they feed. In some cases, output inverters are also an option. Generating the masks needed to implement custom PLAs is commonly done with the aid of computer programs.

To use PLAs efficiently, the number of first-stage gates, i.e., the number of columns, must be minimized. Thus the processes discussed in Chapter 3 for finding minimal SOP and POS expressions, including the extensions to multi-output functions, are applicable here. Note that only the *number* of first-stage gates affects the size of the PLA. Fanin to these gates affects only propagation delay.

Even when a PLA is programmed to realize a minimized set of logic expressions, it may still be far from optimum in its use of chip area. Referring to Figure 4.17, it can be seen that the rows correspond to input lines and output lines, and that each column corresponds to a first-stage gate (a NOR-gate in this case). A key point is that the area occupied by a PLA is proportional to this product, and is *not* affected by the actual number of activated transistors in the array. Let us see how this area can be reduced.

First, we introduce a useful way of portraying a programmed PLA. Omitting the fine detail of Figure 4.17, we can represent it by a rectangular grid, with circles at the intersections with active (as opposed to disconnected) transistors. Figure 4.18 is such a simplified version of Figure 4.17. Consider how the pair of functions $Z_1 = A(B + C)(B + D)$, $Z_2 = (B + C)(A + D)(C + D)$ might be implemented with a PLA. Both expressions are in minimal form for multi-output NOR-NOR logic, with the shared term $(B + C)$. A direct PLA realization is shown in Figure 4.19a, where the columns are numbered for reference. Note that there is no column in which rows A and B both have circles. We shall exploit this fact by rearranging the array so as to make it possible for inputs A and B to share the same row. This is done in Figure 4.19b, in which columns 2 and 3 are interchanged. (This of course does not alter the behavior of the array.)

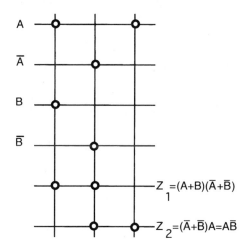

Fig. 4.18. A simplified representation of the programmed PLA of Figure 4.17.

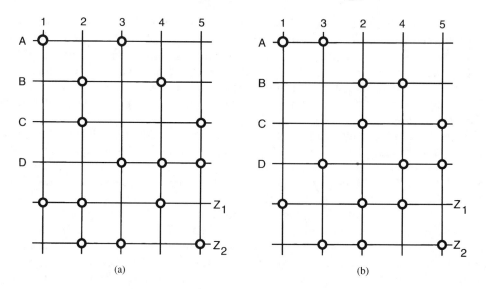

Fig. 4.19. A PLA realization of a pair of functions.

In the modified array, the A-circles are all to the left of all of the B-circles. It is therefore possible to connect input-A to the left side of the line and input-B to the right side of the same line, breaking the wire between them, as depicted in Figure 4.20. The area occupied by the PLA is thus reduced without affecting its operation. The number of activated transistors is unchanged, but the number of *inactive* transistors has been reduced.

This process, commonly called *row folding*, might better be called *input line sharing*. Where the number of columns is large and where the fanout from the inputs is not great, the potential for utilizing this technique may be substantial. In the limit, the PLA area may be almost halved. In the examples presented here, we avoid situations where input line sharing requires an input variable and its complement to appear on opposite

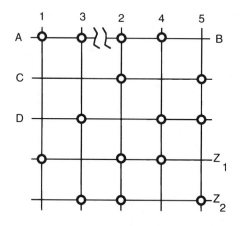

Fig. 4.20. PLA of Figure 4.19b with rows A and B folded together.

sides of the PLA. The resulting complication in wiring often, but not always, makes this undesirable.

Determining how to permute the columns in such a manner as to maximize the number of input lines that may be shared is often a very difficult problem; there are no known, efficient, systematic methods that guarantee optimal solutions. In some cases, the fact that when input line sharing is used not all of the inputs to the PLA go to the same side, may complicate chip layout and increase the area that must allocated to on-chip wiring. But in some cases, where the inputs originate at diverse locations, this may actually be an advantage.

A related technique, commonly called *column folding*, and better described as *gate line sharing*, can be applied where two different output rows are fed by gate lines that have no circles in the same rows. For example, in the PLA of Figure 4.20, developed in the previous example, the gate corresponding to column 1 feeds Z_1, and the gate corresponding to column 5 feeds Z_2. There is no row with a circle in both of these columns; i.e., no input is common to both of these gates. This makes it possible to reorganize the PLA to eliminate a column. First, as shown in Figure 4.21a, we permute the rows so that the rows represented in the column feeding Z_1 appear above the Z_1 row, and so that the rows feeding Z_2 appear below Z_2, which, in turn, is placed below Z_1.

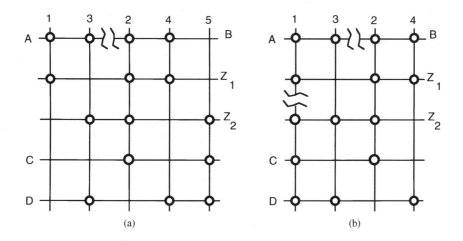

Fig. 4.21. Gate-column sharing for the PLA of previous figure.

Next, as shown in Figure 4.21b, columns 1 and 5 are merged, with the line between Z_1 and Z_2 cut. Thus, two different gates, with disjoint inputs and outputs going to disjoint sets of output lines, share the same vertical line of the PLA. Note that it is necessary to supply distinct pullup resistors to each segment of the merged line (see Figure 4.17), which detracts somewhat from the savings in area. Readers should compare the reduced PLA, with area equal to $4 \times 5 = 20$, with the original PLA of Figure 4.19a, with area equal to $5 \times 6 = 30$, to satisfy themselves that the same two functions are realized. The problem of rearranging an array to maximize opportunities for gate line sharing is similar to that for input line sharing; only suboptimal solutions can be systematically generated.

One more method for optimizing the use of PLAs is worthy of mention. Rather than involving transformations of the PLA itself, as do the row and column merging techniques, this method involves preprocessing some of the PLA inputs. Consider some pair of inputs A and B. It is a common practice to present these to the PLA in both complemented and uncomplemented forms, i.e., the inputs A, $\overline{A}$, B, and $\overline{B}$ are supplied, one to each of four input lines (e.g., see Figures 4.16–4.18). Any NOR-gate can be programmed to receive any subset of the $2^4 = 16$ possible Boolean sums of these terms (among its other inputs). But these 16 possibilities include seven combinations such as $(A, \overline{A}, B)$, all of which have the same effect on the NOR-gate output, namely they force it to 0. Thus only 8 of the 16 combinations can be considered as useful (remember that the 16 possibilities include using *none* of the four signals).

Now suppose that we replace these four very simple functions of A and B by four other functions of the same two variables. Suppose we use the outputs of a 2–4 decoder controlled by A and B. Then, each of four input lines would receive one of the product terms $\overline{A}\overline{B}$, $\overline{A}B$, $A\overline{B}$, AB, and any of the 16 combinations of *these* signals could be connected to any first-stage NOR-gate. Apart from the two trivial functions 0 and 1, which were dismissed as useless in the previous case, we have 14 different functions of A and B, including all of the eight possibilities available by using the simpler inputs. That is, the effect of feeding both AB and $A\overline{B}$ to a NOR-gate is the same as the effect of giving it just an A, the effect of $\overline{B}$ is obtained by using $(\overline{A}\overline{B}, A\overline{B})$, etc. In addition, such combinations as (AB), $(\overline{AB})$, and $(A\overline{B}, \overline{A}B)$ can also be supplied as NOR-gate inputs. With this richer choice of inputs, it is often possible to realize a set of functions with a smaller PLA than would otherwise be necessary.

Consider the function corresponding to

$$Z = (A + C + D)(B + C + D)(\overline{A} + \overline{C} + \overline{D})(\overline{B} + \overline{C} + \overline{D})(\overline{A} + E)(B + E) \quad (1)$$

realized by the 10-row, 6-column, single-output PLA of Figure 4.22. Several input-row mergers are possible, but let us instead multiply pairs of parenthesized pairs of expres-

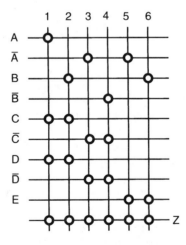

Fig. 4.22. Initial programming of a PLA for a five-variable function.

sion (1), that is, multiply out the pairs

$$(A + C + D)(B + C + D), (\overline{A} + \overline{C} + \overline{D})(\overline{B} + \overline{C} + \overline{D}),$$

and $(\overline{A} + E)(B + E)$. This converts the expression to

$$Z = (AB + C + D)(\overline{A}\,\overline{B} + \overline{C} + \overline{D})(\overline{A}B + E). \qquad (2)$$

Observing that several products of A and B literals appear in this expression, we generate those signals, as discussed above, by decoding the pair of signals A and B (only three of the four possible outputs are required). If these are used along with the C, D, and E signals as inputs to a PLA, the result is as shown in Figure 4.23a. The PLA area has been reduced from $6 \times 10 = 60$ to $3 \times 9 = 27$. Of course, the savings in area of the PLA is partly dissipated by the area consumed by the decoder. By permuting columns 1 and 3, the configuration shown in Figure 4.23b is obtained, which leads to the three row mergers shown in Figure 4.24. Now the PLA area is down to $3 \times 6 = 18$.

The direct effect of the use of decoders with PLAs is to reduce the number of PLA columns. The resulting *area* savings, which is what really counts, is directly proportional

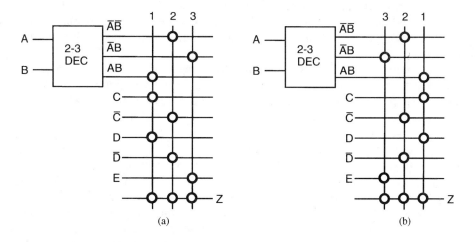

Fig. 4.23. PLA with some decoded inputs realizing the function of Figure 4.22.

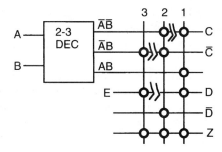

Fig. 4.24. PLA with a decoded input and merged rows.

to the number of rows, or inputs to the PLA. Hence the gains per decoder are greater for larger PLAs. (There may also be a savings in power if the total number of gates, taking into account the gates used in the decoder, is reduced.)

In general, the use of mergers or decoders to reduce PLAs constitutes a trade-off between reduced chip area and simplicity. It is, after all, the simple, regular nature of the PLA that is its chief attraction. Simplicity and regularity help save the time of designers, minimize design error, and accelerate the design process. The area minimization process can be time-consuming and makes the resulting design less regular. Furthermore, it is not always feasible to have inputs to a PLA coming from opposite direction, as is required for line sharing. Thus far, there are no generally known methods for systematically exploiting the decoder technique in an optimal manner.

4.2.4. Gate Arrays

A *gate array* (the IBM term is *master slice*) is a chip with a fixed, regular array of gates and predetermined channels that can be used to interconnect these gates. Logic designers are free to generate logic circuits (combinational or sequential) using as many gates as required up to the number available on the chip. The resulting logic circuits are then mapped onto the gate array, either manually or with the aid of computer programs. Each gate in the logic circuit must be "placed" in the array (i.e., assigned to a specific array gate), and then wiring paths on the chip must be specified to make all of the necessary connections. Constraints on the number of "wires," i.e., conducting paths, in each channel must be satisfied.

The advantage of gate arrays over pure "custom" design, in which each circuit is designed from scratch with no constraints on the placement of gates or wires, is in the manufacturing process. Depending on the technology, production of an IC chip may require that five to ten or even more unique masks be generated. This is a very time-consuming, costly process, with many opportunities for errors to creep in. With the gate-array approach, the set of masks corresponding to the basic gate array is used over and over again for entirely different logic circuits. For each specific application, it is only necessary to generate masks for the wiring together of the gates. Two or three masks usually suffice for this purpose.

Gate arrays are particularly useful for high-performance systems using bipolar technologies such as emitter-coupled logic (ECL), where the number of masks needed for the basic gate is relatively large. It is possible to order gate-array chips from certain vendors, who will fill the order with off-the-shelf components, completed according to the customer's wiring specifications.

A common practice in organizations using gate arrays is to develop a library of designs for commonly used devices such as latches, XOR-gates, and decoders. Wiring layouts are developed that involve neighboring gates, which can then be treated as units for incorporation in larger circuits. Gates with different numbers of inputs are often included in the gate array. Sometimes the designer may be able to choose among several power levels for each gate, thus permitting trade-offs between speed and power consumption in various parts of the overall system implemented on the chip. Gate arrays come in sizes ranging from several hundred to over 50,000 gates per chip, and as IC technology continues to develop, we may expect this number to continue to grow. Placement and wiring is by no means simple, and the development of computer algorithms for handling these chores is an ongoing process.

A more recent development is the *field-programmable gate array* (*FPGA*). This is a gate array that can be programmed by the user, using relatively simple equipment. A number of varieties are commercially available. These include chips where the basic elements are very simple configurations of transistors, and others where the basic elements are more complicated gates. Some such systems are *erasable*, in the sense that the same chips can be reprogrammed. Finally, the most flexible FPGAs have their programs in ordinary random-access memories, so that the logic configurations can be changed while the system is in use.

There is, of course, a price that must be paid for such convenience. FPGAs for a given level of technology may be an order of magnitude slower than custom chips, and the density of gates on a chip is also much lower. There are at least two important uses for FPGAs. One is in the design of systems where performance is not at a premium, and where it is not necessary to pack as much as possible onto each chip. The other is in situations where a complex system is being developed, and FPGAs can be used to construct early versions of all or parts of the system to check out the design logic, to exercise fault detection routines, and perhaps to test software. While FPGAs are, as just mentioned, much slower than customer chips, such "breadboard" models are likely to be a great deal faster than simulated versions of the same systems.

Another approach to the problems discussed here is the *standard cell* concept. The focus here is on the design process. A library of basic elements, similar to that referred to earlier, is developed to the point of specifying detailed mask designs. These standard cells include basic gates with various numbers of inputs. Each of the standard cells is carefully tested via circuit simulators and certified as being correct with respect to logic function, and within tolerance with respect to delays, noise margins, etc. Logic designers are restricted to using library elements as components. If these are not sufficient, designers may design new cells to add to the library. Thus, while the resulting chips are custom made, the masks are produced by assembling pretested components. Placement and wiring of the standard cells remains a significant problem.

Overview and Summary

Certain logic functions have been found to be useful in a wide variety of digital systems, and so they are frequently used as building blocks. In this capacity they are available as distinct IC components, and there are standard designs for them as modules on LSI and VLSI chips. An important example is the *decoder*, which indicates the state of its input variables by activating exactly one of its outputs. A related device is the *multiplexer*, or MUX, which can switch any selected input onto the output line. An interesting application of a MUX is to implement a scaler, which shifts binary strings in either direction. A MUX can also be used to implement any function of its input variables. A *simple encoder* generates a unique state of its output signals that concisely indicates which of its input terminals has been activated. A more sophisticated device is a *priority* encoder, which indicates the active input with the highest priority.

There are a number of techniques for implementing irregular functions with logic circuitry organized in a regular pattern. The *read-only memory* (*ROM*) is a device useful for such purposes (among others). It is implemented as a regular, rectangular array of transistors organized as a multi-output, two-stage logic circuit. There is a first-stage gate corresponding to each state of the input variables. Those gates that do *not* correspond to minterms of the function to be realized are disabled in the "programming" process. A more efficient device is the *programmed array logic* (*PAL*) unit, a device organized in a similar manner, in which there is a limited number of first-stage gates, each connected to exactly one of the second-stage gates. The logic designer can specify which inputs should be connected to each first-stage gate. The programmable logic array (PLA) is a further refinement of the same general scheme, in which the logic designer can specify the inputs both to the first *and* to the second-stage gates. Thus, minimal SOP or POS multi-output logic is generated with the components arranged on the chip in an orderly array. There is still likely to be a significant amount of chip area wasted, due to unused positions in the array. This can be reduced by techniques for merging input rows or output columns. It is also possible to use 2–4 decoders in place of some pairs of inputs to realize many functions with small arrays. ROMs, PALs, and PLAs are all available on individual IC chips as standard parts, in various sizes and configurations. PLAs are commonly incorporated on LSI and VLSI chips to realize logic functions within larger systems.

In a *gate array* chip the gates are fixed in regular arrays, and space is left for specifying wiring to interconnect them. Logic designers are then free to specify whatever logic circuits they find most suitable, and then the assignment of gates in the array and the intergate wiring are usually generated with the aid of computer programs. This means that only a few masks need be developed specifically for each application. Libraries of efficiently designed and carefully tested gate-array subcircuits implementing commonly used modules, such as XOR-gates and MUXs, are usually assembled to aid the logic designer. Such libraries are often compiled even when the chips are *not* organized in array form. This is the *standard cell* approach, in which designers are limited to constructing all of their circuits with standard building blocks. *Field programmable gate arrays* (FPGAs) can be personalized by the user with simple equipment. They are useful for building low performance systems or for emulating parts of complex systems.

SOURCES

Decoders, MUXs, encoders, etc., are treated in standard texts on logic circuits and digital design such as [Fletcher, 1980], [Friedman, 1986], [Hill, 1981], [Lee, 1976], [Mano, 1984], [McCluskey, 1986], [Winkel, 1980]. ROMs are dealt with in [Hodges, 1983], [Millman, 1979], and [Mukherjee, 1986]. PALs are treated by [Roth, 1992], and PLAs, including work on folding, are discussed in [Hachtel, 1982], [McCluskey, 1986], [Mukherjee, 1986], [Wood, 1979], and [Weste, 1992].

PROBLEMS

4.1. Show how to realize a 2–4 decoder with NOR-gates.

4.2. Show how to construct an 8–1 MUX from two 4–1 MUXs having tristate outputs.

4.3. Explain why there is an advantage to placing a 0 on the selected output of a decoder if it is to be implemented with NAND-gates.

4.4. Implement a 4–16 decoder with five 2–4 decoders, four of which have enable inputs.

4.5. Show how to implement a generalized MUX that connects one of n inputs to one of k outputs (an n–k MUX). Use macros discussed in the text.

4.6. Implement a two-wide 4–1 MUX using (a) two 4–1 MUXs; (b) a 2–4 decoder and 10 NAND-gates.

4.7. Design a 6–1 MUX using transmission gates.

4.8. How many transistors are required for a CMOS version of a 2^k-1 MUX built as a two-stage NAND-gate circuit? How many are needed for the same MUX constructed in tree form with 2–1 MUXs built with transmission gates?

*4.9. Design a CMOS pass-transistor circuit implementing a 1–2 signal distributor with the unchosen output = 0. Assume single-rail inputs. This can be done with eight transistors.

4.10. Modify the scaler shown in Figure 4.11 to make it into a circular shifter, as described in the text.

4.11. Assuming A is available double-rail, show how $f(A, B, C) = \Sigma m(2, 3, 4, 7)$ can be realized with a 4–1 MUX and no other logic elements.

*4.12. Implement the five-variable function mapped in Figure 3.23 (p. 69) with just an 8–1 MUX, assuming E is available double rail.

4.13. Implement an 8–3 encoder
(a) With OR-gates.
(b) With NOR-gates.

4.14. Implement with NOR-gates; (a) a 3–2 priority encoder with outputs Z_1, Z_0, and inputs X_i, with $i = 1, 2, 3$. The output 00 is to indicate that none of the inputs are on; (b) a 7–3 priority encoder of the same type, with three outputs and inputs labeled X_i with i ranging from 1 to 7.

4.15. Program the 3-output, 2-input ROM of Figure 4.14 to realize the set of functions $Z_1 = \overline{A} \oplus B$, $Z_2 = \overline{AB}$, $Z_3 = A + B$.

4.16. Draw a diagram of a 3-input, 2-output PAL in the form of Figure 4.15, with enough AND-gates to realize the set of functions $Z_1 = \Sigma m(3, 4, 5, 6, 7)$, $Z_2 = \Sigma m(0, 2, 6, 7)$.

4.17. Show how to realize a function requiring four p-terms using a PAL that connects only three AND-gates to each of its output OR-gates.

4.18. Draw a diagram of a 2-output PLA in the form of Figure 4.18, for three input variables with a sufficient number of AND-gates to realize the set of functions specified in Problem 4.16.

4.19. For the PLA of Figure 4.22, show how three row mergers can be achieved, *without* putting any X on the opposite side of the PLA from $\overline{X}$.

4.20. Using a PLA, in an economic manner, implement the five-variable logic function shown below. Assume that one 2–4 decoder may be used at the input, and that row sharing is permitted (without, however, placing complementary variables or decoder outputs at opposite ends of the PLA). (*Hint:* Consider using the decoder on inputs A and D.) $Z = \Sigma m(0, 3, 4, 6, 7, 10, 11, 13, 21, 24, 25, 26, 31, 33, 36, 37)$.

4.21. Assume a sophisticated, high-performance arithmetic logic unit (ALU), including a set of registers, is to be implemented for a 32-bit word size. Briefly discuss the suitability of using a large ROM chip, a PLA embedded on a VLSI chip, or a VLSI gate-array chip for this purpose.

5
Symmetric and Iterative Circuits

Designers of complex systems always appreciate finding parts of these systems that can be viewed as being regular or repetitive in nature. This simplifies almost every aspect of their work. In Section 4.4, methods for regularizing a part of the implementation of random logic through the use of read-only memories (ROMs), programmable array logics (PALs), or programmed logic arrays (PLAs) were shown to be useful. The subject of this chapter is a class of functions that are *themselves* regular in significant ways, and so are particularly amenable to realization in regular form.

Symmetric functions, treated first, constitute a subset of the broader class of iterative functions, which are then dealt with afterward. Realizations of iterative functions by linear iterative arrays are treated first, and then it is shown how all such functions can be realized in tree form, which greatly reduces propagation delays. There is a strong analogy between iterative functions in the space domain, and sequential functions in the time domain. This makes the study of iterative functions a useful bridge toward an understanding of sequential functions. It is also directly applicable to a number of important problems.

5.1. SYMMETRIC FUNCTIONS

5.1.1. Definition, Notation, and Examples

There are a number of important functions in which the variables are interchangeable. We have already encountered several of these, namely the AND, OR, NAND, NOR, and XOR functions. They are characterized by the fact that the value of each such function is completely determined by the *number* of variables equal to 1; it does not matter *which* particular variables they are. Such functions are said to be *totally symmetric*, and, where no confusion is likely, we shall refer to them simply as *symmetric*. (Functions for which *some*, but not *all*, of the input variables are interchangeable are said to be *partially symmetric*. These are not treated in this book.)

Symmetric functions can be precisely and compactly described by a simple notation that indicates the names of the variables and the numbers that must have 1-values (must be on) in order for the value of the function to be unity. For example, a 3-input AND-gate with inputs A, B, and C would be described by $S_3(A, B, C)$. A 3-input OR-gate would be specified by $S_{1,2,3}(A, B, C)$. An XOR function with four input variables would be described by: $Z = S_{1,3}(A, B, C, D)$, indicating that $Z = 1$ iff (if and only if) exactly one or exactly three of the variables A, B, C, and D are on. More generally, n-variable parity or XOR functions would be described as $Z = S_{1 \text{ MOD } 2}(X_1, X_2, \ldots, X_n)$.

This is a good opportunity to introduce, as an additional example, a very useful function, the *majority* function, M(X) (often referred to as the *majority gate*). It has an odd number of inputs, most commonly 3 and, in symmetric function notation, is described by $M(A, B, C) = S_{2,3}(A, B, C)$. In general, the output value is that of the *majority* of the input variables.

5.1.2. Operations on Symmetric Functions

There are useful operations that can be described in this notation to specify the combining of symmetric functions in various ways to generate other symmetric functions. Readers should have no difficulty in validating the following four manipulations of symmetric function expressions.

1. The Boolean sum of two symmetric functions is found by taking the *union* of sets of subscripts. Thus, we have

$$S_{1,2,4}(A, B, C, D) + S_{0,2,4}(A, B, C, D) = S_{0,1,2,4}(A, B, C, D).$$

2. Similarly, the Boolean product of two symmetric functions is found by taking the *intersection* of the sets of subscripts of the individual functions. For example,

$$[S_{0,1,3}(A, B, C, D)][S_{1,2,4}(A, B, C, D)] = S_1(A, B, C, D).$$

3. The complement of a symmetric function corresponds to taking the complement of the set of subscripts. That is, a nonnegative integer i is in the set of subscripts for an n-variable function $\bar{f}$, iff it does not exceed n and does not appear as a subscript for f. For example,

$$\bar{S}_{0,2,3}(A, B, C, D) = S_{1,4}(A, B, C, D).$$

4. If all of the *arguments* of an n-variable symmetric function are complemented, then each subscript i is replaced by $n - i$. Thus we have

$$S_{1,2,4}(\overline{A}, \overline{B}, \overline{C}, \overline{D}) = S_{3,2,0}(A, B, C, D) = S_{0,2,3}(A, B, C, D).$$

Many, if not most, symmetric functions have many prime implicants, each covering a relatively small number of 1-points. Thus, their minimal sum or product expressions are relatively costly. A supreme example of this is the n-variable XOR-function. An inspection of Figure 5.1, a K-map plot of the four-variable XOR = function, makes this point quite clear.

Although the pattern is completely regular, a perfect checkerboard, it would take eight 4-literal pi's to express this function in minimal sum-of-product (SOP) form. (Readers should satisfy themselves that the situation is no better from the product-of-sum [POS] point of view.) Exactly the same pattern prevails for all XOR functions, regardless of the number of variables. Thus, the number of pi's doubles for every unit increase in the number of variables, and the cost of each pi goes up linearly with the number of variables. Hence the cost of realizing XOR functions with two-stage logic rises at a slightly more than exponential rate with the number of variables.

While this is a worst case, many, if not most, other symmetric functions are almost as bad. (Exceptions are the AND, NAND, OR, and NOR functions.) This complexity

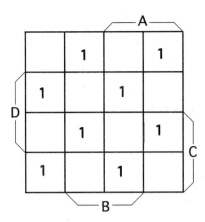

Fig. 5.1. K-map representation of A $\oplus$ B $\oplus$ C $\oplus$ D.

associated with very regular functions seems to contradict the assertion at the start of this chapter that regularity tends to reduce costs. The explanation is that if we do not restrict ourselves to two-stage logic, we can indeed exploit the regularity of *all* symmetric functions to implement them economically. In fact, the methods developed in Section 5.3 enable us to realize all members of a broader class of functions, of which the symmetric functions are a subclass, at costs that rise only linearly with the number of input variables. But first we consider a switching circuit useful for implementing symmetric functions.

5.1.3. Implementing Symmetric Functions with the Tally Circuit

Given any integer n, the full set of $n + 1$ functions of n variables, $S_0, S_1, \ldots, S_n$, can be realized with a very regular switching network (see Subsection 1.2.1 and Section 3.8). The circuit shown in Figure 5.2 is for the case $n = 3$, the variables being A, B, and C. How does it work? Every node not a 0- or 1-input (i.e. ground or, V_{dd} respectively) is connected both to X-branches and $\overline{X}$ branches (for some variable X), each linking to 0, 1 or to another such node. Suppose we start at an output terminal S_i. Each uncomple-

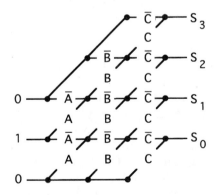

Fig. 5.2. Tally circuit realizing $S_0, S_1, S_2,$ and S_3.

mented variable moves the path one level down and one column left. Each complemented variable moves the path only to the left. In order to reach the 1-input, the
number of level changes must be *exactly i*. If the number of uncomplemented variables
exceeds i, the path will terminate *below* the 1-input, at the 0-node. If the number of
uncomplemented variables is *less* than *i*, the path terminates *above* the 1-input at the
0-node.

This circuit is a refinement of an earlier circuit devised for realization with relay
contacts. The total number of switches required for *n* variables is $n(n + 3)$, assuming
double-rail inputs. For three, or perhaps for four variables, the tally circuit is well suited
for implementation with *n*-channel MOS (nMOS) transistors. The regularity of the
geometry facilitates chip fabrication. The limiting factor is the number of transistors in
the longest path (which is equal to the number of variables). Coping with this problem
entails the use of more complex circuits with buffers to break up the longest transistor
chains.

In cases where not all of the outputs are needed, various parts of the circuit can be
omitted. As is suggested in one of the problems at the end of this chapter, the circuit can
be modified to realize symmetric functions with multiple subscripts.

5.2. ITERATIVE FUNCTIONS

Iterative functions are introduced not by defining them in terms of some abstract
property, but, rather, by starting at the other end and discussing implementations. That
is, they are presented initially by defining them as the class of functions realizable in a
particular form. Subsequently, a more basic way of defining and specifying them is
developed, and then realizations in a different form are discussed.

5.2.1. Initial Definition and Some Examples

An iterative function is defined as an *n*-variable function realizable by a circuit in the
form of Figure 5.3. (In this diagram, the convention is followed that a diagonal slash
through a signal line indicates that the line represents a multiplicity of signals, *k* in the
present case. The underscored symbols $\underline{Y}$ and $\underline{y}$ represent vectors, i.e., ordered sets of
signals, such as Y_1, Y_2, etc.) Such a circuit is called a *linear iterative circuit*. The boxes
labeled "typical cell" represent combinational logic circuits with inputs X_i and

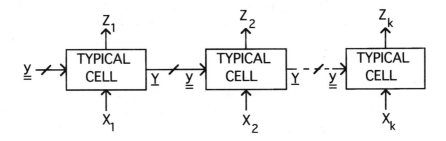

Fig. 5.3. A linear iterative circuit.

$y = y_1, y_2, \ldots, y_k$, and outputs Z_i and $Y = Y_1, Y_2, \ldots, Y_k$. All of the typical cells are exactly the same. The Xs are the external inputs and the Zs are the external outputs. The Ys and y's are *internal* variables. Note that each cell generates internal variables Y_j that are fed to the next cell on the right under the title y_j. The leftmost cell receives an externally generated, fixed y-input.

The overall circuit is strictly combinational in nature, and there are no feedback paths. In some cases, only the output from the rightmost cell, Z_k is of interest. An important characteristic of the functions realizable in this manner is that the number of input variables is not important; it must be possible to describe the function without this being a limitation.

Example 5.1

Consider now an iterative circuit with $k = 2$, with left end input $y_1 = y_2 = 0$, and with typical cell logic described by $Y_1 = X$, $Y_2 = Xy_1$, and $Z = \overline{X}y_2$. A detailed examination reveals that the Z-output from any cell is 1 only if the X-input to that cell is 0 and the y_2-input to that cell is 1. But the latter condition requires that the Y_2-output from the *previous* cell (i.e., the cell to the left) be 1, which is true only if the X-input to that cell is 1 and the X-input to the cell on *its* left have an X-input of 1. Hence we can conclude that the Z-output of the ith cell is 1 iff the sequence of three X-inputs terminating with the X-input to the ith cell, i.e., $X_{i-2}X_{i-1}X_i$, is 110. Thus, for the X sequence shown below, we would expect the circuit to produce the Z sequence shown immediately under it:

$$X: 0\,1\,0\,0\,0\,1\,1\,0\,1\,0\,1\,1\,1\,1\,0\,0$$
$$Z: 0\,0\,0\,0\,0\,0\,0\,1\,0\,0\,0\,0\,0\,0\,1\,0.$$

In this example, we see that the y-signals entering a cell summarize the sequence of X-signals to the left of that cell. This is a simple example of an iterative circuit capable of detecting the presence of a particular pattern of 0s and 1s in the spatial sequence of input signals.

We shall shortly see that iterative circuits can realize many interesting functions, including all symmetric functions. Another example of a *non*symmetric function is an *incrementer*, a device that accepts a binary number b as an input and generates output $b + 1$. As will become evident shortly, the class of circuits characterized by Figure 5.3 can be expanded by permitting more than one X-input to, and/or more than one Z-output from, the typical cell. With this generalization we can add other interesting functions, such as the binary adder to our category of iterative circuits. Next we consider a systematic way of describing functions realizable by iterative circuits, and a systematic procedure for using this description as a starting point in implementing such functions.

5.2.2. Flow Table Representations

Let us define the *state* of a prefix of a sequence of input signals as a summary of certain significant properties of that subsequence. The circuit introduced in Example 5.1 detects the subsequence 110. In that case, what information about input signals $X_1, X_2, \ldots, X_{i-1}$ is relevant for determining the values of outputs $Z_j, Z_{i+1}, \ldots$? Only the values of the last two members of the sequence are of interest; in particular whether they are 11, 01, or -0. All we need know about the sequence prefix is which of those three descriptions is satisfied (note that they cover all possibilities). Each of those three possibilities would be defined as a *state*. In an iterative circuit, the ith typical cell receives the state information from the y-input. This is combined with the value of X_i to generate the next member of the Z-output sequence and to produce an updated version of the state, in the form of the Y-signals, that is passed on to the next cell.

In order to understand fully the concept of a state and of how to make use of it, it is necessary to examine a few examples in detail. For the case just discussed, it is possible to represent the summarizing information precisely and compactly in the form of a *flow table* as shown in Figure 5.4. Each of the three rows corresponds to one of the states, as indicated by the comments on the right side (note that # stands for the left boundary of the sequence). The left and right columns, respectively, represent the 0 and 1 values of the X-input signal. There is a pair of entries in each position of the table, i.e., at the intersection of each row and column. The first entry indicates what the state corresponding to the row changes to *after* the X-signal corresponding to the column is added to the prefix. The second entry specifies the value of the Z-output in the position corresponding to the X-signal. Thus, if the sequence prefix corresponds to row 1 (which describes not only the situation where the rightmost member of the prefix is a 0, but also the situation at the left end of the sequence, where the prefix is null), and if the addition to the prefix is a 1-signal, then the entries in row 1, column 1 indicate that the state generated by the extended prefix is 2, and that the Z-output should be 0. If the prefix generates state-3, which occurs if it terminates with 11, then, if the X-signal is a 0, the row 3, column 0 entries specify a next-state value of 1 and a Z-output of 1, i.e., that an instance of the 110 pattern has been found terminating at this point in the sequence. Readers should examine the table carefully to ascertain that both the next-state and output entries are all consistent with the row comments and with the objective of identifying 110 subsequences.

Suppose that, instead of having a set of n binary input variables, $X_1, X_2, \ldots, X_n$, representing an ordered sequence of input values, all available at the same time, we had only *one* input variable, X, whose value changed at regular intervals (determined perhaps by some regular source of clock pulses). In n units of time, n values of X would arrive. Then we could talk about a *time* sequence of inputs in the same way that we have just been discussing *spatial* sequences of inputs. We might be interested, for example, in designing a circuit that would indicate if the input sequence included the subsequence 110, or if it contained exactly one or exactly two 1s. Our verbal description of the sequences of interest would be essentially the same, whether we were talking about sequences of values of n different variables (i.e., *spatial* sequences) or about sequences of values of single variables over a period of time (i.e., *temporal* sequences). But any other description of a desired pattern could also be used in either case, *including flow tables*.

X

	0	1	
1	1,0	2,0	0 or # on left
2	1,0	3,0	01 or #1 on left
3	1,1	3,0	11 on left

Fig. 5.4. Flow table for function that detects the subsequence 110.

This is a very important observation, since it means that most of what we know about sequential functions and their implementation can also be applied to iterative functions and their implementations. This applies, in particular, to the use and manipulation of flow tables. More will be said about this in the chapters dealing with sequential circuits, but the reader might meanwhile get accustomed to thinking about the strong analogy between spatial and temporal sequences, particularly with respect to their flow table descriptions. In the next example, a flow table specification of a symmetric function is analyzed.

Example 5.2

Let $S_{1,2}$ represent the subclass of a symmetric functions, $S_{1,2}(X_1, X_2, \ldots, X_n)$, for all positive values of n. A circuit realizing this function, when presented with the sample input sequence shown below, should generate the output sequence shown beneath it:

$$X: 0\,0\,1\,0\,0\,0\,1\,0\,1\,0\,0\,1\,1$$
$$Z: 0\,0\,1\,1\,1\,1\,1\,1\,0\,0\,0\,0\,0.$$

In the flow table shown as Figure 5.5, which corresponds to this function, the columns again represent values of the ith X-input, i.e., of X_i. The rows, numbered 1 through 4, as in the previous case, represent the states produced by the prefix of the X-input sequence terminating with X_{i-1}.

Let us now analyze this table in detail. Starting at the left end of the input sequence, the state of the system is 1, since there are obviously no 1-inputs to the left. If $X_1 = 0$, then the entry in row 1, column 0 specifies that the state remains 1 and that output $Z = 0$. Moving to the right, as long as the X_i are 0s, the state remains 1 and the Z-outputs remain 0. When a 1-input is encountered, the relevant table entry is in the 1 column of row 1; it specifies that the state at this point should be changed to 2, and that the Z-output corresponding to this X-input should be set at 1. Moving on to the right in the sequence, the next specification is taken from row 2 of the table, which indicates, in accordance with the comment on the right side, that, at this point, there is exactly one X-input to the left with the value 1. Depending on whether the next X-input is 0 or 1, the specification is taken from the 0 or 1 column of row 2. Should a second 1-input be encountered, the state will change to 3, and if additional 1-inputs appear later in the X sequence, the state changes to 4. Note that there is no escape from state-4; it is what is termed a *trap state*, with only 4 as next-state entries. This is as it should be, since obviously once a certain number of 1s has been found in our left-to-right inspection of the X-inputs, there is no way that the cumulative number can be reduced. Readers

	X		
	0	1	
1	1,0	2,1	no 1 to the left
2	2,1	3,1	one 1 to the left
3	3,1	4,0	two 1s to the left
4	4,0	4,0	more than two 1s to the left

Fig. 5.5. Flow table describing $S_{1,2}$.

should satisfy themselves that the table generates, for each value of i, the correct value of Z_i, which is a 1 iff the cumulative number of 1s among all of the inputs X_j for $j \leq i$ is either 1 or 2.

An important characteristic shared by this function and the previous one that makes it possible to describe them with such tables, is that, regardless of the number of inputs, the total number of states necessary to characterize any prefix of the X sequence is bounded. Consider any other symmetric function where the maximum value of its subscripts is t. In that case, the values of the inputs of any prefix of the sequence of X-inputs are adequately summarized by a statement of exactly how many 1-values it includes up to and including t, or if the number exceeds t. This corresponds to $t + 2$ states. Hence, any such function does indeed correspond to a flow table. In the next subsection, it is shown how such tables can be used to generate iterative circuits realizing the function described. But first let us consider some additional examples. In the next one we examine a symmetric function in which the subscripts do *not* have any upper limit.

Example 5.3

Recall that the XOR function can be described as $S_{1 \, MOD \, 2}(X_1, X_2, \ldots, X_n)$. This is equivalent to $S_{1,3,5\ldots}(X_1, X_2, \ldots, X_n)$. An example of an output generated by a particular input is shown below:

$$X: 0\,0\,0\,1\,0\,0\,1\,1\,1\,0\,0\,1\,0\,1\,0\,0$$
$$Z: 0\,0\,0\,1\,1\,1\,0\,1\,0\,0\,0\,1\,1\,0\,0\,0.$$

Since, for a sufficiently large value of n, any odd number might be included among the subscripts, this function does not satisfy the requirement stated earlier as a sufficient condition for correspondence to a flow table. Does this mean that no flow table representation of XOR exists? By no means—in fact, this is a very easy function to describe in flow table form. The reason is that the only pertinent information about the input sequence prefix is whether the number of 1s is odd or even. This leads to the very simple two-row flow table shown in Figure 5.6. At the left end of the sequence the initial state corresponds to the "even" case, since 0 is an even number, and obviously no 1s appear in a prefix of 0 length.

In the next example, an important iterative function is discussed in which the input sequence consists of *pairs* of binary signals, so that the flow table has four columns, one for each possible input combination.

X
	0	1	
1	1,0	2,1	even number of 1s
2	2,1	1,0	odd number of 1s

Fig. 5.6. Flow table for the XOR function.

Example 5.4

The arithmetic operation, binary addition, can be thought of as an iterative function. Let us represent the operands as A and B, and the sum as S. Then the input signals consist of pairs of binary-valued signals representing the bits of the binary numbers A and B. Assume that the input sequence begins with the least significant, i.e., the rightmost, bits, and that the signal flow is leftward. Below is an example of the summation process in which the operands A and B are shown along with the bits carried in C, and the sum-bits, S. This may be helpful in following the subsequent discussion, in which a flow table describing addition is derived.

$$A: \quad 1\,1\,0\,1\,0\,1\,1\,0$$
$$B: \quad 0\,1\,0\,1\,1\,1\,0\,0$$
$$C: 1\,1\,0\,1\,1\,1\,0\,0\,0$$
$$S: 1\,0\,0\,1\,1\,0\,0\,1\,0$$

The carry-bits convey information about the input sequence prefixes. They correspond to the states represented by the flow table rows. Suppose we assume that row 1 of the flow table represents the situation in which the carry out from the prefix is 0. Then the current value of the sum-bit S is found from the A and B bits by taking their sum modulo-2 (i.e., $S = A \oplus B$). If *both* A and B are equal to 1, then a carry is generated, and so the next-state entry in the flow table for the column corresponding to AB = 11 is 2. Otherwise, in the other columns, the next-state entry is 1, signifying that no carry has been generated. This accounts for the entries in row 1 of the flow table shown in Figure 5.7.

		AB		
	00	01	11	10
1	1,0	1,1	2,0	1,1	no carry from the right
2	1,1	2,0	2,1	2,0	carry from the right

Fig. 5.7. Flow table for a binary adder.

When a carry-in occurs in the course of a binary addition, the effect is to add a 1 to the sum of the A and B bits. This makes the sum-bit S the complement of the modulo-2 sum of the A and B bits. Hence, the output entries in row 2 are 1s in columns 00 and 11, and 0 in the other two columns. If either A or B, or both, are 1s, there will be a carry to the next position, so that the row 2 next-state entries are 2s in all columns but 00.

Example 5.5

As a final example of a flow table for a nonsymmetric function, consider the input-output relation typified by the following pair of sequences:

$$X: 1\,1\,0\,1\,1\,1\,0\,0\,1\,0\,0\,0\,1\,1\,0\,1\,1\,1\,1\,0\,1\,0$$
$$Z: 0\,0\,1\,0\,0\,0\,0\,0\,0\,0\,0\,0\,0\,1\,0\,0\,0\,0\,1\,0\,0.$$

If we define a 1-*block* as a consecutive subsequence of 1s bounded on the left by a 0 or by the left end of the sequence (#), and bounded on the right by a 0, then a Z-signal is to appear in any position where an $X = 0$ signal terminates a 1-block of even length, i.e., a block of 2, or 4, or 6, ... 1s.

The first row of the flow table represents the state at the start of the sequence, or immediately to the right of an $X = 0$ signal. The second row corresponds to the state in which an unterminated odd-length 1-block is immediately to the left. Finally, the third row corresponds to the state in which an unterminated *even*-length 1-block is immediately to the left. These constitute a complete set of possible states, in that every possible prefix would fall into exactly one of these categories. It is evident from the flow table shown in Figure 5.8 that it is also sufficiently refined for the purposes of realizing the function.

	X		
	0	1	
1	1,0	2,0	# or 0 on left
2	1,0	3,0	odd 1-block immediately on left
3	1,1	2,0	even 1-block immediately on left

Fig. 5.8. Flow table for detecting even-length 1-blocks.

It is evident from the preceding discussion that flow table descriptions can be found for all symmetric functions and for all other functions for which the ith output can be generated from a knowledge of the ith input and a statement as to which of a finite number of categories characterizes the first $i-1$ inputs. Although this class of functions is substantial and includes many interesting and useful members, there are also many other well-defined, easily described functions that can*not* be specified in flow table form. Consider, for example, the function for which $Z_i = 1$ iff the cumulative number of 1s included in the first i inputs exceeds the cumulative number of 0s in that same prefix. In order to represent this function with a flow table, a unique state would have to be assigned for *every* number that, at any point in the sequence, could be the difference between the numbers of 1s and 0s. But since the flow table would have to be valid for arbitrarily long input sequences, it follows that there is no bound on the number of states required. Hence the function cannot be described by a flow table. Next, we consider how to generate linear iterative circuits for functions that can be described by flow tables.

5.2.3. Synthesis of Linear Iterative Circuits

When a flow table description has been obtained for the function to be realized, it may be necessary to eliminate redundant states. Procedures for accomplishing this are presented in Chapter 6. For the examples in this section, this step can be omitted, as the tables are already in minimum-state (or row) form. Take another look at Figure 5.3, which depicts the general form of a linear iterative circuit (except that, as noted previously, there may be several X-inputs to and several Z-outputs from each typical cell). We shall associate the states of the X-inputs to a cell with the columns of the flow table to be realized, and associate the states of the y-variables with the rows of the table.

Thus, given a reduced flow table, the first step in the synthesis procedure is to code the rows in terms of y-states. Next we generate K-maps describing the Z- and Y-functions to be realized. The problem then becomes on of combinational logic synthesis, as discussed in Section 3.5. The procedure is illustrated below by means of several examples.

Example 5.6

Consider Figure 5.7, the flow table describing a binary adder. Since there are only two rows, we can use a single y-variable to encode them. Let us assign $y = 0$ to row 1 and $y = 1$ to row 2; y plays the role of the carry bit. The result is shown as Figure 5.9a, an example of what is called a flow *matrix*, i.e., a flow table with rows encoded in terms of the states of y-variables.

The next step is to specify the logic functions that must be realized by the typical cell, which is, of course, simply a combinational logic circuit. In this case, two functions must be specified, that for the sum-bit S, and that for the *next* value of y, which is Y. Since Figure 5.9a is already in the form of a three-variable K-map, a K-map for S is obtained from it by simply copying the values of the output entries for each position in the flow matrix. This yields Figure 5.9b. To get the corresponding map for Y, we observe that Y should be set to 1 in every position where the next-state entry corresponds to a row for which $y = 1$. In this case, where $y = 1$ only in state-2, a 1 should be placed in the K-map for Y in precisely those positions where the next-state entries are 2s. This yields Figure 5.9c.

At this point the problem has been reduced to a standard multi-output logic synthesis problem. It is interesting to note that for this example, both Y and S are symmetric functions of A, B, and y. $S = A \oplus B \oplus y$ and $Y = S_{2,3}(A, B, y)$, which is the majority function referred to at the end of Subsection 5.1.1. Thus a binary adder can be implemented in linear iterative form, with signals propagating from right to left, as shown in Figure 5.10 (where the gate labeled M is the majority gate). A module performing the two functions of its typical cell is often referred to as a *full adder*. Note that, depending on such matters as the kinds of gates used and the number of logic stages allowed, there are various ways of combining the logic for the XOR and majority gates. (The common name for the linear iterative form of the adder is *ripple-carry adder*.)

Because of its great importance as a basic component of virtually all computers, the binary adder has received a great deal of attention, and many techniques have been used to speed its operation. A high-speed adder is discussed in Subsection 5.3.6. Our next example concerns the XOR function, or parity circuit, another important device.

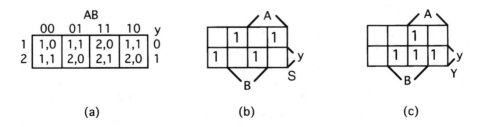

Fig. 5.9. Synthesizing a binary adder as a linear iterative circuit.

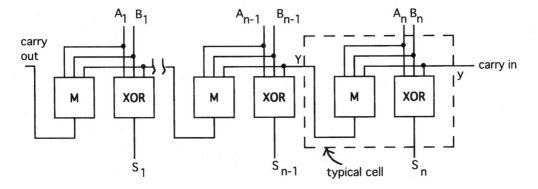

Fig. 5.10. A linear iterative realization of the binary adder.

Example 5.7

Refer now to Figure 5.11*a*, which describes the multi-input XOR-function (better known as a *parity-check* circuit). It too has only two rows, so again the coding problem is trivial; we assign $y = 0$ to row 1 and $y = 1$ to row 2. This leads to the flow matrix shown in Figure 5.11*a*. The K-maps for Z and Y follow immediately, as shown in Figure 5.11*b* and Figure 5.11*c*.

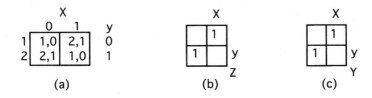

Fig. 5.11. Derivation of a linear iterative realization of a parity-check circuit.

The resulting logic expressions are simply $Z = Y = X \oplus y$, resulting in the iterative circuit shown in Figure 5.12*a*, in which the signals flow from left to right. This is a case where we are interested only in the output from the last cell, which gives the modulo-2 sum of all of the inputs. Since the output of the leftmost XOR-gate in Figure 5.12*a* is clearly just X_1, it may be eliminated, resulting in the circuit of Figure 5.12*b*, in which the number of XOR-gates is $n-1$. Although the linear iterative circuit is quite simple, it is also slow, because the X_1-signal must propagate through a long chain of logic to affect the output. The same problem afflicts the adder. Faster implementations of both are developed in Section 5.3.

Example 5.8

For the flow table in Figure 5.8, describing the function of Example 5.5, the coding problem is less trivial. At least two y-variables are needed to code the three rows. All that is needed to generate a valid circuit is that a distinct y-state be assigned to each row. There are many ways to do this. The complexity of the logic for the typical cell is

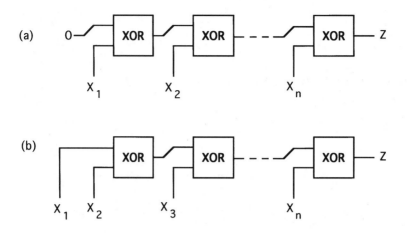

Fig. 5.12. Linear iterative circuit for a parity function.

dependent on the choice made. Much work has been done to develop ways of assigning y-states to flow table rows so as to optimize the resulting logic. Although some useful results have been achieved, the problem is far from being solved. State assignment methods are not treated in this book, but some appreciation of the variations possible may be gained by trying out two different assignments for the current example.

In Figure 5.13, two y-variables are used. The flow matrix, the Y-matrix (showing the values of $Y_1 Y_2$), and the Z-matrix are shown in Figure 5.13a, b, and c, respectively. Since row 2 is the only row for which $y_1 = 1$, the 1-entries for Y_1 in the Y-matrix are placed in just those positions where the next-state entries in the flow matrix are 2s. Similarly, the

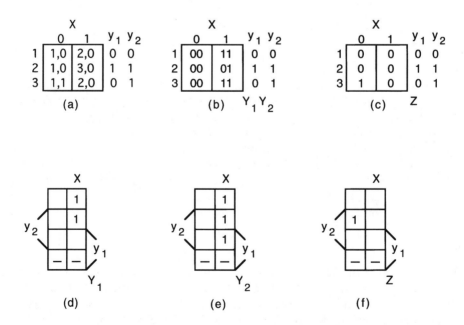

Fig. 5.13. Two-variable state assignment for the flow table of Example 5.5.

1-entries for Y_2 are placed wherever the next-state entries are either 2 or 3, since $y_2 = 1$ in both rows 2 and 3. The Z-matrix is formed simply by copying the Z-entries from the flow matrix. (An alternative technique for generating the Y-matrix is to substitute for each next-state entry in the flow matrix a $Y_1 Y_2$-value corresponding to the $y_1 y_2$-value of the next-state. For example, wherever 2 appears a next-state entry in the flow matrix, we place 11, the code assigned to row 2, in the corresponding position of the Y-matrix.)

K-maps for Y_1, Y_2, and Z are shown as Figure 5.13d, e, and f. Note the don't care entries corresponding to the unused y-state, 10. Logic expressions for the typical cell are easily found from the K-maps to be $Y_1 = X\bar{y}_1$, $Y_2 = X$, $Z = \bar{X}\bar{y}_1 y_2$.

An alternative (by no means the only one) state assignment for the same flow table is shown in Figure 5.14. This is an example of what is called a 1-*hot code*, a term derived from the fact that, for each row of the table, exactly one y-variable is "hot," i.e., has the value 1. One advantage of this type of coding is that it simplifies the task of generating logic expressions. Observe that $y_1 = 1$ in row 1, and that all of the next-state entries in the $X = 0$ column are 1s. Hence $Y_1 = \bar{X}$. To produce the Y_2-expression, we must identify the flow matrix positions for which the next-state entries are 2s. One of these is in the first row of the 1 column. Since $y_1 = 1$ uniquely identifies row 1, this corresponds to the *p*-term Xy_1. The other *p*-term for Y_2 is obtained from the row-3 entry of the 1 column, corresponding to Xy_3. Thus we have $Y_2 = Xy_1 + Xy_3$. (Since *both* of the rows involved here are uniquely characterized by 0s in the y_2-column, this expression can be simplified to $y_2 = X\bar{y}_2$.) In a similar manner, it can be seen that $Y_3 = Xy_2$, and that $Z = \bar{X}y_3$.

$$
\begin{array}{c}
\quad\quad\;\; \mathbf{X} \\
\begin{array}{c|c|c|ccc}
 & 0 & 1 & y_1 & y_2 & y_3 \\
\hline
1 & 1,0 & 2,0 & 1 & 0 & 0 \\
2 & 1,0 & 3,0 & 0 & 1 & 0 \\
3 & 1,1 & 2,0 & 0 & 0 & 1 \\
\end{array}
\end{array}
$$

Fig. 5.14. A 1-hot code for the table of Example 5.5.

Linear iterative circuits have some very nice characteristics. Their regular nature facilitates design, manufacture, and testing. They tend to be economical in terms of numbers of logic elements (these rise only linearly with the number of input variables), and wiring patterns tend to be simple. Their principal, perhaps only, drawback is the fact that, in some cases, signals must traverse long paths through many typical cells. (This, as pointed out earlier, is true for the important cases of the binary adder and the parity circuit. Note that it is not true for all linear iterative circuits; for example, the signal paths for a linear iterative realization of the function for detecting the pattern 110, treated in Example 5.1 and Figure 5.4, is not in this category.) A consequence of long signal paths is delay, the enemy of high-performance systems. Major reductions in signal-path lengths can be achieved by realizing iterative functions in tree form, the subject of the next section.

5.3. REALIZATIONS IN TREE FORM

5.3.1. Introduction to the Tree Form

Consider the linear iterative realization of the parity function shown in Figure 5.12. The X_i-signal is not used until all the X_j-signals, where $j < i$, have been processed. The process is strictly serial in that the partial result flows rightward, each cell producing effective Y- and Z-outputs, only after its y-input becomes valid. We could view this from the point of view of a nested algebraic expression, as shown below for $n = 8$:

$$((((((X_1 \oplus X_2) \oplus X_3) \oplus X_4) \oplus X_5) \oplus X_6) \oplus X_7) \oplus X_8). \tag{1}$$

The linear iterative parity circuit evaluates this expression, starting with the innermost subexpression and working its way out, adding one term at each step. But XOR is an associative operation, which means that we can rearrange the parentheses in any convenient manner without altering the value of the expression. In particular, it can be arranged as below:

$$((X_1 \oplus X_2) \oplus (X_3 \oplus X_4)) \oplus ((X_5 \oplus X_6) \oplus (X_7 \oplus X_8)). \tag{2}$$

From (2) it is evident that circuitry could *simultaneously* perform the operations $(X_1 \oplus X_2)$, $(X_3 \oplus X_4)$, $(X_5 \oplus X_6)$, and $(X_7 \oplus X_8)$. The results of the first two of these operations could then be processed simultaneously with the results of the third and fourth operations, corresponding to the second level of parentheses. Finally, in a third stage, the outermost XOR could be performed. A logic circuit performing in this manner would be in the form of a *tree*, as depicted in Figure 5.15. The advantage of this form is that the longest signal path is $\log_2 n$ stages, as opposed to $n-1$ stages for the linear realization. For the parity circuit, the building blocks used are exactly the same, and exactly the same number ($n-1$) is used. Since parity circuits often have a large number of inputs, the reduction in delay is often very substantial (5 stages as opposed to 31 for $n = 32$, a common case). Hence, parity circuits are almost always realized in tree form. Unlike the parity function case, in general, the building blocks (or cells) required for realizing iterative functions in tree form are more complex than those required for the linear form. Next, the concepts necessary for treating the general case are developed by means of a running example.

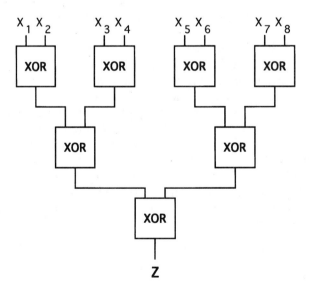

Fig. 5.15. Tree realization of a parity circuit.

5.3.2. State Mappings and Their Products

Example 5.9

A key point is that the set of next-state entries of a column of a flow table can be thought of as a mapping of the set of states of the table. That is, under a specific input, each row of the table is mapped into another row (possibly itself). In our example, shown in Figure 5.16, the 0 column corresponds to the mapping $(1, 2, 2)$, in that it maps 1 into 1, 2 into 2, and 3 into 2. Similarly, the mapping corresponding to the 1 column is $(3, 1, 1)$. It maps 1 into 3, and both 2 and 3 into 1.

The mapping concept can be extended to apply to a *sequence* of inputs. That is, any finite sequence of inputs can also be thought of as generating a mapping of the states. In our example, the input sequence 10 maps 1 into 2 (first the 1-input maps 1 into 3, then the 0-input maps 3 into 2). Similarly, 10 maps both 2 and 3 into 1. Hence 10 performs the mapping $(2, 1, 1)$. If the 10 sequence is extended to 101, the resulting mapping is $(1, 3, 3)$.

Given any pair of mappings M_1 and M_2, it is useful to define the *product* $M_1 M_2$ as the mapping produced by applying first M_1 and then M_2. Thus, for the example treated in the previous paragraph, we can write

$$(3, 1, 1)(1, 2, 2) = (2, 1, 1), \text{ and } (2, 1, 1)(3, 1, 1) = (1, 3, 3).$$

This multiplication operation is associative, but not commutative. By repeated multiplications of the mappings corresponding to the input columns, mappings corresponding to any input sequence can be obtained. How many different mappings might there be for any particular table? For an n-row table, the mappings consist of n-tuples, each element of which can be independently chosen from the n-rows. The number of different n-tuples is n^n. Thus, for an n-row table, the maximum number of different mappings that could be generated by any set of finite sequences is n^n. This is, of course, a finite number. For many examples, it appears that the actual number of different mappings generated is much less than this maximum. (All associative operations define what are known as *semigroups*. In this case the semigroup is finite.)

For any flow table, we can generate the complete semigroup of mappings as follows: Start with the set of mappings corresponding to the input columns. Multiply each member of the set by each of the input mappings, adding to the set any new mappings that result. Continue until no new mappings are produced. It is convenient to organize this process in tabular form, with columns corresponding to the mappings, essentially generalizing the flow tables by adding columns corresponding to mappings that correspond to multiple-member sequences. At the same time that this table is produced, it is useful to generate a multiplication table showing, for each ordered pair of mappings, what the product mapping is. This process is carried out below for the current example of Figure 5.16.

	X	
	0	1
1	1,0	3,0
2	2,1	1,0
3	2,0	1,1

Fig. 5.16. Flow table of a function to be realized in tree form.

The mapping table is shown as Figure 5.17a, with columns M_0 and M_1 corresponding, respectively, to the 0 and 1 columns of the flow table. Multiplying M_0 by M_0 yields M_0, and $M_0M_1 = M_1$, adding nothing new to the table. These operations account for the first two columns of row M_0 of the multiplication table, shown as Figure 5.17b. Continuing to multiply each column of Figure 5.17a, first by M_0 and then by M_1, we find $M_1M_0 = M_2$ and $M_1M_1 = M_3$, giving us two new mappings. This adds two columns to Figure 5.16a, and adds the row-M_1 entries of columns M_0 and M_1 of the multiplication table. Since multiplying M_2 and M_3 by M_0 and M_1 yield no new mappings (although they do fill in the next two rows of the M_0 and M_1 columns of Figure 5.17b), Figure 5.17a is now complete, as are the first two columns of the multiplication table, Figure 5.17b. The total number of mappings for this flow table is thus four, out of the $3^3 = 27$ possible mappings for a 3-row table.

The multiplication table can be completed by simply carrying out all the operations for the remaining columns. For example, the row-M_0, column-M_2 entry (corresponding to M_0M_2) can be found by evaluating $(1, 2, 2)(2, 1, 1) = (2, 1, 1) = M_2$. Or the associative nature of the multiplication can be exploited to generate all the remaining entries from the information contained in the first two columns. For example, to find M_1M_2, we replace the second operand, M_2, by a product that generates it, which can always be found in the first two columns. This gives us $M_1M_2 = M_1(M_1M_0) = (M_1M_1)M_0$. The result of the first product can be found in the first two columns of the table, so we have $M_1M_2 = (M_1M_1)M_0 = M_3M_0 = M_0$.

5.3.3. Specifications of Tree Circuit Cells

We now return to the problem of realizing iterative functions in tree form. Two kinds of combinational logic cells must be designed for each function. The first, called a *primary-level* or IZ cell, is shown in Figure 5.18a. It has as its inputs one of the primary inputs of the circuit (i.e., one of the X_i), and the state of the system (suitably encoded)

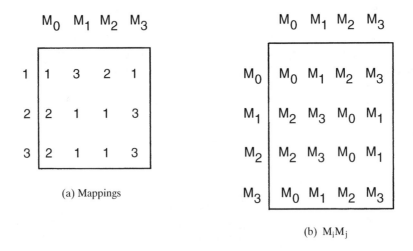

Fig. 5.17. Mappings and multiplication table for the flow table of Figure 5.16.

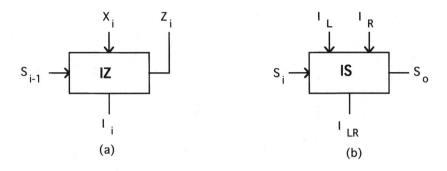

Fig. 5.18. Cells for tree realizations. (a) Primary-level cell, (b) interior cell.

following the previous input. This is exactly the same input information received by a typical cell of a linear iterative circuit. One of the outputs of this cell is the same as one of the outputs of the typical linear cell, namely the external, or Z-output, corresponding to the X-input to that cell. This output is a function of both inputs. The other output specifies (again in suitably coded form) which of the mappings corresponds to the cell input X. This output is a function only of the X-input.

The other cell, shown in Figure 5.18b, called an *interior* or IS cell, has three inputs. Input I_L, the *left* input mapping, specifies one of the mappings such as the M_i of Figure 5.17a. This represents the mapping performed by some consecutive subsequence of inputs. The second input, I_R, the *right* input mapping, represents a similar mapping, but for a subsequence immediately to the right of the subsequence represented by I_L. The third input, S_i, specifies the state of the system just before the input corresponding to I_L. The interior cells each have two outputs. I_{LR} specifies the mapping corresponding to the product $I_L I_R$. S_o indicates the state that I_L maps S_i into. (Note that S_o is a function only of S_i and I_L; it is independent of I_R.) The design of these cells is discussed after we show how they are assembled to realize the function.

5.3.4. How the Cells Are Interconnected in the Tree Circuit

Refer next to Figure 5.19. The IZ cells are shown along the top of the figure, receiving, in order from left to right, the X-inputs and generating the corresponding Z-outputs. They also feed I-signals, indicating the mappings performed by their inputs to the first level of IS cells. Each IS cell generates a mapping for the subsequence of inputs that contributes to its inputs. Thus, for example, the IS cell receiving signals from the IZ cells with inputs X_3 and X_4 (the second cell from the left in the top row of IS cells) generates output I_{3-4}, specifying the mapping performed by the input sequence $X_3 X_4$. The leftmost IS cell in the second row of these cells generates I_{1-4}, the mapping performed by the input sequence $X_1 X_2 X_3 X_4$, which is the set of inputs that flow into this cell. (In topological terms, considering this diagram as a tree, we would say that this IS cell *spans* the *leaves* of the tree represented by the IZ cells fed by inputs $X_1 X_2 X_3 X_4$.)

Each IS cell also generates a signal specifying the state of the system following the X-input that is the rightmost element of the string whose mapping is indicated by the left input I_L. Thus, the first IS cell just referred to (the one spanning $X_3 X_4$) carries out the mapping on S_2 specified by its I_L-input, which corresponds to X_3. It generates the state S_3, which it feeds to the IZ cell producing Z_4. The second cell described (spanning

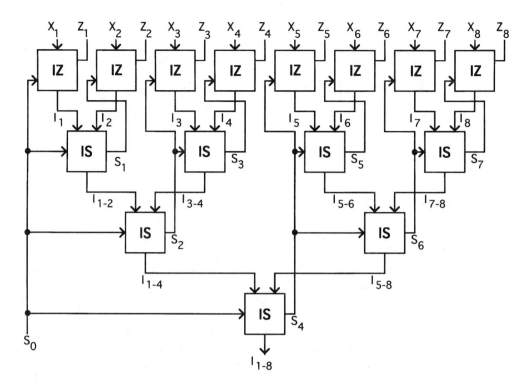

Fig. 5.19. Tree circuit for function of Figure 5.16 with eight X-variables.

$X_1 X_2 X_3 X_4$) calculates the result of the mapping corresponding to $X_1 X_2$ operating on state S_0. It generates S_2, which it feeds to the IZ cell generating Z_3, as well as to the aforementioned IS cell that generates S_3. The IS cell at the bottom (root) of the tree receives state-input S_0 and an I-input from the left providing the mapping produced by X_1–X_4. It generates S_4, which it feeds to two IS cells and to the IZ cell generating Z_5.

The longest signal path through the circuit is from an input to the leftmost IZ cell (either S_0 or X_1) through the three IS cells along the left edge of the tree, then back up the tree again from the S_4-output of the root cell through the IS cells on the *right* edge of the tree to the rightmost IZ-cell to generate Z_8. A total of seven cells is traversed in this case—the same number that would have been traversed in a linear iterative circuit realizing the same function with the same inputs. But in terms of the tree parameters, the path length is one less than twice the depth of the tree, or $2(1 + \log_2 n) - 1 = 2\log_2 n + 1$. For larger numbers of inputs, the path length for the linear circuit is clearly greater. For $n = 32$, the path in the linear case is 31, while the path for the tree circuit is only 11. (It should be clear how this configuration can be generalized to deal with 16, 32, or more inputs.)

5.3.5. *Logic Design of the Tree Cells*

We turn now to the question of how to generate the logic for the IZ and IS cells. Consider first the IS, or interior cells. The S_o-output of the IS cell is a realization of a state-to-state mapping. Consider Figure 5.17*a*, the list of mappings corresponding to the table being realized in our example. Suppose the I_L-input to the cell specifies M_2 as the mapping carried out by the input sequence spanned by I_L, and that the S_i-input to the cell indicates that the state immediately prior to this sequence is 3. Then, from the table we can deduce that the S_o-output of the cell should correspond to state-1. The situation is exactly the same as if Figure 5.17a were a flow table with inputs $M_0, M_1, M_2,$ and M_3 and we were concerned with the next-state function. As in the case where the flow table is being realized with a linear iterative circuit, it is necessary to encode the states in terms of binary variables. We have the additional task of similarly encoding the mappings, M_i. Once these tasks have been completed, we can generate the logic for the S_o-output in a routine fashion. Let us encode the states in terms of y-variables as shown in Figure 5.20a, and the mappings in terms of w-variables, as shown in Figure 5.20b. The mappings chosen are arbitrary, meeting the only essential requirement: each state must have a unique code, and each mapping must have a unique code.

Given the assignments of Figure 5.20a, we can now proceed to design the logic for the state output of the interior cell. Following the convention used for linear iterative circuits, the signals defining the states generated in a cell are designated by uppercase Ys, while the signals conveying state information to a cell are labeled as lowercase y's. The process of generating the K-maps for the Y-output of the interior cell precisely parallels that for the case of Example 5.5, in which the development of the Y-logic for a linear iterative circuit is depicted in Figure 5.13. The steps are shown in Figure 5.21 in which the mappings (Figure 5.17a) are repeated in Figure 5.21a. Figure 5.21b is analogous to Figure 5.13b, with the w-signals playing the part of the Xs. This table was generated by replacing the M_i with the w-encodings of Figure 5.20b, and replacing both the row labels and the table entries with the y-encodings of Figure 5.20a. The w's used as column labels refer only to the I_L or *left* inputs to the I cells—only these affect the Y-outputs. The K-maps of Figure 5.21c and d follow immediately from the Y-matrix of Figure 5.21b. (Note the don't cares resulting from the fact that the $y_1 y_2$ code 10 is unused.)

Readers should have no trouble verifying that minimal SOP expressions for Y_1 and Y_2 are

$$Y_1 = \overline{w}_{L1} w_{L2} \overline{y}_2 + w_{L1} \overline{w}_{L2} y_2 \quad \text{and} \quad Y_2 = \overline{w}_{L2} y_2 + w_{L2} \overline{y}_2.$$

The other output of the IS cell, I_{LR}, specifies how the mappings corresponding to its I_L- and I_R-inputs combine, i.e., the *product* of these mappings. The necessary information for specifying the logic is contained in the multiplication table of Figure 5.17b, and in

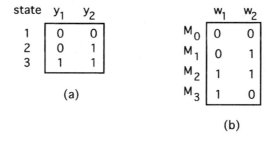

Fig. 5.20. State and mapping encodings for tree circuit example.

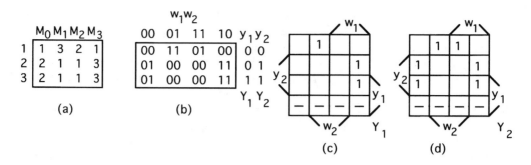

Fig. 5.21. Derivation of K-maps for the Y-outputs of the interior cell.

the encodings of the mappings shown in Figure 5.20b that were used earlier to generate the Y-expressions. As in the case of state signals, lowercase symbols, w_i, are used to specify mappings I_L and I_R serving as cell inputs, and the cell output, I_{IR}, is described in terms of uppercase symbols, W_i. The development of the K-maps for the W_i in our example is shown in Figure 5.22. First, a matrix of W_i-values, shown in Figure 5.22a, is produced from the multiplication table (Figure 5.17b), by replacing row labels, column labels, and entries of that table with the corresponding encodings of Figure 5.20b. The row entries correspond to the I_L-signals and so w_L-variables are used, while the column entries are replaced with w_R, corresponding to I_R-signals. The information in the W-matrix is then reorganized in the form of the K-maps shown in Figure 5.22b and c.

From the K-maps we obtain the minimal SOP expressions

$$W_1 = \overline{w}_{R1}w_{L2} + w_{R1}\overline{w}_{L2} \quad \text{and} \quad W_2 = w_{L2}\overline{w}_{R2} + \overline{w}_{L2}w_{R2}.$$

This completes the design of the interior cells. The primary level, or IZ cells, are designed next, a relatively simple task.

The logic for the Z-outputs is exactly the same as for a linear iterative cell for the same function and state coding. Refer to the flow table shown in Figure 5.16. With the state assignment of Figure 5.20a, we obtain the Z matrix shown in Figure 5.23a, which leads directly to the K-map of Figure 5.23b. This yields the expression for the output logic: $Z = \overline{X}\overline{y}_1y_2 + Xy_1$. The design of the I-logic for the primary cells is even less daunting.

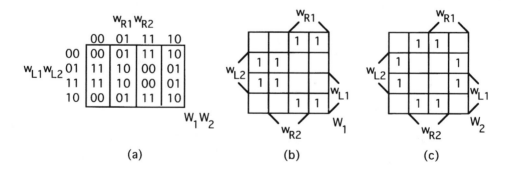

Fig. 5.22. Derivation of K-maps for the W-outputs of the interior cell.

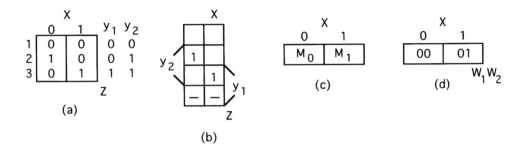

Fig. 5.23. Derivation of tables specifying the primary cell logic.

The I-output of a primary cell simply specifies which of the mappings corresponds to the input signal. In our example, as shown in Figure 5.17a, the mappings for $X = 0$ and $X = 1$ are M_0 and M_1, respectively. This information is conveyed in Figure 5.23c. Replacing M_0 and M_1 by their w codes produces the W_1W_2-matrix of Figure 5.23d. It is not necessary to generate the K-maps in order to generate the "logic expressions": $W_1 = 0$ and $W_2 = X$. This completes the design of the tree circuit for our example.

5.3.6. Some Refinements, Derivation of the Full Carry Lookahead Adder

There are a number of possible refinements of the preceding method that are not treated here. It is possible to group the inputs to the primary cells in pairs, triples, etc., so as to eliminate one or more levels of the tree. Along similar lines, we could increase the fanin to the interior nodes, reducing the number of tree levels by a multiplicative factor. As is also the case with linear iterative circuits, some of the logic formally generated for tree circuits is not needed. For example, the I-outputs of the primary and interior cells along the right boundary are not used, and so need not be produced. More important, it is possible to refine the general structure of these circuits so as to reduce the maximum number of cells that must be traversed to about half of the value for the design shown.

The methods described here (and the refinements mentioned, but not discussed) can be applied very effectively to the case of the binary adder. The linear iterative realization of this device, as previously stated, suffers from the fact that, in the worst case, carry signals for an n-bit adder must propagate through n cells to generate the highest order output bit. The speedup factor available through the use of tree realizations of this very important device is therefore substantial. Long before the development of the general theory presented here, a tree circuit implementation of the binary adder was developed that has found very wide application. In the next example, this device, called a *full carry lookahead adder* (*FCLA*), is derived using the concepts just presented.

Example 5.10

The flow table for the binary adder, shown in Figure 5.9a is reproduced in Figure 5.24a. There are three different mappings specified by the primary inputs. These are shown in Figure 5.24b, as M_G, M_P, and M_K (for reasons that will become evident, these are commonly referred to as carry *generate*, carry *propagate*, and carry *kill*, respectively).

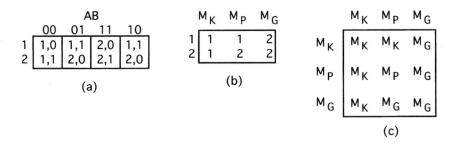

Fig. 5.24. Tables for a full carry lookahead adder.

Note that, with respect to the block diagram of a tree circuit shown in Figure 5.19, the least significant bits of the number being added are on the *left*. Applying the methods introduced earlier produces no additional mappings. The multiplication table is shown in Figure 5.24c.

The carry generate mapping, M_G, corresponds to the input state where $A = B = 1$. Regardless of whether or not there is a carry in to that position, there *will* be a carry out, i.e., a carry is generated. Conversely, for M_K, the carry kill mapping, $A = B = 0$. Even if there is a carry in to that position, there will be no carry out. The input states for which $A = 0$ and $B = 1$ or vice versa give rise to M_P, the carry propagate mapping. For this case, a carry out occurs iff there is a carry in. As shown in Figure 5.25a, the state is encoded in terms of the variable c, corresponding to the word *carry*. We set $c = 0$ for the first row of the flow table and let $c = 1$ for the second row. An economical encoding for the mappings is shown in Figure 5.25b. The variables are named p and g, with $g = 1$ only for M_G, $p = g = 0$ for M_K, $p = 1$ and $g = 0$ for M_P, with p left unspecified for M_G. (The variable names and terminology employed in this example conform to what is used in the general literature to describe the FCLA, and should therefore make it easier for readers to comprehend that literature.) With these assignments, readers should be able to verify that the logic expressions for the primary cell are $Z = A \oplus B \oplus c$, $P = A + B$, $G = AB$. For the interior cell, substituting the encodings of Figure 5.25b in the multiplication table of Figure 5.24c leads to the logic expressions $P = p_L p_R$, $G = g_R + g_L p_R$, and a similar substitution in the mapping table of Figure 5.24b results in $C = g_L + c p_L$.

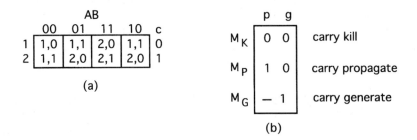

Fig. 5.25. Variable assignments for the full carry lookahead adder.

Because of the simplicity of the logic expressions derived above, the FCLA can be, and usually is, implemented by merging tree nodes. A common practice is to process four bits of each operand per primary-level cell (as opposed to one in our example), and to have each interior cell receive signals from four other cells (as opposed to two in our example.) This reduces the signal path lengths by a factor of about 2. The refinements referred to earlier result in adders with even shorter propagation paths than in the FCLA.

Overview and Summary

Totally symmetric functions are characterized by the fact that all of the variables are interchangeable. The values of such functions depend only on *how many*, not on *which* of the variables are on. Common examples of symmetric functions are the AND, OR, NAND, NOR, XOR, and majority functions. Adding or multiplying symmetric functions produces other symmetric functions, as does complementing a symmetric function or its arguments. Two-stage logic realizations of symmetric functions are often very uneconomical and the fanin may also be quite large. One more attractive approach, well suited for nMOS technology, is the *tally* circuit, a very regular, multi-output switch network.

Iterative functions are those combinational functions realizable in the form of a linear array of identical combinational logic modules (or typical cells), each receiving a fixed number of external X-inputs and a fixed number of y-inputs generated by the cell immediately on the left. Each cell also generates a fixed number of external Z-outputs. The cost of such implementations is linear in the number of variables. Examples of iterative functions include *all* symmetric functions, detectors of fixed patterns, and the binary adder.

All iterative functions can be described by *flow tables*, which have columns for X-input states and rows for classes of input prefixes. These classes, corresponding to the states of the y-signals, represent summaries of information about the inputs to the left that are pertinent to the problem. Each flow table entry specifies the next state and the output at this point. Flow tables describe functions whose arguments are *time* sequences as well as the *spatial* sequences that we have been discussing. They are thus a basic tool for dealing with sequential logic circuits.

Given a satisfactory flow table description, the next step in the design is to assign to each row a unique state of the internal y-variables. The complexity of the resulting logic circuits is strongly affected by the row assignment. Techniques useful for generating good assignments are known, but are not discussed in this book. Given the state assignment, the next step in design is to generate K-map specifications for the Ys and Zs of the typical cells. This is a simple process that reduces the problem to one of combinational logic design.

The basic drawback of the linear iterative circuits described earlier is that the signal paths may have to pass through many, sometimes *all*, of the cells. This means delays linear in n, the number of inputs. All iterative functions can also be realized in *tree* form, with the worst case delay proportional to $\log n$. The number of cells is linear in n, but they are generally more complex than cells for linear iterative circuits. A simple example is the tree realization of the parity circuit. Another very important example is the binary adder; the methods described here lead directly to the standard full carry lookahead adder.

The concept on which tree realizations are based is that each input can be regarded as a specification of a transformation, or mapping, of the set of flow table rows onto itself.

This constitutes an associative operation. We can define the product of two such mappings as the mapping that results if the mappings are applied in sequence. There are only a finite number of such mappings ultimately obtainable starting with any set of input columns. Refinements, not treated here, can further reduce the maximum lengths. (This is apart from the straightforward approach of simply increasing the fanin of the tree nodes.)

SOURCES

Symmetric functions and their realization with regularly structured switching circuits, including the tally circuit realized with relay contacts, were first described in [Shannon, 1938]. The more general class of iterative circuits appears to have been first recognized, also in terms of relay contact networks, in [Keister, 1951], which also treats symmetric circuits, including the tally circuit, in some detail. The version of the tally circuit for nMOS technology is found in [Mead and Conway, 1980] and in [Mukherjee, 1986]. Material on both symmetric and iterative functions and their realizations can be found in [Kohavi, 1978]. Applications to transistor switching circuits are treated in [Mead, 1980], and in [Mukherjee, 1986]. The analogy between iterative and sequential functions was first pointed out in [Huffman, 1955]. More general types of iterative circuits are treated in [Hennie, 1961]. The material on tree realizations, including applications to adders, is based on [Unger, 1977b], which also include the refinements mentioned, but not discussed here. The full carry lookahead adder is treated in [Morgan and Jarvis, 1959] and in [Flores, 1963] (which also deals with other related types of fast adders). See also [Koren, 1993]. Methods, not treated here, for increasing the effective speed of iterative circuits by having them generate completion signals are presented in [Unger, 1977a], which carries further earlier work of [Gilchrist et al., 1955] and [Waite, 1964]. Useful material on the state assignment problem is in [Armstrong, 1962], [Dolotta and McCluskey, 1964], [Hartmanis and Stearns, 1966], and [Kohavi, 1978].

PROBLEMS

5.1. Describe 3-input NAND- and NOR-gates in symmetric function notation.

5.2. Write SOP expressions for $S_2(A, B, C)$ and for $S_{0, 2, 3}(A, B, C, D)$.

5.3. $M(A, B, C, D, E)$ is a majority function. Describe, in symmetric function notation.
   *(a) $M(A, B, C, D, E) + \bar{S}_{2, 3}(A, B, C, D, E)$
   (b) $M(\bar{A}, \bar{B}, \bar{C}, \bar{D}, \bar{E}) + \bar{S}_{2, 3}(A, B, C, D, E)$
   (c) $[M(A, B, C, D, E)][S_{2, 3}(A, B, C, D, E)]$

5.4. Describe in symmetric function notation, with uncomplemented arguments, the dual of $S_{0, 2, 3}(A, B, C, D)$.

5.5. Specify in symmetric function notation the n-variable function equal to 1 iff the number of variables equal to 1 is exactly 2 modulo-3.

5.6. (a) Show that for any integer n, and any integer $k \le n$, that

$$S_k(X_1, X_2, \ldots, X_n) = X_1 S_{k-1}(X_2, \ldots, X_n) + \bar{X}_1 S_k(X_2, \ldots, X_n).$$

   (b) Express $S_{0, 2, 3}(A, B, C, D)$ in the form $f(D, g)$, where f is a Boolean function, and g is a symmetric function of A, B, and C.

5.7. (a) Give examples of positive and negative four-variable symmetric functions.
   (b) Give an example of a four-variable symmetric function that is *not* unate.

5.8. Show how to modify the tally circuit of Figure 5.5 by deleting switches and adding one or more connections to realize the function $S_{2,3}$ (A, B, C).

5.9. Construct a flow table describing an iterative function for detecting odd-length 1-blocks in the input sequence. Z should be 1 for every position in which an input 0 terminates an odd-length 1-block.

5.10. Construct a flow table describing an iterative function for detecting whether there is an odd number of 1-blocks to the left. That is, Z should equal 1 in all positions where an $X = 0$ signal terminates the kth 1-block in the sequence, where k is odd. An example of the specified input–output relation is shown below:

$$X: 1\,1\,0\,0\,0\,1\,1\,1\,0\,1\,0\,1\,1\,0\,0\,0\,0\,1\,1\,1\,0\,1\,0\,1\,1\,1\,1\,0\,1\,1$$
$$Z: 0\,0\,1\,0\,0\,0\,0\,0\,0\,0\,1\,0\,0\,0\,0\,0\,0\,0\,0\,0\,1\,0\,0\,0\,0\,0\,0\,1\,0\,0$$

5.11. A three-way comparator is a circuit with two binary numbers, A and B, as inputs that indicates if $A > B$, $A = B$, or $A < B$. Write two flow tables for such a device.

*(a) First assume that the signal flow in the iterative circuit is from the most significant bit to the least significant bit.

(b) Then assume the flow is in the reverse direction.

5.12. Write a flow table for an incrementer, described at the end of Subsection 5.2.1 (p. 132).

5.13. Suppose the function $S_{29,31}(X_1, X_2, \ldots, X_n)$ is to be realized by a linear iterative circuit, where $n = 32$. Transform the expression to simplify the problem and write the flow table. You need not proceed further with the design.

5.14. Realize the function described by the following flow table (Figure 5.26) with a linear iterative circuit.

	AB			
	00	01	11	10
1	1,010	2,001	1,010	3,100
2	2,001	2,001	2,001	2,001
3	3,100	3,100	3,100	3,100

Fig. 5.26.

5.15. Adjacent pairs of 1-input cells of a linear iterative circuit can be merged, leading, in some cases, to savings in components, as the combined cell may be less than twice as costly as the original cell. Transform the flow table of Figure 5.27 into a table reflecting such a merger. Hint: The new table will have four columns.

	X	
	0	1
1	2,0	3,1
2	1,0	3,0
3	3,0	2,1

Fig. 5.27.

5.16. Modify the state assignment shown in Figure 5.13 (p. 140) by interchanging the y-states assigned to rows 2 and 3. Then generate new SOP logic expressions and compare them with those resulting from the original assignment.

5.17. Suppose that the complement of a two-input XOR-gate could be implemented with one less transistor than the XOR-gate (without any disadvantage). How could this best be exploited in the design of a 32-input parity check circuit?

5.18. Implement, in tree form, the iterative function described by the flow table of Figure 5.4.

5.19. Implement, in tree form, the iterative function described by the flow table of Figure 5.26.

5.20. Specify logic for a FCLA with merged primary cells and 4-to-1 fanin to interior cells as described at the end of Subsection 5.3.6. Each interior cell must produce an I-output specifying the mapping performed by the X-inputs that it spans, as well as three outputs indicating the states following the inputs spanned by each of the three least significant I-inputs that it receives.

6
Sequential Logic Circuits

Whereas combinational logic circuits generate outputs only on the basis of current inputs, sequential logic circuits take into account the effects of past inputs as well. This means that they must incorporate some form of memory. The information contained in that memory is referred to as *state* information. Thus, the output of a sequential circuit is a function of the current input and the state. (Note on terminology: The terms *finite state machine*, *sequential machine*, *state machine* or, in the proper context, simply *machine*, are often used to refer to sequential circuits or to the functions they realize.)

In one class of sequential logic circuits, time is quantized; all actions take place at discrete intervals of time, determined by a regular source of pulses, called a *clock*. These are referred to as *synchronous* or *clocked* circuits. For the other, more general class, *asynchronous* circuits, events may take place at any time, provided that input changes are not too closely spaced. Since circuits that must be designed to operate in the asynchronous mode are necessary components of synchronous circuits, and since clocked circuits are more directly related to the iterative circuits discussed in the previous chapter, the discussion here will alternate between the two types, beginning with a treatment of an idealized version of synchronous circuits.

6.1. Synchronous Circuits with Ideal Unit Delays

In this section it is assumed that logic elements and the wires that connect them operate with zero delay. Furthermore, it is assumed that we have at our disposal ideal delay elements with magnitudes exactly equal to the unit of time between consecutive input signals.

6.1.1. Flow Table and State Diagram Descriptions of Sequential Functions

Flow tables were introduced in Subsection 5.2.2 as a tool for handling iterative combinational functions, i.e., functions of spatial sequences of input symbols. It was pointed out there that flow tables are equally useful for dealing with sequences in *time*. Hence, much of the material of Section 5.2 is applicable to sequential logic circuits. Nevertheless, the treatment here will not presume a prior acquaintance with the aforementioned material.

The process of formulating a flow table and then using it to generate the logic for realizing the corresponding function is illustrated by means of an example.

Example 6.1

Suppose that a digital system is being constructed to process data in the form of algebraic expressions. Assume that preliminary processing has already verified that the syntax of the expressions is correct and has mapped the symbols of the expression into one of three symbols, L, R, and C, where L and R represent left and right parentheses, respectively, and C represents a numeral, letter, or operation such as $+$. The input to our circuit is a sequence of symbols drawn from the alphabet, L, R, and C representing a valid algebraic expression. A 1-output is to be produced on output-terminal Z whenever the input symbol is an R that terminates an innermost parenthesized expression, i.e., a parenthesized expression that does not include any other parenthesized expressions. For example, the algebraic expression

$$A + (B(C + DE)(A(B + C)) + D)(E + F) + K$$

would have been mapped into the first sequence shown below, and the desired Z-output sequence is shown immediately below it:

```
C C L C L C C C C R L C L  C C C R R C C R L C C C R C C
0 0 0 0 0 0 0 0 0 1 0 0 0  0 0 0 1 0 0 0 0 1 0 0 0 0 1 0 0
```

An examination of the problem statement reveals that there should be a 1-output coincident with the last symbol of subsequences of the form LCC...CR, that is, whenever an R appears following a string of Cs preceded by an L. Otherwise, the output should be a 0. If the system could remember whether or not an R has appeared since the last occurrence of an L, then it could decide if the appearance of an R should now cause the Z-output to be set to 1.

The specification can be described precisely by a flow table, as shown in Figure 6.1a. In form, the table has a column corresponding to each input state, and a row corresponding to each internal state—or condition that it must distinguish. The table entries are ordered pairs, consisting of a specification as to what the next state should be, and a specification of the value of Z. At any time, if the system is in a state corresponding to row *i*, and if an input corresponding to column *j* appears, we consult the entry pair in the row *i*, column *j* position to determine the next state and present output, respectively.

In this example (see comments to the right of each row) the first row corresponds to the state in which no L has as yet appeared, or in which no L has appeared since the last occurrence of an R. State-2 corresponds to the condition that no R has appeared since the last occurrence of an L.

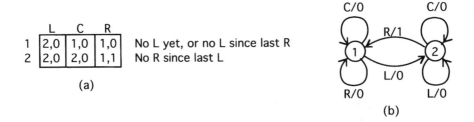

	L	C	R
1	2,0	1,0	1,0
2	2,0	2,0	1,1

1 No L yet, or no L since last R
2 No R since last L

(a)

(b)

Fig. 6.1. Function that detects innermost parenthesized subexpressions.

Let us now see if the Figure 6.1*a* table fits the problem under consideration. Row 1 clearly describes the initial state prior to any inputs. If a C-input occurs, the basic condition described remains the same, i.e., no L has yet appeared, and so the next-state entry is properly a 1. The output of course is 0, as it must be for all states in the C and L columns. Only after the occurrence of an L-input should the state change. And this indeed does happen, as indicated by the next-state entry of 2 in column L of row 1. Once the system is in state-2, any number of L- and C-inputs leaves the state unchanged, as specified by the next-state entries in the C and L columns of row 2. When an R-entry finally occurs, the state reverts to 1 and a $Z = 1$ output signal is produced. Readers should satisfy themselves that the table specifies correct operation under all conditions.

Figure 6.1*b* is an example of what is called a *state diagram* or *graph*. It conveys exactly the same information as does the accompanying flow table. The nodes of the graph correspond to the internal states of the system (or the rows of the flow table). The labeled arcs of the graph indicate what should happen under each input condition. For example, starting in state-1, if a C-input occurs, the path taken is via the arc labeled $C/0$ originating at node-1. This arc terminates at node-1, indicating that the next state is 1. The output value is given as the second part of the arc label, 0 in this case. The result of an R-input when the system is in state-2 is indicated by the arc labeled $R/1$. The next state is the destination of the arc, namely node-1, and the output is the second entry on the arc, which is 1. State graphs sometimes provide added insight into the nature of the function depicted. Flow tables are more compact and usually are more compatible with important synthesis methods. There are, however, many situations where the number of input variables is large, but where only one or two input variables are significant for any internal state. In such case (one is presented later in this chapter) a flow table would have an inordinate number of columns, with most entries unspecified, while the state graph representation would be relatively compact.

The process of constructing flow tables from verbal descriptions or other informal specifications cannot be an algorithmic process. That is, there can be no set of precise rules for doing it. One consequence of this observation is that one can acquire skill at this task only by examining a number of examples, and by practice. A second consequence is that, in the course of constructing the table, one is forced to make many decisions as to what one wishes to happen under various circumstances that might not have been considered originally. By the time the flow table has been completed, every possible set of inputs will necessarily have been considered either explicitly or implicitly. This is an important benefit derived from the construction of the table, even before any other use is made of it.

Yet another consequence of the informal nature of the process is that an excessive number of states may have been specified. It may be that a smaller table could have been constructed describing exactly the same function. Since the flow table itself *is* a formal description, it is possible to manipulate it after it is constructed so as to eliminate redundant states. This is the subject of the next subsection.

6.1.2. Reducing Flow Tables

In order to eliminate redundant states of a machine, we must determine which states are interchangeable. We could then replace each set of interchangeable states by a single state, and the resulting table would still describe the same function. Two states S_i

and S_j are defined as interchangeable or *equivalent* (i.e., $S_i \approx S_j$) if, for all input sequences, the outputs are the same regardless of whether the circuit starts in S_i or in S_j. According to this definition, $S_i \approx S_j$, and $S_j \approx S_k$ implies that $S_i \approx S_k$. This *transitivity* property means that the state equivalence relation partitions the states of the table into disjoint sets, each consisting of states that are equivalent to one another. Each of these sets (called *equivalence* classes) corresponds to a state of a reduced table equivalent to the original table. We now proceed to develop a procedure for determining if two states are equivalent and for finding the equivalence classes. A running example is used to illustrate what is going on.

Example 6.2

Rather than determine directly which pairs of states are equivalent, our approach is to identify all pairs of states that are *not* equivalent. Pairs not so identified are clearly equivalent. To determine that a pair is not equivalent, it is sufficient to find any input sequence that distinguishes the members by producing different output sequences when applied to each of them. Consider now the flow table of Figure 6.2a.

Are states 1 and 2 equivalent? Certainly not, since the input sequence consisting of just A produces a 0 when the initial state is 1, and a 1 if the initial state is 2. In general,

	A	B	C
1	5,0	7,0	6,0
2	3,1	4,0	1,0
3	4,0	4,0	6,0
4	3,1	5,0	6,0
5	3,1	2,0	1,0
6	2,0	7,0	1,0
7	4,1	4,0	6,0

(a) Flow table

1						
X	2					
45,47	X	3				
X	16,45	X	4			
X	24	X	16,25	5		
25	X	16,24,47	X	X	6	
X	16,34	X	34,45	16,24, 34	X	7

(b) Initial implication chart

Fig. 6.2. A flow table and implication charts.

1						
XX	2					
45,47	XX	3				
XX	16,45	XX	4			
XX	24	X	16,25	5		
25	XX	16,24,47	X	X	6	
XX	X16,34	X	X34,45	X16,24, 34	X	7

(c) Intermediate implication

1						
XX	2					
XX45,47	XX	3				
XX	16,45	XX	4			
XX	24	XX	16,25	5		
25	XX	XX16,24, 47	XX	XX	6	
XX	XX16,34	XX	XX34,45	XX16,24, 34	XX	7

(d) Final implication graph.

Fig. 6.2. *Continued.*

if for some input column an output variable Z_k is specified at 0 for one row and 1 for the other, then the rows are said to be *output distinguishable*, and are therefore not equivalent. That this condition is only sufficient, not necessary, is illustrated by the pair of rows 5 and 7 in the same table. They are not output distinguishable, since both have output patterns 100 for the three inputs. But if we apply the input sequence AA to states 5 and 7 (i.e., to the machine when it is in each of these states), the resulting output sequences are 10 and 11, respectively. Hence, they are not equivalent.

Suppose an input sequence I maps states S_i and S_j into a pair of states S_p and S_q, respectively, that are known to be distinguishable. Then it follows that S_i and S_j are also distinguishable, since if we append to I a sequence that distinguishes S_p and S_q, the result will be a sequence that produces different outputs when applied to S_i and S_j. This observation serves as the basis of an efficient procedure for identifying distinguishable pairs of states.

A preliminary definition is useful in describing it. If an input sequence of length 1 maps S_i and S_j into S_p and S_q, respectively then we say that S_iS_j *implies* S_pS_q ($S_iS_j \Rightarrow S_pS_q$). Thus, in Figure 6.2$a$, 57 implies 34 (under input A), 24 (under B), and 16 (under C). Implications of a pair of rows are easily found, as before, by noting the pairs of next-state entries for these rows in each column. (Ordering of the pairs is not significant, so we shall adopt the convention of ordering the pairs lexicographically in increasing order; e.g., we write 15 rather than 51. This facilitates the comparison of pairs.) The degenerate cases typified by $ij \Rightarrow kk$ or $ij \Rightarrow ij$ are ignored.

Observe now that if $S_i S_j \Rightarrow S_p S_q$, and if S_p and S_q are distinguishable, then it follows that S_i and S_j are also distinguishable. It is not hard to prove that S_i and S_j are distinguishable iff either $S_i S_j$ is output distinguishable or there is a chain of implications from $S_i S_j$ to a pair of states $S_p S_q$ that are output distinguishable. This is equivalent to saying that if S_i and S_j are distinguishable, then either $S_i S_j$ is output distinguishable or $S_i S_j \Rightarrow S_p S_q$, and S_p and S_q are distinguishable.

From the preceding discussion, it can be seen that the set of all distinguishable pairs can be constructed as follows: (1) Begin with the set of pairs that are output distinguishable; (2) repeatedly add to this set pairs that imply any that are already members. This task can be efficiently organized by using a *pair chart*, as shown in Figure 6.2b. Note that the same numbers are used to label rows and columns. Thus, row 2 has just one cell (containing an X), row 4 has three cells. Column 2 has five cells, with a cell containing X at the top. There is no row 1 and there is no column 7. For each unordered pair of states in the table, there is precisely one position in the chart. For example, in Figure 6.2b, the cell in column 2, row 4 corresponds to the pair 24. The entry in that position is a set of pointers, {16, 45}, indicating that $24 \Rightarrow 16$ (under input C) and $24 \Rightarrow 45$ (under input B). These can be verified by inspection of the flow table in Figure 6.2a. All of the implications are shown in Figure 6.2b, except for those positions corresponding to pairs that are output distinguishable, where Xs are placed. (Trivial implications such as $45 \Rightarrow 33$, and $34 \Rightarrow 34$ are omitted.) Since, in this context, we are using the pair charts as *implication charts*, they are so designated in the figures.

The next step is to add to the set of known distinguishable pairs (by marking their cells with Xs), all of the pairs that imply pairs already in that set. This is done by considering each X-marked cell in turn and looking for other cells that contain its coordinates. Starting with the 12-cell, we look for other cells with entries 12. Upon not finding any, we put a second X in the 12-cell, and consider it no further. The same result is obtained for X-marked cells 14, 15, 17, 23, and 26. When we get to 34, we find that cells 27, 47, and 57 each have a 34-entry. So we place an X in each of them to denote the fact that it too is distinguishable, by virtue of the fact that it implies a distinguishable pair. We then place a second X in 34. This brings us to the point shown in Figure 6.2c.

Continuing the process, we check for appearances of 27, and find none. So we put a second X in the 27-cell. Similarly, we dispose of X-marked cells 35, 37, and 46. Arriving at cell 47, we find a 47 in cell 13, and so place an X in cell 13. Since there are no pairs implying 13, we simply place a second X in that cell. The process is completed with Xs added to the 56, 57, and 67 cells, without any additional distinguishable pairs being identified. The final pair chart is shown as Figure 6.2d. One verification of the accuracy with which the previous steps have been executed is the fact that all of the entries in the uncrossed cells, i.e., the implications of the pairs that have *not* been identified as distinguishable, correspond to other cells in this same category. The next step is to generate the equivalence classes from Figure 6.2d. (The implications are ignored from now on. All that is of interest is whether a cell is uncrossed or crossed, i.e., whether it represents an equivalent pair or not.)

This is easily accomplished by starting at the leftmost column of the final implication chart and adding to state-1 all of the rows of column 1 that are not crossed. These correspond to all of the rows equivalent to row 1. This yields the set 16. We then move to the next column on the right that does not correspond to a row already included in an equivalence set; in this case, column 2. We add to 2 the set of all rows corresponding to uncrossed rows in this column to form the next equivalence class: 245. Repeating this process, we conclude the exercise by adding the one-member classes 3 and 7 to the collection. The set of equivalence classes is thus: $\{16, 245, 3, 7\}$. The last step is to construct the reduced table from the original table and the set of equivalence classes.

The original flow table is reproduced in Figure 6.3a, and the reduced table is shown in Figure 6.3b. The construction of the latter is initiated by drawing the skeleton of the table, with one row for each equivalence class. (The classes are listed at the left of the rows, arranged in increasing order of the first members from top to bottom.) Row 1 of the reduced table is equivalent to rows 1 and 6 of the original table. Hence, it must have the same outputs as these rows (which, being equivalent, must of course have the same output entries in each column). This yields the output pattern 000 for row 1. The next-state entry in row 1, column A must correspond to a row that is equivalent to the next-state entries for *both* rows 1 and 6 in column A of the original table. These are rows 5 and 2, respectively. Inspection of the row labels indicates that these are both included in the set corresponding to row 2 of the reduced table. Hence, the next-state entry in row 1, column A is 2. The column B next-state entry in row 1 is found by noting that it must point to a row equivalent to 7 of the original table. This is row 4 of the new table. It should now be clear how to complete the table. The fact that no contradictions are encountered in this process with respect to finding either output or next-state entries is a partial verification that the process has been carried out correctly.

	A	B	C
1	5,0	7,0	6,0
2	3,1	4,0	1,0
3	4,0	4,0	6,0
4	3,1	5,0	6,0
5	3,1	2,0	1,0
6	2,0	7,0	1,0
7	4,1	4,0	6,0

(a) Original flow table.

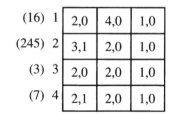

		A	B	C
(16)	1	2,0	4,0	1,0
(245)	2	3,1	2,0	1,0
(3)	3	2,0	2,0	1,0
(7)	4	2,1	2,0	1,0

(b) Reduced flow table.

Fig. 6.3. Original flow table and the minimal-row equivalent.

A USEFUL VARIATION

(The following is not essential for understanding subsequent topics.) Although the method presented earlier for finding the equivalent pairs of states is quite efficient for hand computation, it can be improved with respect to machine computation. Consider the part of the process in which we are looking for state-pairs that imply a nonequivalent (i.e., distinguishable) pair. In Example 6.2, at one point we focused on the X in position 34 of the pair chart and scanned the chart (Figure 6.2b) for other cells with the implication 34. Very quickly we saw that such an entry appears in positions 27, 47, and 57. We then entered Xs in those positions. (This is based on the idea that a pair that implies a nonequivalent pair, must also be a nonequivalent pair.) Because people are very good at pattern recognition, it did not take long for us to find the cells with 34 entries. However, when the same process is executed by a computer, it must sequentially examine each position of the pair chart. For an n-state flow table it takes order n^2 time for such a search (proportional to the number of entries in the pair chart). Since the upper bound on the number of such searches is also of order n^2 (again corresponding to the number of cells in the chart), the total time of this part of the process is of order n^4.

Consider now a variation in the way we describe the implications (refer to Figure 6.4; the flow table is repeated in Figure 6.4a). Suppose that we construct the initial implication chart by showing the implications in the reverse direction. That is, instead of showing that 27 implies 34 by putting a 34 in cell 27, as is done in Figure 6.2b, we put a 27 in cell 34 (see Figure 6.4b). In general, instead of listing in cell ij the pairs that state-pair ij implies, we list in cell ij those pairs that imply ij. It is then easy to extend a chain of distinguishable pairs back up the implication chain. If cell ij contains an X, then we put Xs in all cells listed in ij, and then add a second X to ij to indicate that we are finished with it. No searching is necessary to identify the pairs that imply ij.

We obtain Figure 6.4b by first entering Xs in cells corresponding to output distinguishable pairs, just as in the original procedure. Next, consider in turn each of the cells *without* Xs. Start with 13. From the flow table we see that 13 implies 45. We therefore put a 13-entry in cell 45 (instead of a 45 in 13). Similarly, we put a 13 in 47, reflecting the fact that 13 also implies 47. (The trivial implication, 13 $\Rightarrow$ 66 is ignored.) Next, moving on to cell 16, we see that 16 implies 25. So we place a 16-entry in cell 25. (The trivial implications 16 $\Rightarrow$ 77 and 16 $\Rightarrow$ 16 are ignored.) In similar fashion, we process 24, 25, 27, 36, 45, 47, and 57 to complete the initial implication chart shown in Figure 6.4b.

Next, we generate the final implication graph by tracing backward from the entries known to correspond to nonequivalent pairs. We consider only entries that have both Xs and other entries. (There is no need to consider entries that have Xs without implications.) In this example, the only candidate is 34. Inspection of cell 34 leads us to add Xs in 27, 47, and 57, since each of these pairs implies 34, which is a nonequivalent pair. We then add a second X in 34, to indicate that we are done with it. We now have a new candidate cell, 47, which has an X and two pointers to it. This causes us to add an X to 36. (There is already an X in 13.) Since there are no further candidates, Figure 6.4c is the final implication chart. The set of cells without Xs (i.e., the equivalent pairs) in the final reverse implication chart (Figure 6.4c) is the same as that in the final direct implication chart (Figure 6.2d). The remainder of the reduction procedure is the same as for the previous approach.

This procedure leads to a minimal-row equivalent table that is unique except for a possible reordering and relabeling of the rows. The corresponding problem for incompletely specified flow tables, i.e., those for which some of the next-state or output entries are don't cares, is far more difficult. The procedure is considerably more complex, and the solutions are not always unique. Some insight into the nature of the difficulty may be obtained by considering the flow table of Figure 6.5a, which has a don't care entry for one output entry. One might initially imagine that the following procedure would be

	A	B	C
1	5,0	7,0	6,0
2	3,1	4,0	1,0
3	4,0	4,0	6,0
4	3,1	5,0	6,0
5	3,1	2,0	1,0
6	2,0	7,0	1,0
7	4,1	4,0	6,0

(a) Flow table

(b) Initial reverse implication chart.

1					
X	2				
	X	3			
X	25,36,57	X27,47,57	4		
X	16,45	X	13,24,47	5	
24,27,36, 45,57	X		X	X	6
X		X	13,36		X	7

(c) Final reverse implication chart.

1						
X	2					
X	X	3				
X	25,36,57	XX27,47, 57	4			
X	16,45	X	13,24,47	5		
24,27,36, 45,57	X	X	X	X	6	
X	X	X	XX13,36	X	X	7

Fig. 6.4. Using reverse implications.

effective in finding a minimal-row solution:

1. Assign a 0 to the don't care entry, thus making the table fully specified.
2. Using the procedure just discussed, find the minimal-row equivalent.
3. Assign a 1 to the don't care entry and repeat step 2.
4. Select whichever of the solutions found in steps 2 and 3 is minimal.

	A	B
1	1,–	3,0
2	3,1	1,0
3	2,0	1,0

(a)

	A	B
(12)1	2,1	2,0
(13)2	1,0	2,0

(b)

Fig. 6.5. An incompletely specified flow table and a smaller covering table.

Readers carrying out this procedure on the Figure 6.5a table will quickly find that no reduction at all results. However, the two-row table shown in Figure 6.5b *covers* the table of Figure 6.5a in that, for any state S_i of Figure 6.5a, and for any input sequence, there is a state of Figure 6.5b (two in the case of state-1) that yields an output sequence that matches that produced by S_i wherever that sequence is specified. In effect, the don't care entry is not replaced by either a 0 or a 1, but is allowed to take on different values for different parts of an input sequence. Methods for accomplishing such reductions are not presented in this book.

6.1.3. Generating Logic Expressions from Flow Tables

Once a reduced flow table has been obtained, the next step in the synthesis process is to generate specifications for the logic realizing it. The general form of these circuits is shown in Figure 6.6. State information is stored in the delay elements located in the feedback paths. In this section, we assume that these are perfect one-unit delays. In a subsequent section it is shown how more practical devices can be used to achieve the same effects.

Synchronous (clocked) systems operate as follows. Assume that the y_i-signals emerging from the delay elements at the start of the first time interval correspond to a code for the initial state of the system. The y-signals, along with the X-signals representing the first input, are processed by the combinational logic to generate the outputs $\underline{Z}$, and a set of Y-signals that encode the *next* state of the system. The Y-signals serve as inputs to the unit delay elements, emerging at the start of the next time interval (as the next set of

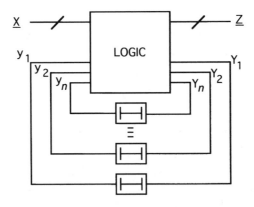

Fig. 6.6. Model for a synchronous sequential circuit with ideal delay elements.

y-signals), in time to combine with the next set of X-signals to generate the output and next-state signals for the second time interval, and so forth.

In the next example, the synthesis procedure for this type of circuit is continued with a reduced three-row table. It is essentially the same process treated in Subsection 5.2.3 in connection with iterative functions.

Example 6.3

First it is necessary to find a suitable state assignment for the table, i.e., to assign a unique state of a set of internal variables to each row of the flow table. At the same time, we must decide how to encode the input variables. One possibility for the inputs is to use a single variable for each input state (in which case we would say that the input is *decoded*). Another option is to encode the input states in terms of the states of two X-variables. In many cases, the form of the inputs is a matter that is decided at some other point in the design process. In this example, let us represent inputs A, B, and C respectively, by the X_1X_2-states 00, 01, and 11, respectively. This decision and a reasonable choice for the encoding of the rows is illustrated in Figure 6.7a.

Referring to the circuit model of Figure 6.6, we can see that the Y-signals constitute the *next* values of the corresponding y-signals. We can obtain the *Y-matrix* (Figure 6.7b), which specifies the Ys as functions of the Xs and y's, by replacing in the *flow matrix* (Figure 6.7a) occurrences of each row number with the codes assigned to that number. For example, since the y-state assigned to 2 is 01, the four occurrences of 2 as a next-state entry are all replaced by 01 in the corresponding positions in the Y-matrix. Taking into account the fact that the X-state 10 and the y-state 10 are both unassigned, thus leading to columns and rows, respectively, of don't cares, we produce the K-maps for Y_1 and Y_2 shown as Figure 6.7c and d. Finally, the K-map for Z, Figure 6.7e, is obtained directly from Figure 6.7a by simply copying the output ones into the corresponding positions (actually there is only a single 1-point).

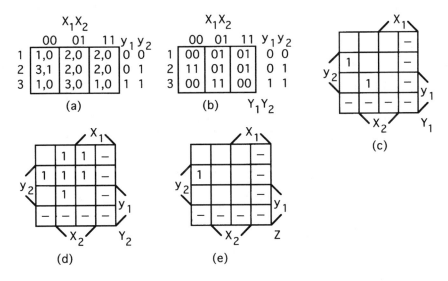

Fig. 6.7. Development of logic specifications for Example 6.3.

Readers who have mastered the techniques, presented in Subsection 3.5.1, for multi-output logic synthesis should have no trouble in deriving from this set of three maps the following optimal sum of product (SOP) expressions. (Recall that terms appearing more than once are underscored after the first appearance, to indicate multiple use in the circuit, via fanout.)

$$Y_1 = \overline{X}_1 X_2 y_1 + \overline{X}_2 \overline{y}_1 y_2, \qquad Y_2 = X_2 \overline{y}_1 + \underline{\overline{X}_1 X_2 y_1} + \underline{\overline{X}_2 \overline{y}_1 y_2}, \qquad \text{and}$$

$$Z = \underline{\overline{X}_2 \overline{y}_1 y_2}.$$

The resulting circuit, assuming double-rail inputs, is shown in Figure 6.8. Note that the design derived here would be valid for a linear iterative realization (see Subsection 3.5.1) of the same flow table. For that application, the Figure 6.8 circuit can be modified to serve as a typical cell, if we simply omit the delay elements and connect the Y_i-outputs of one cell to the corresponding y_i-input of the next cell in the chain. Each cell would be fed different X_1- and X_2-inputs, which would be ordered in a spatial, rather than in a temporal sequence.

6.2. ASYNCHRONOUS CIRCUITS

6.2.1. Some Basic Concepts

The condition of a *synchronous* sequential circuit (SSC) is normally described entirely by its internal state. During periodic intervals the input is scanned to determine, in conjunction with the internal state, an output and the next internal state. By contrast, for an *asynchronous* sequential circuit (ASC), the inputs play a continuous role in determining the output and internal state. Another way to put this is that for an SSC, the operating point is the internal state (or *row* of the flow table), whereas for an ASC, the

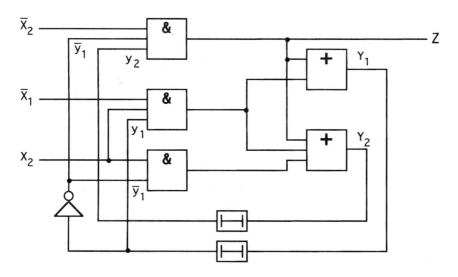

Fig. 6.8. Sequential logic circuit realizing the flow table of Figure 6.7a.

operating point is the *total* state, determined by both the input and internal states (or the row–column position in the flow table). The flow table shown in Figure 6.9 illustrates these concepts.

The particular device described by this flow table, the *positive-edge-triggered D-flip-flop* (PETDFF), is itself of considerable interest. Its output changes only when the C-input (clock-input) changes from 0 to 1, i.e., it changes only on the positive-going edge of the C-pulse; hence the designation "positive-edge-triggered." When this event occurs, the output of the flip-flop (henceforth abbreviated FF) assumes the value that the D-input (data-input) held just prior to this change. More is said about this device later in this chapter. For the present, we are interested only in using the flow table to illustrate certain ideas.

The form of the flow table for an ASC is the same as that for an SSC, except for the circles that appear around those next-state entries where the next-state is the same as the present state. These serve only to emphasize that the total states so distinguished are *stable*, in the sense that no change in the internal state is about to occur. Suppose that, initially, the internal state is 1 and the input state is 00 (i.e., the *total* state is 1-00). This is a stable state, since the next-state entry, 1, corresponds to the present state. Thus, the output will remain at 0 and no internal state change will occur until the input is changed. Should the input now change to 01 (i.e., if D goes on), the operating point reflects this change by moving horizontally to total state 1-01. Here the output remains at 0, but since the next-state entry is 2, the internal state changes, so that the operating point moves vertically to the position 2-01. This new total state is stable, so the output remains at 0, and nothing else happens until another input change occurs.

Suppose now that C goes on. Then again there is a horizontal move, this time to column 11 (total state 2-11), which is an unstable state. As dictated by the next-state entry of 3, there is now a vertical move of the operating point to total state 3-11, where the output is 1. Note that the output in the 2-11 was left unspecified. This is because we usually would not care whether, when the output changes, the change occurs in the unstable state or in the succeeding stable state. (The output change, as indicated in the explanation of the behavior of this FF, occurred when C changed in the positive direction.) Should D now revert to 0, the operating point would move horizontally to the stable state 3-10.

This example typifies the most important type of ASC, in that, for any single change of the input state, the output signal changes at most once. Functions of this type are called *single-output-change* (or SOC) functions. Related to the SOC concept is the notion of *fundamental mode* operation. This means that no input change is permitted unless the circuit has reached a stable state following any previous changes. Two other types of

	CD			
	00	01	11	10
1	①,0	2,0	①,0	①,0
2	1,0	②,0	3,–	–
3	4,1	③,1	③,1	③,1
4	④,1	3,1	–	1,–

Fig. 6.9. Flow table for a positive-edge-triggered D-FF.

ASC functions, *multiple-output-change* (MOC) and *unbounded-output-change* (UOC) functions are illustrated in Figure 6.10*a* and *b*, respectively.

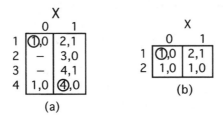

Fig. 6.10. Tables for an MOC and for a UCO function.

If, for the ASC described in Figure 6.10*a*, the initial state is 1-0, then the circuit is stable with output Z = 0. When X is turned on, the next total state is 1-1, where Z goes to 1. Since the next-state entry is 2, the operating point moves down to 2-1, causing Z to return to 0. But 2-1 is also unstable, so a further move is made down to 3-1, causing Z to change again to 1. Once again the next-state entry differs from the current internal state, so the operating point moves downward to row 4, where Z is turned off. Since 4-1 is stable, nothing more happens until X is turned off, whereupon, with no change in Z, the circuit returns to its initial state 1-0. Thus, one change in the input state resulted in an output signal changing four times (producing two pulses).

We now turn to Figure 6.10*b* to show how a single change in the input state can produce an unbounded number of changes in an output variable. Starting in the stable state 1-0, with Z = 0, a change in X brings the system to 1-1, turning Z on. Since the next-state entry is 2, the operating point moves to 2-1, where Z is turned off. But the next-state entry in 2-1 sends the system back to 1-1. Hence, as long as X remains on, the total state oscillates between 1-1 and 2-1, with Z going on and off with each cycle. The process stops when X is returned to 0. (This is precisely how a door buzzer works!)

In many situations, the inputs to an ASC are constrained so that the only one variable at a time may change; this is called *single-input-change* (SIC) operation. Certain basic problems arise in connection with allowing multiple-input changes that are addressed later in this chapter. For the present, note that for the D-FF of Figure 6.9, it is not permissible for the D-input to change at the same time that the C-input is going from 0 to 1 (i.e., on the triggering edge of the C-input). This accounts for the don't care entries in total states 2-10 and 4-11.

6.2.2. *Primitive Flow Tables and Flow Table Reduction for Asynchronous Functions*

When developing a flow table for an asynchronous function, it is best to start with what is called a *primitive flow table*, which is a flow table that has at most one stable state per row. This allows the specification to be most general. Subsequently, the primitive table is reduced to a minimal-row form. The process of developing and using a primitive flow table is illustrated below for a simple, but very important device.

Example 6.4

Our object is to design a storage element with two inputs, S and R, and one output, Q. At most, one of the inputs is permitted to be on at any time. When S goes on, Q goes on (if it is not already on), and it remains on when S goes off. When R is turned on, Q returns to 0 (if it is not already at 0), and remains off when R reverts to 0.

The flow table (see Figure 6.11) has three columns corresponding to the three possible input states (recall that SR = 11 is not permitted). We start developing the primitive flow table in row 1 at an obvious stable state; in the 00 column, with Q = 0. Then, in the SR = 01 column of row 1, there must be a next-state entry pointing to another state where Q is still 0. Let this state be in row 2. The stable state for row 2 is thus in the 01 column. Returning to row 1, we complete its entries by noting that if S is set to 1, the system must go to a state with Q = 1. Therefore, we put a next-state entry of 3 in this position and add row 3, with a stable state in which Q = 1, in the 10 column. We have now reached the point shown in Figure 6.11a.

Since only one input at a time can change, there is no way to reach total state 2-10, and so a don't care entry is placed there. If, from stable state 2-01, R is turned off, the device should revert to its initial state in row 1. From stable state 3-10, if S is turned off, a new stable state, with Q = 1, must be added in the 00-column. This requires the addition of a fourth row. It should now be clear how the table is completed as shown in Figure 6.11b.

The next step is to reduce the flow table to a minimum-row equivalent using the general procedures presented in Subsection 6.1.2. These are applicable to ASC flow tables, but optimal results for SOC-type functions result when the flow tables are in the standard form described as follows: next-state entries in unstable states correspond to the stable states ultimately to be reached, and the output entries in these unstable states are either unspecified or are the same as in the aforementioned stable states. (See [Unger 1969] for a proof of this assertion.) It is always easy to transform a table into this form without changing the function specified. As is usually the case for primitive flow tables, the Figure 6.11b table is already in standard form. If we treat the don't care entries as being equal to whatever are most convenient, it is easy to reduce the Figure 6.11b table to the table shown in Figure 6.11c.

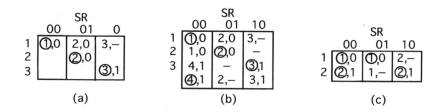

Fig. 6.11. Developing and reducing a primitive flow table for Example 6.4.

6.2.3. Types of Delays

Delays appear in logic circuits in either of two ways. First there are delays inevitably associated with wiring and logic devices. Not there at the invitation of the designer, these are referred to as *stray* delays. Logic designers can expect only general information as to their magnitudes, typically upper and lower bounds. In contrast to the stray delays are delay *elements* that are deliberately inserted by designers for various purposes that are discussed further along in this chapter. Their minimum values are specified by the designer, and, given the tolerances on such elements, their maximum values can then be calculated. In many technologies, delay elements are not available as specific devices, but rather are constructed out of cascaded pairs of inverters.

There are two important models of delays. The first is the *pure* delay, which simply shifts the input signal in time by D, the magnitude of the delay, as illustrated in Figure 6.12. (The signals involved are assumed to be two-valued. Note that the waveforms are idealized as rectangular. In reality, their shapes are likely to be rather irregular, and, in particular, the rising and falling edges must be sloping rather than vertical.)

The other type of delay model is the *inertial* delay (or *ID*). An inertial delay with magnitude D behaves the same way as a pure delay, except that it not only delays the input signal by D, but also filters out positive or negative pulses of duration less than D. In other words, when an input change occurs, the output changes D units of time later, *unless* the input reverts to its original value *before* the output responds. In this case, the output does not reflect either change. The waveforms in Figure 6.12 illustrate the difference between the responses of pure and inertial delays to the same input waveform.

Neither pure nor inertial delays can be perfectly realized in the real world. Furthermore, real delays are often somewhat better modeled as combinations of the two types. A simple transmission line behaves roughly like a pure delay. An ID is fairly well approximated by an electromagnetic relay. There is always a delay between the time a voltage across a relay coil is changed and the time the contacts respond by opening or closing. If a sufficiently short voltage pulse is applied across a relay coil, a combination of inductive reactance and mechanical inertia prevents the contacts from responding at all. At the other extreme, a simple RC circuit does tend to suppress short pulses, but it also significantly reduces the slope of the leading and trailing edges of the pulses that do get through. Electronic circuits of varying degrees of complexity can realize IDs reasonably well. In certain extreme cases, however, such as when the input pulse width is very close to the delay value, realizable IDs may not even *approximate* ideal behavior. This issue is discussed at greater length in Subsection 6.7.2.

An important practical consideration is that, for many circuits, signals changing from high to low values are delayed by significantly different amounts than are signals changing in the opposite direction (see Appendix A.2.3).

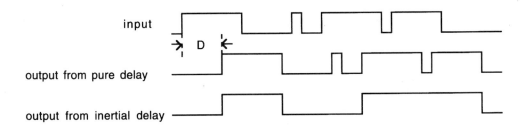

Fig. 6.12. Waveforms illustrating the behavior of pure and inertial delays.

6.2.4. Generating Logic Expressions for Asynchronous Functions

The general model for an ASC is the same as that shown in Figure 6.6 for synchronous circuits with ideal delay elements, and the design process is essentially the same once the flow table states have been encoded. Consideration of that important problem is left for the next subsection. The problem treated next is that of generating logic from a flow matrix (i.e., from a flow table with a state assignment).

Example 6.5

We now continue the process of synthesizing the storage element that was the subject of Example 6.4. Encoding the two-row flow table that was produced in that example and shown in Figure 6.11 is not a very imposing task. Let us assign $y = 0$ to row 1 and $y = 1$ to row 2, as shown in Figure 6.13a. Note that the SR = 11 column corresponding to the forbidden input state appears as part of Figure 6.13a with all of its entries unspecified. Precisely as in Example 6.3, dealing with the synthesis of a synchronous circuit (refer to Figure 6.7) we produce a Y-matrix and then a pair of K-maps describing the functions for Y and Q. These are shown Figure 6.13b, c, and d.

From the K-map for Q (Figure 6.13d), it is clear that we can make $Q = y$. The expression $Y = S + \bar{R}y$ is easily obtained from Figure 6.13c. Suppose, however, that we are interested in realizing the device with NOR-gates. Then, as was shown in Section 2.6, it is best to start with a minimal product-of-sums (POS) expression. As illustrated in Example 3.2 of Subsection 3.3.4, this can best be found by complementing the map of the function, generating a minimal SOP expression, and applying DeMorgan's law. The complement of the Y-map is shown in Figure 6.13e. From this we obtain $\bar{Y} = R + \bar{S}\bar{y}$. Complementing both sides and applying DeMorgan's law then gives us $Y = \bar{R}(S + y)$, corresponding to the circuit shown in Figure 6.14a. It is redrawn in Figure 6.14b in a more popular form. (Note that no delay element appears in the Y branch. The question of when delay elements are necessary in ASCs is considered later in this section.) This device is the well-known *set–reset flip-flop* (SR-FF), the simplest type of storage device. (Terminology in this area is not uniform. The term "SR-latch," or even simply "latch" is often applied to the SR-FF.)

This device is important enough to justify some additional analysis. The don't-care column corresponding to $S = R = 1$ in the flow table, and ultimately in the K-map for Y, has been specified by the implicit choices that led to the Figure 6.14 circuit. Working backward from the logic expression for Y that essentially determined that circuit, we can see that the Y-matrix is as shown in Figure 6.15a, and that this leads to the flow matrix

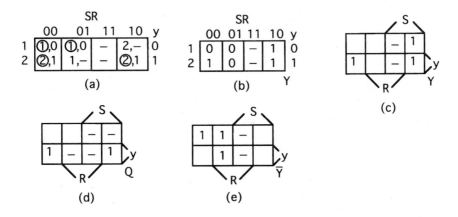

Fig. 6.13. Synthesis of an ASC realizing the flow table of Figure 6.11.

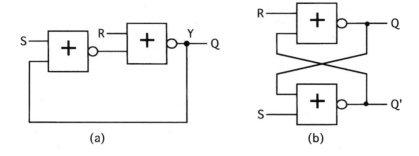

Fig. 6.14. Circuit for the SR-FF, derived from Figure 6.11.

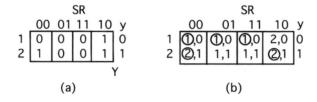

Fig. 6.15. The Y-matrix and complete flow matrix for the SR-FF of Figure 6.14.

shown in Figure 6.15*b*. Inputs S and R are commonly referred to as *set* and *reset*, respectively. The row-1 and row-2 states are commonly called the *reset* and *set* states, respectively.

The output signal labeled Q′ in Figure 6.14*b* is described by the logic expression $Q' = \overline{S + y} = \overline{S}\,\overline{y} = \overline{S}\,\overline{Q}$. When S = 0, clearly $Q' = \overline{Q}$. If S = 1, then Q′ = 0. If S = 1 and R = 0, then in the corresponding stable state of the flow table we can see that Q = 1, so again $Q' = \overline{Q}$. Thus, for stable states in the three columns of the flow table for which S and R are not both on (i.e., for the inputs for which the device was originally defined by the Figure 6.11 flow table), $Q' = \overline{Q}$. Now let us examine what happens if, for the circuit derived earlier (Figure 6.14*b*), we allow the condition S = R = 1. Clearly, Q′ = 0, and, in the stable state in the 11 column, Q is also equal to 0. No problems result from using the S = R = 1 input if this is clearly understood *and* if, with S = R = 1, both inputs are not simultaneously turned off. Violation of the latter constraint can lead to indeterminate operation, as discussed in Subsection 6.7.3. The SR-FF implemented as in Figure 6.14 is commonly called a *reset-dominated* SR-FF, since, with S and R *both* on, the FF goes to the reset state.

Next we look more closely at the problem of assigning y-states to flow table rows. It is shown that the fact that we cannot make sets of delay elements with precisely equal values significantly complicates the design of ASCs.

6.2.5. Critical Races and How to Avoid Them: The State Assignment Problem

Consider again the flow table (Figure 6.9) for the PETDFF discussed at the start of this section. If we wished to synthesize this device, then the next step would be to encode the rows of the flow table. Suppose this is done as shown in Figure 6.16*a*. The resulting

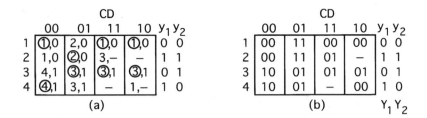

Fig. 6.16. A state assignment that leads to trouble.

Y-matrix is shown in Figure 6.16b. Without even carrying out the rest of the design process to obtain the logic circuits, it can be shown that serious problems would result from this encoding.

Assume the system starts in total state 1-00. This is a stable state, and so the inputs to the two y-branch delay elements (the values of Y_1 and Y_2) are both 0. These are the same values as y_1 and y_2, the outputs of the delays (refer here to the block diagram of Figure 6.17). If the D-input is turned on, bringing the system to 1-01, then, since the encoding for the next state (2), is 11, Y_1Y_2 changes to 11. Then we can imagine that 1-signals are simultaneously moving through both delay elements. However, since we have dropped the assumption that the delay elements are exactly equal in value, this process must be viewed as a *race*, with any of three outcomes possible. The least likely outcome is a tie, i.e., that y_1 and y_2 will both change values from 0 to 1 at the same instant. It is far more likely that either y_1 or y_2 will win the race.

Suppose y_2 wins. Then the y-state becomes 01, corresponding to row 3, a stable state in this column. Since the Y_1-value in 3-01 is 0, the signal change at the input to the y_1-branch delay has persisted for less than the magnitude of that delay. Hence, assuming it is inertial, there will be no change in the value of y_1, and the system will remain in stable state 3-01—the wrong state. If the delay element is not inertial, then y_1 may change at some later time, but while the system is in state 01, the Y_1-value will be 0. Hence, some time after the system state changes to 2-01 as a result of y_1 changing to 1, the 0 will get through the delay and the state will revert to 3-01. This oscillation might continue indefinitely. Thus, regardless of whether the delays are inertial, pure, or

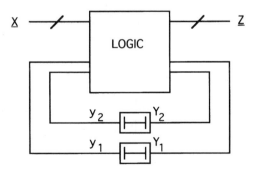

Fig. 6.17. Block diagram for ASC with two y-variables.

something in between, a malfunction is likely as a result of the outcome of the race. (If y_1 had been the winner, the result would have been similar.)

A situation in which several y-variables become unstable simultaneously is called a *race* condition. If for any outcome of the race, the state reached is not the one specified in the flow table, then the race is said to be *critical*. Thus, for ASCs, state assignments must satisfy more stringent criteria than simply distinguishing among the flow table rows. The problem of finding state assignments free of critical races is not trivial. It is sometimes necessary to use more than the minimal number of y-variables needed to provide a unique y-state for each row. But a great deal of work has been done on this problem, with the result that there exist a number of good solutions.

For the Figure 6.16 example, we note that there are transitions between the following (unordered) pairs of rows: $(1, 2), (2, 3), (3, 4), (1, 4)$. If the rows are plotted on a K-map as shown in Figure 6.18a, it can be seen that the members of each pair are in adjacent pairs of cells, i.e., that the members of each pair are separated by only one y-variable. If the y-states are assigned this way, as is done in Figure 6.18b, then only one y-variable changes for each transition; there are no races at all. The next example poses a more difficult problem.

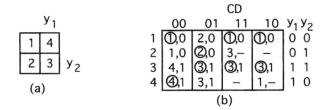

Fig. 6.18. A race-free state assignment for the flow table of Figure 6.16.

Example 6.6

Consider the three-row flow table of Figure 6.19a. The pairs between which transitions occur are $(1, 2), (2, 3), (1, 3)$. Attempts to place the three rows on a K-map so that the members of each pair are adjacent, will inevitably fail. Why? Observe that we would be trying to find a closed path on a K-map linking $1, 2, 3$ (see Figure 6.19b). Each step along a path linking adjacent nodes corresponds to a change in exactly one y-variable. If a path is closed, then each variable must be restored to its original value upon

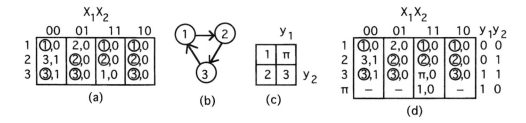

Fig. 6.19. A harder problem and a solution.

completion of the circuit; hence, each variable must have changed an even number of times (possibly 0 times). Then the total number of changes—and therefore the total number of steps—must be *even*. Thus, no closed path containing an *odd* number of nodes can exist on a K-map, which explains our failure in this case of a three-node circuit.

There are several ways to escape from this impasse. One is to permit one or more of the flow table transitions to involve a sequence of two or more y-variable changes. In this case, as shown in the Figure 6.19c K-map, the rows 1, 2, and 3 are assigned to a connected chain of y states. A row labeled π is introduced to enable the chain to be continued from 3 to π and then back to 1. A flow matrix utilizing this scheme is shown as Figure 6.19d. The transitions $1 \Rightarrow 2$ and $2 \Rightarrow 3$ each involve a change in one y-variable (i.e., they are one-step changes). But the transition $3 \Rightarrow 1$ (see Figure 6.19a) has been broken up into two transitions: $3 \Rightarrow \pi$ and then $\pi \Rightarrow 1$, i.e., $3 \Rightarrow 1$ is now a two-step transition. This technique of inserting intermediate states to make it possible to fit the rows on a K-map with the necessary adjacencies is very powerful, and can be applied in all cases. However, there are no known simple algorithms for applying it. Furthermore, the *shared-row method*, as it is known, has the drawback that multistep transitions are sometimes undesirable because they slow down circuit operation. An approach that leads to *single-transition-time* state assignments is illustrated next, for the same flow table.

Example 6.7

Instead of trying to eliminate races altogether, which, as was just shown, may require that some flow table transitions be made in several steps, we can make all transitions direct, allowing some, or all, of them to involve *noncritical* races. Such a solution is shown in Figure 6.20 for the flow table of the previous example. There are three y-variables, and each of the three transitions in the flow matrix involve races between two y-variables. The key to making the races noncritical is to ensure that, for all states that may be passed through in the course of a transition, the next-state entry corresponds to the destination of the transition. As is the case here, this often requires the addition of rows to the flow matrix.

Consider the transition in column 01 from row 1 to row 2. This entails a race between y_2 and y_3, both of which are required to change. If the race is a tie, there is of course no problem. Suppose y_2 wins. Then the system enters the state 010-01. Here (see Figure 6.20) the next-state entry is 2, so the signal at the Y_3-terminal remains at 1 and y_3 will

$$X_1 X_2$$

	00	01	11	10	y_1	y_2	y_3
1	①,0	2,0	①,0	①,0	0	0	0
2	3,1	②,0	②,0	②,0	0	1	1
3	③,1	③,0	1,0	③,0	1	1	0
–	–	2,0	–	–	0	0	1
	3,1	2,0	1,0	–	0	1	0
	3,1	–	–	–	1	1	1
–	–	–	1,0	–	1	0	0

Fig. 6.20. A single-transition-time state assignment for the flow table of Example 6.6

eventually change when that signal gets through the delay in that branch. Similarly, if y_3 is the winner, the system will enter the state with $y_1y_2y_3 = 001$, where the next-state entry in the 01 column is also 2. There are also races for the other two transitions that are noncritical for the same reason.

There are systematic methods for generating such *unicode* (because there is *one* y-state per row) single-transition-time (USTT) state assignments for any flow table, but they are not presented in this book. The principal approaches, called the Liu and the Tracey methods, are reasonably tractable, and lead to results that are usually not excessively costly to implement. The next example illustrates the application of a general procedure enabling the Y- and Z-variable logic expressions to be generated by inspection.

Example 6.8

The state assignment used in this example is known as the *1-hot code*, a term derived from the fact that, for an *n*-row flow table, *n* y-variables are used and that for each row exactly *one* of the y-variables has the value 1. Transitions are all two-step. The method is illustrated with the flow matrix of Figure 6.21, where the 1-hot state assignment is shown.

In the circuit corresponding to this assignment, a transition from row i to row j is executed by y_j turning on and *then* y_i turning off. The logic expression for Y_i has two parts. First there are one or more *turn-on* terms, each corresponding to an unstable state representing a transition *to* row i. These are the terms responsible for causing the first part of the transition to row i to occur, i.e., they turn on Y_i, causing y_i to follow suit after a delay. An example of a turn-on term for Y_5 is Dy_3, corresponding to the total state 3-D, where the next-state entry is 5. A turn-on term for Y_2 is Cy_1, corresponding to total state 1-C.

Once y_i is on, it should remain on until a transition to another state j has reached the point where y_j has been turned on. This is taken care of by the second part of the Y_i-expression, which is called the *hold* term. It is a single product term added to the sum of turn-on terms. The hold term consists of the product of y_i and the *complements* of all of the y's corresponding to the rows that can be reached directly from row i. In our example, the rows reachable directly from row 1 are 2 and 5. Hence the hold term for Y_1 is $y_1\bar{y}_2\bar{y}_5$. The hold term for Y_2 is $y_2\bar{y}_3$, as the only unstable state in row 2 has 3 as its next-state entry.

	A	B	C	D	y_1	y_2	y_3	y_4	y_5
1	①,0	①,0	2,0	5,0	1	0	0	0	0
2	②,0	3,0	②,0	②,1	0	1	0	0	0
3	③,0	③,0	4,0	5,0	0	0	1	0	0
4	3,0	④,1	④,0	5,0	0	0	0	1	0
5	1,0	3,0	⑤,0	⑤,0	0	0	0	0	1

Fig. 6.21. Flow matrix with 1-hot code.

Putting together the turn-on and hold terms for the five y's gives us

$$Y_1 = Ay_5 + y_1\bar{y}_2\bar{y}_5, \qquad Y_2 = Cy_1 + y_2\bar{y}_3, \qquad Y_3 = Ay_4 + By_2 + BY_5 + y_3\bar{y}_4\bar{y}_5,$$

$$Y_4 = Cy_3 + y_4\bar{y}_3\bar{y}_5, \qquad Y_5 = Dy_1 + Dy_3 + Dy_4 + y_5\bar{y}_1\bar{y}_3.$$

The Z-expression is easily obtained by summing all of the terms corresponding to total states where $Z = 1$. For our example, these states are 4-B and 2-D, leading to the expression: $Z = By_4 + Dy_2$. The method described here is applicable to all SOC functions. It can be elaborated to encompass MOC functions, but this is not done here.

The 1-hot code certainly minimizes designer time and effort. Since the transitions are all 2-step, it leads to circuits slower than could be built employing a single-transition-time assignment. The cost in logic complexity is usually somewhat higher than for some other methods, but is generally not far out of line. For a flow table with a relatively small number of transitions it is sometimes quite competitive with respect to gate-input count.

6.2.6. Combinational Hazards and What to do about Them

Quite apart from the critical race problem addressed in the previous subsection, there is another class of problems related to delays in logic circuits. Although these problems involve the transient behavior of *combinational* circuits, their consequences are most significant when these circuits are embedded in ASCs. They are introduced in the next example, which also introduces another important type of storage element.

Example 6.9

The flow table shown in Figure 6.22a describes what is known as a *latch* (*D-latch*, *polarity hold latch*, *transparent latch*, and *Earle latch* are alternative names for it). The Q-output of a latch copies the value of the D-input as long as the C-input is on. When the C-input goes off, the Q-output becomes constant, independent of any changes in D. Latches are widely used in modern computers.

Using the state assignment shown in Figure 6.22 (obviously no races are possible with only one contestant!), the K-map for Y shown in Figure 6.22c is easily obtained (no map is necessary for the output, since Q is obviously equal to y). The expression $Y = CD + \bar{C}y$ follows immediately, leading to the circuit realization of Figure 6.23a (the two AND-gates are numbered for identification in the discussion to follow). Readers should satisfy

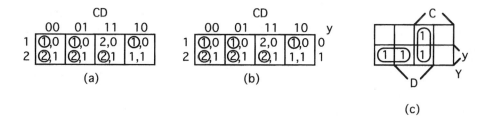

Fig. 6.22. Tables for the design of a latch.

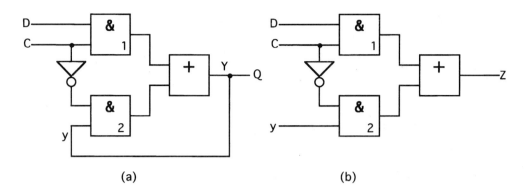

Fig. 6.23. Circuit corresponding to the design developed in Figure 6.22.

themselves, by tracing through various input sequences, that this circuit behaves according to the verbal description of a latch presented at the outset of this example.

Consider what happens if, with the latch in the 1-state (Q = 1) and C = D = 1, C is
turned off. Initially, both inputs to AND-gate-1 are 1s, so the output of that gate is 1.
The output of the inverter is 0, and this signal keeps the output of AND-gate-2 at 0.
Hence, the OR-gate output (and, hence, Y = Q) is kept on by the 1 at its upper input.
When C goes to 0, AND-gate-1 is turned off, which turns off the upper input to the
OR-gate. But, through the lower path from the C-input, the inverter output is turned on,
which causes AND-gate-2 to be turned on. This sends a 1-signal to the lower input of the
OR-gate, keeping its output at 1. But suppose that the stray delays in the path from C
through the inverter to the AND-gate-2 input are large relative to the delays in the path
from C through AND-gate-1. Then there will be an interval during which *both* inputs to
the OR-gate are off. If this interval is sufficiently long, y will go to 0 and this, in turn, will
keep the output of AND-gate-2 off. The circuit will then be in the 0-state instead of in
the 1-state.

The basic problem resides in the combinational circuit generating Y, which is isolated
in Figure 6.23*b* (with its output relabeled Z). As indicated earlier, with D = y = 1, a
change in C from 0 to 1 (abbreviated as C: 0 ⇒ 1) can cause the output to change from 1
to 0 to 1. That is, a 0-pulse can occur. In cases such as this, where, as a result of stray
delays in the logic, there is a *possibility* of a false output pulse (sometimes called a
glitch) accompanying an input change, there is said to be a *combinational hazard*. If the
output is supposed to remain the same following the input change, the hazard is *static*. If
the value of the output is specified to be 1, then we have a 1-*hazard*. Thus, the Figure
6.23*b* circuit has a static 1-hazard. Some light can be shed on the problem by fixing
D = y = 1 in the circuit, and then examining the resulting simplified circuit. With
D = 1, the C signal passes right through AND-gate-1, which can therefore be eliminated. In a similar manner, we can dispose of AND-gate-2 as a result of y being 1. This
reduces the circuit to what is shown in Figure 6.24*a*, where the C-signal reaches the
OR-gate through two branches, one of which contains an inverter. The corresponding
logic expression is $Z = C + \overline{C}$.

As shown by the waveforms in Figure 6.24*b*, when C drops to 0, there may be a time
lag before $\overline{C}$ rises to 1. During that interval, *both* of the inputs to the OR-gate are 0, so
its output falls to 0 for a short interval. If, on the other hand, the delay is greater in the

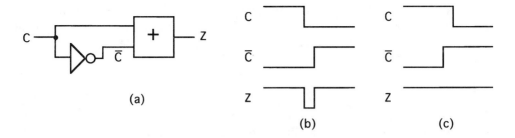

Fig. 6.24. Circuit showing the essence of a 1-hazard.

upper path from C, so that the signal through the inverter arrives at the OR-gate *before* the uncomplemented C-signal, then the 1-signals to the gate overlap in time and the output remains constant, as shown in Figure 6.24c.

There are two ways of using delay elements to deal with the hazard problem. The first is suggested by Figure 6.24. If we are concerned with the possibility of a glitch when C is changing from 1 to 0, as in the Figure 6.24b situation, then we could insert a delay element (pure or inertial) in the path of the uncomplemented version of C. This produces the situation shown in Figure 6.24c, where the overlapping inputs to the OR-gate keep its output on. But then we have ensured that for the reverse transition. C: $0 \Rightarrow 1$, there *will* be an interval during which both OR-gate inputs are off and the output will dip to 0. There are situations in which this is acceptable.

The other approach is to accept the possibility of the glitch being generated and to insert an inertial delay element (ID) at the output of the circuit to filter it out. The drawback of this approach, in addition to the cost of the ID, is the fact that it slows the operation of the circuit. Furthermore, the ID will also distort the waveform of the output signals by reducing the slopes of the leading and trailing edges unless it incorporates costly active devices designed to prevent this.

Another view of the problem, which leads to an elegant solution that does not entail the use of delay elements can be obtained by examining the K-map of the function, i.e., Figure 6.22c. The subcubes corresponding to the *p*-terms feeding the OR-gate are ovalled. The 1-point, 111, corresponding to the start of the input transition, is in a different subcube from the adjacent 1-point, 011, corresponding to the endpoint of the transition. Hence, following the C-change, the AND-gate realizing one of these subcubes goes off and the AND-gate realizing the other transition goes on. If these events occur in the order just stated, then we have a glitch; otherwise not.

In a situation in which only one input variable is changing, (i.e., *single-input-change* or *SIC* operation) the two 1-points involved are necessarily adjacent and, hence, can be covered by a single subcube. Suppose that an AND-gate realizing this "bridging" or "consensus" subcube is added to the circuit, with its output going to the OR-gate. Since the subcube corresponds to a *p*-term that does *not* contain the variable that is changing, it remains on during the transition. This ensures that the OR-gate stays on, thereby eliminating the hazard. In the latch example depicted in Figure 6.22, the added *p*-term would be Dy, changing the Y-expression to $Y = CD + \overline{C}y + Dy$.

Any logic function can be realized by a two-stage AND-OR circuit without 1-hazards for SIC operation, iff the SOP expression corresponding to the circuit includes *p*-terms covering all boundaries between adjacent 1-points as well as all 1-points. Clearly, it is advantageous to use pi's to cover the boundaries for the same reasons of greater coverage at lower cost that dictate their use simply for 1-point coverage (see Section 3.2). The next example serves to illustrate the process.

Example 6.10

Consider the logic function treated in Example 3.6 (p. 62). The K-map with the optimal selection of subcubes for a minimal SOP expression is repeated in Figure 6.25a. There are four uncovered boundaries between adjacent 1-points, marked with Xs. Note that we need not consider boundaries between 1-points and don't care points because the latter are treated as though the corresponding input state will never occur. In Figure 6.25b, three additional subcubes are designated to cover the marked boundaries. The 1-cube corresponding to $A\overline{B}C$ covers the lower right member of this set, $\overline{A}\,\overline{D}$ covers the two left members, and BD covers the remaining X-marked boundary (just below the middle of the map). The process can be carried out systematically for any function using a simple extension of the same general approach developed in Sections 3.3 and 3.4.

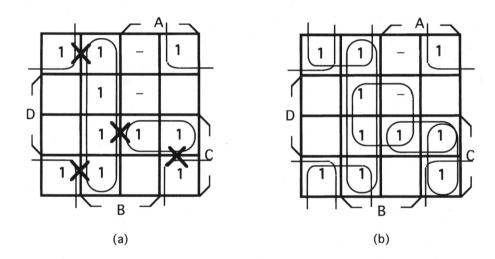

(a)   (b)

Fig. 6.25. Finding a minimal hazard-free realization of a function.

Example 6.11

In addition to static 1-hazards, there are also static *0-hazards*. For two-stage AND-OR circuits, which we consider first, they are present only under the rather peculiar circumstances illustrated in Figure 6.26a, a logic circuit corresponding to the expression $Z = A\overline{B} + AC + BC\overline{C}$. The output is specified at 0 when A = 0 and B = 1, independent of the value of C (i.e., Z = 0 for the input states ABC = 010 and 011). But a C-change from either of these states may cause Z to go on briefly. The mechanism for this is made evident by Figure 6.26b, the equivalent of Figure 6.26a when AB is fixed at 01. This circuit, directly corresponding to the expression $Z = C\overline{C}$, is the dual of Figure 6.24a, just introduced to reveal the anatomy of the 1-hazard.

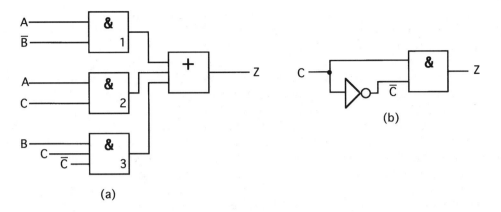

(a)

(b)

Fig. 6.26. An example of a circuit with a 0-hazard.

During a C: $0 \Rightarrow 1$ change, if the delay in the path from C to the AND-gate through the inverter exceeds the delay in the direct path from C to the AND-gate, there will be a brief interval during which *both* AND-gate inputs are 1s, thus producing a spurious 1-pulse at the output. The possibility of such behavior, i.e., an output change from 0 to 1 to 0 following a single change in the input state, constitutes a 0-hazard.

Avoiding 0-hazards in two-level AND-OR logic is easy. A necessary and sufficient condition for their existence is that:

1. The corresponding SOP expression include a *p*-term with inputs X and $\overline{X}$;
2. For some state of the other input variables, the other elements of this term be 1;
3. For the same input state all other *p*-terms be 0.

In our current example, the *p*-term of (1) is $BC\overline{C}$, with C serving as the X, and with the state of the other variables being $AB = 01$. Since there is no reason to use such strange *p*-terms, nothing is sacrificed by omitting them. Thus, if we drop the $BC\overline{C}$ term from the expression for Z, which means that AND-gate-3 in Figure 6.26a is eliminated, the resulting circuit will no longer have the 0-hazard. (A circuit corresponding to the logic expression derived in connection with the table of Figure 6.25 is free of 0- as well as 1-hazards.) Incidentally, it is convenient to refer to the presence or absence of hazards in a *logic expression*, when we really are talking about the logic circuit *corresponding* to that expression.

In addition to static hazards, a circuit might also have a *dynamic* hazard, which is a possibility of a multiple output change such as Z: $0 \Rightarrow 1 \Rightarrow 0 \Rightarrow 1$, or Z: $1 \Rightarrow 0 \Rightarrow 1 \Rightarrow 0$, following a single change in the input state that is supposed to cause Z to change once. In two-stage AND-OR logic, this also implies the existence of a gate with complementary inputs. A characteristic algebraic form of a dynamic hazard is $X + X\overline{X}$, where X is the changing input variable. (The other characteristic form is the dual expression, $X(X + \overline{X})$.) Thus, by avoiding terms with complementary literals to ensure that there are no 0-hazards, we are also ensuring that there are no dynamic hazards for SIC operation.

An example of a circuit with a dynamic hazard for a change in A when $B = C = 1$ is that corresponding to the expression $A\overline{B} + AC + A\overline{A}BC$. For $B = C = 1$, this reduces to the characteristic form $A + A\overline{A}$.

The situation is considerably more complex when multiple input changes are permitted. Dynamic hazards are not treated further here. Next we extend the discussion of combinational hazards to encompass multistage logic.

When certain Boolean algebra transformations are carried out on logic expressions, the combinational hazards are either unaffected or are changed in a precisely known way. These transformations can be used to reduce an expression corresponding to a multistage logic circuit to an equivalent SOP expression with the same hazards. Then we can apply the ideas just discussed to identify these hazards, or to establish that there are none. Conversely, we can generate a SOP expression free of hazards and then apply one or more of these transformations to obtain a multistage logic circuit that is also hazard-free. The most important of these transformations are briefly introduced below, without proofs (see Figure 2.3 for a listing of the laws):

1. The associative laws preserve combinational hazards.
2. The distributive law preserves static hazards when used in either direction. When used to factor it can eliminate, but never add, dynamic hazards.
3. The DeMorgan laws preserve all combinational hazards.
4. The duality transformation converts hazards to their duals and moves the *locations* to dual points (e.g., it converts a 0-hazard between 011 and 010 to a 1-hazard between 100 and 101).

An example of a transformation that is *not* hazard preserving is the consensus theorem. That is, $AB + \overline{A}C + BC = AB + \overline{A}C$ does not preserve hazards. Deleting the consensus term *adds* a 1-hazard. The transformation $A + \overline{A}B = A + B$ *removes* a 1-hazard.

A simple application of transformation (2) converts the hazard-free expression found for the function of Example 6.11 from $Z = A\overline{B} + AC$ to $Z = A(\overline{B} + C)$. In this case, we still have two-stage logic, but now it is OR-AND, which is appropriate for a NOR-gate realization.

A more elaborate application of factoring would be to transform the function of Example 6.10. Here we generated (see Figure 6.25*b*) the hazard-free expression $Z = \overline{A}\overline{D} + BD + \overline{A}B + \overline{B}\overline{D} + ACD + A\overline{B}C$. This can be factored to produce an expression corresponding to a more economical multistage circuit that is also hazard-free, namely, $Z = AC(\overline{B} + D) + (\overline{A} + \overline{B})\overline{D} + B(\overline{A} + D)$. Next we apply the technique in reverse to analyze a circuit.

The first step in analyzing the circuit corresponding to the expression

$$Z = (A + BC)(\overline{A}D + E)$$

is to multiply it out (i.e., apply the distributive law) to obtain

$$Z = A\overline{A}D + AE + \overline{A}BCD + BCE.$$

We do *not* eliminate the $A\overline{A}D$ term because $X\overline{X} = 0$ is *not* a hazard-preserving transformation. With $D = 1$, $E = 0$, and $BC = 0$ the SOP expression reduces to $Z = A\overline{A}$, so there are three 0-hazards, corresponding to the three ways in which BC can be 0. (Note that SOP expressions with p-terms including complementary literals, which seemed absurd when first presented, appear quite routinely as a result of transformations of reasonable multistage logic expressions.) Readers should verify that there are no 1-hazards for the expression. Hence, the original circuit (not shown here) has no hazards other than 0-hazards for A-changes in the regions defined by the subcubes $\overline{B}CD\overline{E}$, $\overline{BCDE}$, and $B\overline{C}D\overline{E}$.

Another analysis example, involving the use of a DeMorgan transformation, involves the expression $Z = \underline{(A + B)}C\overline{D} + (A + B)\overline{C}D + ABC$. (The underscoring of the first parenthesized term indicates that the output of the gate generating $A + B$ fans out to two places, one of which is the input to an inverter.) Applying a DeMorgan law converts the expression to $Z = \overline{A}\,\overline{B}C\overline{D} + (A + B)\overline{C}D + ABC$. Multiplying out yields $Z = \overline{A}\,\overline{B}C\overline{D} + A\overline{C}D + B\overline{C}D + ABC$. There are no 0-hazards, but, after plotting the expression on a K-map (readers are urged to do this), it becomes evident that a C-change in the region defined by ABD entails a 1-hazard.

The discussion here of combinational logic hazards has been confined to SIC operation. Multiple input change combinational hazards are of somewhat lesser interest. Readers are referred elsewhere for information about them. In subsection 6.2.8, we consider hazards that are fundamentally associated with sequential rather than combinational logic. But first let us examine an interesting device.

6.2.7. Synthesis of a Toggle (or Trigger) Flip-Flop

Synthesizing the toggle flip-flop (T-FF) will serve as a good review of asynchronous circuit concepts. The T-FF (Figure 6.27a) has one input, T, and one output, Q. On the trailing edge of each positive T-pulse, Q changes value. One way to think of this is as a pulse frequency divider. For every two T-pulses, there is one Q-pulse (Figure 6.27b). This indeed is one application of the device. The widths of the T-pulses can be any value exceeding some minimum dictated by the characteristics of the circuit components.

The construction of the primitive flow table (Figure 6.28a) is straightforward. Starting in state 1-0 (with $Q = 0$), which can be considered the initial stable state, changing T to 1 must lead to a new stable state, 2-1, with the same output. Changing T from 1 to 0 (the trailing edge of the pulse) brings the system to a stable state in the 0-column, which cannot be 1-0, since the output is required to be 1. The remaining steps are similar. The flow table is clearly not reducible, so we have a four-row, two-column table to synthesize.

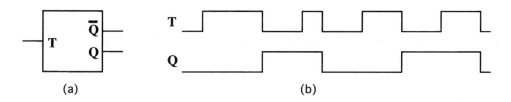

(a)   (b)

Fig. 6.27. The toggle flip-flop and its ideal behavior.

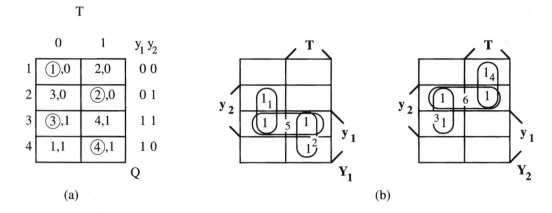

Fig. 6.28. Flow matrix and K-maps for the T-FF.

The state assignment is obvious, and it does not take any heavy machinery to determine that the simplest solution for Q is $Q = y_1$. The Y_1 and Y_2 K-maps of Figure 6.28*b* (rotated ninety degrees to correspond to the flow table) can be readily derived, without the need for a Y-matrix. Observe that, since $y_1 = 1$ for rows 3 and 4, the values of Y_1 must be 1 wherever the next-state entry in the flow table is a 3 or a 4. Elsewhere, $Y_1 = 0$. Similarly, the K-map for Y_2 is obtained by writing in 1-entries in positions corresponding to those where the next-state entry is a 2 or a 3.

Using the ovalled subcubes shown on the maps, we generate the SOP expressions below. Note the inclusion of terms in each expression ($y_1 y_2$ and $\bar{y}_1 y_2$) to eliminate 1-hazards:

$$Y_1 = \overline{T}y_2 + Ty_1 + y_1 y_2, \qquad Y_2 = \overline{T}y_2 + T\bar{y}_1 + \bar{y}_1 y_2, \qquad Q = y_1.$$

Factoring these expressions leads to

$$Y_1 = \overline{T}y_2 + (T + y_2)y_1, \qquad Y_2 = \underline{\overline{T}y_2} + \left(T + y_2\right)\bar{y}_1, \qquad Q = y_1.$$

The underscored terms represent shared gates. The total number of gates in this realization is thus six. However, rather than pursue this line further, let us compare the original SOP expressions for Y_1 and Y_2 with the logic expression for the latch (see p. 179), which is

$$Y = CD + \overline{C}y + Dy.$$

In the expression for Y_1, $\overline{T}$ plays the role of C and y_2 plays the role of D. Similarly, in the Y_2 expression, T behaves as C, and $\bar{y}_1$ behaves as D. We can therefore realize the T-FF with a pair of latches and an inverter, as shown in Figure 6.29.

Having derived the latch implementation of the T-FF via our formal design process, let us now examine it directly to verify that it works as specified. Consider first the situation with T = 0. The C-input to the lower latch is 0, so that y_2 is stable. Since the C-input to the upper latch is 1, the upper latch is in the transparent state, i.e., its output,

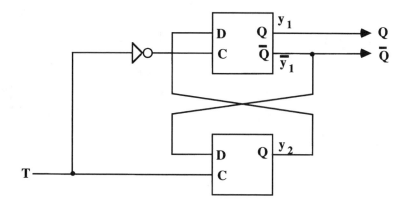

Fig. 6.29. Realizing a T-FF with latches.

y_1, copies the D-input. This input is y_2, so we have a stable state with $Q = y_1 = y_2$. When T changes to 1, the upper latch output, y_1, becomes stable (Q remains constant), and the lower latch becomes transparent. Since the D-input to the lower latch is $\bar{y}_1$, this means that y_2 changes value. Now, when T goes off, $Q = y_1$ will copy the new value of y_2. Hence, we have Q changing on the trailing edge of the T-pulse, which is the desired behavior.

In some practical applications of T-FFs it is necessary to add a second input, a clear signal that brings the device to the $Q = 0$ state, regardless of its initial state. This is easily accomplished by passing each of the y-signals through an AND-gate controlled by the clear signal (Problem 6.14).

6.2.8. Essential Hazards and the Role of Delay Elements

In the introduction to ASCs (Subsection 6.2.1) it is explained that when an ASC is in a stable state, and an input change occurs, the operating point moves within a current row to the new input column and then, if the state arrived at is unstable, a vertical move is made within that column to the next internal state. In this subsection the role of delay elements in ensuring that circuits behave according to this model is explained. It is assumed, unless otherwise stated, that we are dealing with SOC functions operating under SIC restrictions.

Example 6.12

A design for a SOC ASC function described by a three-row flow table is shown in Figure 6.30. Based on the K-maps of Figure 6.30b and c, we can generate the logic expressions $Y_1 = \bar{X}y_2 + y_1$ and $Y_2 = X + y_2$. From the flow matrix, it is obvious that we can set $Z = y_1$. The state assignment is clearly all right (for each of the two specified flow table transitions, only one y-variable changes). Then, since the logic expressions are clearly hazard-free, we might reasonably expect the resulting circuit to behave as specified by the flow table. Let us now see how that circuit, shown in Figure 6.30d might actually function.

The fact that no delay elements have been specified in the y-branches is a key factor. Suppose that the system starts in stable state 1-0, where $X = y_1 = y_2 = 0$. When X is

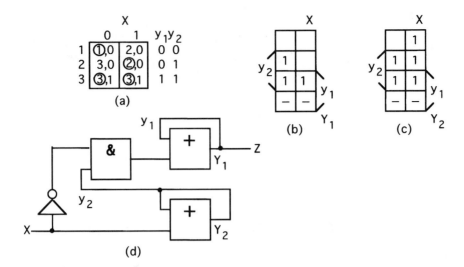

Fig. 6.30. Implementation of a simple flow table.

turned on, signals from the X-terminal flow out along two paths. Assume that the stray delays are such that the X-change propagates swiftly through the lower OR-gate turning on Y_2 and continuing around the feedback path to the AND-gate, arriving there before the output of the inverter changes to 0. Then the AND-gate turns on, and this signal continues through the upper OR-gate to turn on Y_1. The y_1-signal then, through the second feedback path, holds the Y_1 OR-gate on, even after the 0-signal from the inverter finally turns off the AND-gate. The circuit is now in stable state 3-1, the wrong state!

The explanation for this disappointing outcome is that the *direct* effect of the input-change, through the path with the inverter, did not reach Y_1 until *after* it was affected by the internal state change that was caused by the input-change. From the point of view of an observer at the Y_1-terminal, the system started in 1-0 (see the flow matrix in Figure 6.30a), and then, since the first signal to reach Y_1 was from y_2, there was a *vertical* move in the flow matrix down to 2-0. Since this is an unstable state calling for a change in y_1, this causes Y_1 to go on. If the resulting y_1-change reaches the Y_1-terminal before the input change, then the observer there will see a second internal state change, bringing the system to 3-0. When the input-change does finally arrive, the system will be seen as, and will actual be in, state 3-1. (Note that there are both inverted and uninverted paths from X to the AND-gate. This characteristic feature of essential hazards resembles what we saw in connection with combinational hazards—see Figure 6.24—except that here a feedback path is involved.)

Had there been a sufficiently large delay element in the y_2-feedback branch, the news of the X-change would have reached Y_1 *before* the news of the y_2-change, and proper operation would have resulted. The reason for the use of delay elements in y branches is now evident. They are there to ensure that events occur in the proper order, i.e., that circuit operation conforms to our assumption that input changes cause horizontal moves in flow tables followed by vertical moves.

For SIC operation, and even for cases where some multiple-input changes are permitted, it is not always necessary to use delay elements. An example of this is the circuit for an SR-FF shown in Figure 6.14 (p. 172). For some sequential functions it is possible to find state assignments that allow designs without delay elements (called *delay-free* realizations). The function treated in Example 6.12 has a property called an *essential hazard* that makes *any* circuit realizing it vulnerable to the type of trouble encountered with the Figure 6.30*d* circuit. An essential hazard exist in a flow table for a transition involving a change in variable X starting from a stable state S if, starting from S, three consecutive changes in X bring the system to a stable state different from the stable state reached after the first change in X. For example, in Figure 6.30*a*, if we start in the stable state 1-0 and change X once, we reach stable state 2-1. Two more changes in X bring us to a different stable state, namely 3-1. Thus, a change in X from 1-0 entails an essential hazard, and indeed this transition can produce the malfunction described. For the PETDFF, whose flow table is shown in Figure 6.9 (p. 167), there is an essential hazard for a C: $1 \Rightarrow 0$ change starting in 3-10. Note that for the T-FF (p. 184), there is an essential hazard for *every* transition.

In addition to the *steady state* essential hazards just described, there are also *transient* essential hazards. Consider the function of Figure 6.21 (p. 176). Starting from stable state 3-B, an input change from B to C should cause the total state to change to 3-C and then to stable state 4-C, with Z remaining fixed at 0. But if stray delays cause Z to see the internal state change first, then it sees the system pass through 4-B, where Z is specified to be 1. Although the correct state is finally reached, this results in a spurious *output* pulse. Delays in the feedback paths can prevent both transient and steady state essential hazards.

As in the case of combinational hazards, the existence of an essential hazard does not mean that a circuit *will* malfunction. Rather, it means that there exist distributions of stray delay values that would cause faulty behavior. Now let us determine how large the delay elements in the feedback paths must be in order to ensure that for the worst case there will be no malfunction due to signal changes being received in the wrong order, as in the previous examples.

Assume that we are given upper and lower bounds on the time it takes for a signal change to get through the combinational logic from an X- or y-input to a Z- or Y-output. Let the maximum and minimum values of these delays (sometimes called the *long-path* and *short-path* delays), be D_{LM} and D_{Lm}, respectively, and let D_{Em} be the minimum satisfactory value for a delay element in a y branch. Then, starting with an X-change, it would take at least $D_{Lm} + D_{Em} + D_{Lm} = 2D_{Lm} + D_{Em}$ for the signal to penetrate the logic network, change a Y-value, continue through the delay element to change the corresponding y, and then go through the network again to affect another Y. At maximum, it would take D_{LM} for the same X-change to go through the network *directly* to the second Y-terminal. If the former time exceeds the latter time, then no y-change can be seen before the X-change that precipitated it, and so there cannot be the kind of malfunction we are concerned about here. Hence, we require that D_{Em} be large enough to ensure that $2D_{Lm} + D_{Em} \geq D_{LM}$, or that

$$D_{Em} \geq D_{LM} - 2D_{Lm}. \tag{1}$$

Clearly if the short-path delay is not less than half the long-path delay, no delay elements are needed. In particular cases, rather than rely on overall worst-case figures, one might investigate these paths that might cause trouble to determine if and where delay elements need be added. In the circuit of Figure 6.30*d*, for example, we might compare the delays in the two paths leading from the X-input to the AND-gate to determine if a delay element is needed to prevent that gate from going on when the essential hazard transition takes place. In calculating bounds on delays one must consider the many factors that cause variations, such as wiring path lengths, fanin to gates, fanout from gates, process variations, power supply voltage fluctuations, and operating temperature.

To conclude our discussion of delays, let us calculate a conservative lower bound on g, the time that should elapse between consecutive input changes. The calculation is based on the idea that an input change should not reach a Y- or Z-terminal until all effects of the previous input change have reached all terminals. Assume the state assignment is single-transition-time, and that an input change at $t = 0$ causes a change in the internal state. Then, at the latest, the results of that input change will reach all outputs of the combinational logic block by $t = D_{LM} + D_{EM} + D_{LM} = D_{EM} + 2D_{LM}$. The *earliest* time that *any* effect of an input change at $t = g$ can reach any such output terminal is $t = g + D_{Lm}$. Hence, there will be no problem if the second time is later than the first time, i.e., if $g + D_{Lm} > D_{EM} + 2D_{LM}$, or if

$$g > D_{EM} + 2D_{LM} - D_{Lm}. \tag{2}$$

The rate at which a circuit can be operated is governed by g, so that, for high-performance systems, we would like to minimize g. From (1) we can see that the closer the short-path delay, D_{Lm} is to the long-path delay, D_{LM}, the smaller the delay elements needed. From (2) it is evident that a relatively large value of D_{Lm} helps reduce g in two ways. One is by virtue of its appearance in (2) with a minus sign, and the other is through its aforementioned effect on D_E. The larger the short-path delay, the smaller g can be. Thus, we can draw the interesting and important conclusion that faster operation is facilitated by *maximizing* the *minimum* values of logic path delays as well as by *minimizing* their *maximum* values. Another way to state this is that it is very desirable to minimize the *variation* in path delays. That is, we should try to minimize $D_{LM} - D_{Lm}$.

6.2.9. Analysis of Asynchronous Sequential Circuits

Sometimes it is necessary to analyze an ASC to find a flow table that describes it, and, perhaps, to check for the existence of critical races or hazards. For example, if changes are made in the late stages of a circuit design, there might be some doubt as to whether it will work as desired. Another example is when a circuit is encountered, perhaps in a manufacturer's data book, with an incomplete explanation of what it does. Or analysis might be used as an independent check on the validity of an important design. The analysis procedure consists, basically, of running the synthesis procedure in reverse. A widely used circuit is analyzed in the following example to illustrate the procedure.

Example 6.13

The outputs of the Figure 6.31 circuit are labeled Z_1 and Z_2, and the inputs are labeled X_1 and X_2. (Assigning more meaningful variable names might ruin the suspense by giving clues as to what the circuit does.) Ignore the y-labels for the moment.

Because of the feedback paths, it is not possible to express Z_1 and Z_2 as functions of X_1 and X_2 alone. Additional variables, the state variables, must be introduced to represent signals at a sufficient number of points to break all closed loops. A careful examination of the circuit indicates that at least three such cuts are needed, and that they may be chosen in a number of different ways. (Minimizing the number of cuts is not

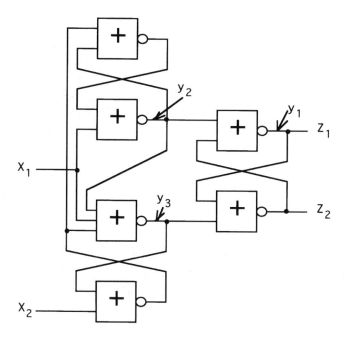

Fig. 6.31. An ASC to be analyzed.

essential, but it simplifies the amount of work that must be done.) One such minimal solution is shown in Figure 6.31 (refer to the points labeled y_1, y_2, and y_3). It is now possible to find expressions for the Zs and Ys as functions of the Xs and y's. (Note that, since no delay elements are shown, $y_i = Y_i$ for each i.) Thus

$$Z_1 = y_1, \qquad Y_1 = (y_1 + y_3)\bar{y}_2, \qquad Y_2 = \overline{X}_1(y_2 + \overline{X}_2\bar{y}_3), \qquad Y_3 = \overline{X}_1\bar{y}_2(X_2 + y_3).$$

(Z_2 is left for Problem 6.28.)

An eight-row, four-column Y-matrix can now be generated from these expressions, as shown in Figure 6.32a. The easiest way to do this is to deal with one Y-variable at a time. Taking Y_1, for example, fill in 0s for all Y_1-entries in all rows where $y_2 = 1$ (due to the $\bar{y}_2$-factor). In the remaining total states, set the Y_1-entries to 1 wherever y_1 or y_3 have 1-values.

Given the Y-matrix, we can produce a flow matrix by simply replacing the Y-entries by the row numbers for the corresponding y-states. (The output entries correspond to $Z_1 = y_1$.) This gives us Figure 6.32b. Examining this table, we see that starting from any stable state and changing one input-variable causes at most one y-variable to become unstable. Following a change in this y-variable, the system either reaches another stable state (e.g., an X_1-change from 1-10 causes y_2 to change, bringing the system to 4-00) or else a second y-variable changes to bring the system to a stable state (e.g., starting in 1-11, an X_1-change causes a y_3-change, bringing the system to 2-01, whereupon y_1 changes, taking the system to 7-01). There are no race conditions originating in any state reachable from a stable state.

If we replace the next-state entries for all two-step transitions with the names of the stable destination states, and delete the rows with no stable-state entries, we obtain

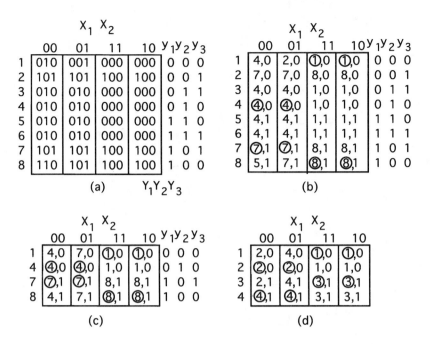

Fig. 6.32. Y-matrix, flow matrix, and flow table for the circuit of Figure 6.31.

Figure 6.32c. Finally, renumbering the rows from 1 to 4 and reordering them yields the flow table of Figure 6.32d. An examination of this flow table indicates that it corresponds to a negative-edge-triggered D-FF (NETDFF). (It cannot, however, be obtained from the Figure 6.9 flow table for a PETDFF by simply replacing C with $\overline{X}_1$. The functions involved differ in the way they respond to multiple input changes; a difference of no practical importance.)

Readers can verify from the Z- and Y-expressions that there are no combinational hazards in the Figure 6.31 circuit. This leaves only essential hazards to concern us. It has been pointed out in Subsection 6.2.8 that there are indeed essential hazards inherent in the ETDFF function. In this case, by tracing through the relevant logic paths, it can be shown that, even without the insertion of delay elements, malfunctions due to essential hazards are very unlikely (see Problem 6.29). Thus, we can conclude that the Figure 6.31 circuit is a valid realization of the Figure 6.32d flow table. It is widely used.

Example 6.14

As an additional example of the analysis process, a storage element popular in small-scale integrated circuits is considered. This is the master–slave JK flip-flop (MS JK-FF, often referred to simply as a JK-FF). As shown in Figure 6.33, this device consists of a connected pair of SR-FFs (see Figure 6.14, p. 172). The leftmost FF, called the master, is controlled by input signals J and K, as well as by a clock signal C, and feedback signals from the output of the rightmost (slave) FF. When C = 0, the state of the master FF is frozen and the slave FF copies the state of the master. When C = 1, the state of the slave FF is frozen and the master FF state is determined by the external inputs. Roughly speaking, the J-signal causes the JK-FF to be set (Q becomes 1), and the

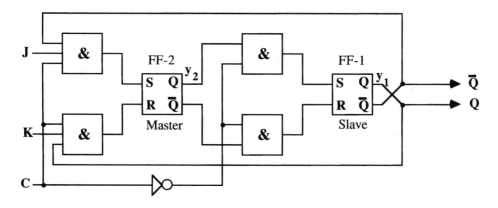

Fig. 6.33. Logic circuit for a JK-FF.

K-signal causes it to be reset (Q becomes 0). If *both* J and K are on while C = 1, then the JK-FF toggles, that is, it changes state. This behavior is described concisely by the following characteristic equation, which gives the new value of Q, designated as Q_n, as a function of J, K, and the value of Q prior to the clock-pulse:

$$Q_n = \overline{K}Q + J\overline{Q}.$$

A more precise description of the device's behavior will emerge from our analysis.

As in the previous example, we begin the analysis by choosing a set of state variables. These, along with the input variables, must be sufficient to determine all logic signals in the circuit. In this case, the outputs of the SR-FFs are chosen. The Q-output of the slave FF (label it FF-1) is assigned the value y_1, and the Q-output of the master FF (label it FF-2) is assigned the value y_2. It is not hard to show that, in the steady state, the $\overline{Q}$-outputs of the FFs in this circuit are always the complements of the Q-outputs, so that the $\overline{Q}$-entries can be considered as equal to $\overline{y}$-signals.

Now we must generate logic expressions for Y_1 and Y_2, the next-state values of y_1 and y_2, respectively. Consider Y_1 first. If FF-1 receives a set signal, or if it is already in the set state (i.e., $y_1 = 1$) and its R-input is 0, then $Y_1 = 1$. This leads to the following expression, in which the first term corresponds to the set condition, and the second term to the absence of a reset condition with y_1 already equal to 1:

$$Y_1 = \overline{C}y_2 + (C + y_2)y_1 = \overline{C}y_2 + Cy_1 + y_1y_2.$$

In a similar manner, we can derive the more complex expression for Y_2 as below:

$$Y_2 = CJ\overline{y}_1 + \left(\overline{C} + \overline{K} + \overline{y}_1\right)y_2 = CJ\overline{y}_1 + \overline{C}y_2 + \overline{K}y_2 + \overline{y}_1y_2.$$

Since the overall output is taken from FF-1, it is clear that $Z = y_1$. There is no need to construct a matrix showing Z-values.

Now we can generate the Y-matrix for the circuit (see Figure 6.34). From the Y-matrix, it is a straightforward matter to generate the flow matrix (Figure 6.35). As mentioned earlier, if the Y_1Y_2-entry in a cell of the Y-matrix is pq, then the next state entry in the corresponding cell of the flow matrix is the row to which pq is assigned. For example, the Y_1Y_2-entry for row 3, column 101 (abbreviated as 3-101) of the Y-matrix is

CJK

	000	001	011	010	110	111	101	100	$y_1\,y_2$
1	00	00	00	00	01	01	00	00	0 0
2	11	11	11	11	01	01	01	01	0 1
3	11	11	11	11	11	10	10	11	1 1
4	00	00	00	00	10	10	10	10	1 0

$$Y_1\,Y_2$$

Fig. 6.34. Y-Matrix for master–slave JK FF.

CJK

	000	001	011	010	110	111	101	100	$y_1\,y_2$
1	①,0	①,0	①,0	①,0	2,0	2,0	①,0	①,0	0 0
2	3,0	3,0	3,0	3,0	②,0	②,0	②,0	②,0	0 1
3	③,1	③,1	③,1	③,1	③,1	4,1	4,1	③,1	1 1
4	1,1	1,1	1,1	1,1	④,1	④,1	④,1	④,1	1 0

Q

Fig. 6.35. Flow matrix for master–slave JK FF.

10. Since 10 is assigned to row 4, the next-state entry for cell 3-101 of the flow matrix is therefore 4.

From the completed flow matrix, we can now see clearly how the master–slave JK FF works. When C = 0, changing J or K has no effect on the internal state (or on the output, which is equal to y_1). Consider now what happens when, with the FF initially in state-1 (the reset state, where Q = 0), C goes on. If J = 1 at the time of the C-change, *or any other time while C = 1*, the internal state changes to state-2. Once in state-2, no further changes in J or K can alter the internal state while C is on. State-2 is stable in all columns with C = 1. When the clock-pulse terminates, i.e., when C returns to 0, the next stable state will be state-3 (the set state, where Q = 1), regardless of the values of J and K. The behavior of the device is analogous when it is in the set state prior to the turning on of C.

Although the *output* changes occur on the trailing edge of the c-pulses, this is *not* an edge-triggered element. Input changes can alter the *internal state* any time while C = 1. But this is also not the way a latch behaves, since, while C = 1, the state can change in only one direction. That is, starting in state-1, a change can occur only from 1 to 2, but not in the reverse direction. Similarly, if the element is originally in state-3, the state can change only from 3 to 4 while C = 1, but never back to 3 again. This mode of operation is called "pulse triggering," and the one-way transitions are often referred to as *0- or 1-catching*.

By tying together the J- and K-terminals, we get a T-FF, useful, for example, in building binary counters (see p. 201, Subsection 6.4.3). As is shown in Subsection 6.5.2, p. 212, the JK-FF is a versatile element, useful as a general building block for synchronous sequential circuits. The master–slave JK FF is rather old technology, now largely confined to small-scale integrated (SSI) circuits. It is considerably more costly than a latch in terms of component counts. The D-FF is a bit simpler and, due to edge triggering, is less sensitive to noise. A somewhat more modern version of the JK-FF is edge triggered, but it too is seldom used in modern very-large-scale integrated (VLSI) circuits.

6.3. DUALITY FOR SEQUENTIAL CIRCUITS

In Section 2.7, the concept of duality of combinational logic functions is used to clarify such topics as negative logic and transformations between NAND and NOR logic. Here the concept of duality is extended to sequential logic.

Consider the sequential circuit diagram of Figure 6.36a, consisting of a combinational logic block L with some of its outputs fed back to its inputs through delays. For our present purposes, this may be considered as either an ASC or a SSC. In Figure 6.36b, the logic block is transformed to realize the duals of the functions realized by L, and both the outputs and the inputs of L^D are complemented through the introduction of inverters. In accordance with the concepts presented in Section 2.7, the function performed by the overall circuit is the same as for the Figure 6.36a circuit. But each feedback path contains two inverters. If we eliminate both inverters, the behavior of the sequential circuit is unchanged; it still behaves as does the Figure 6.36a circuit. Hence, just as in the case of a combinational logic circuit, inverting only the *external* inputs and outputs of a sequential circuit suffices to transform the sequential function realized to

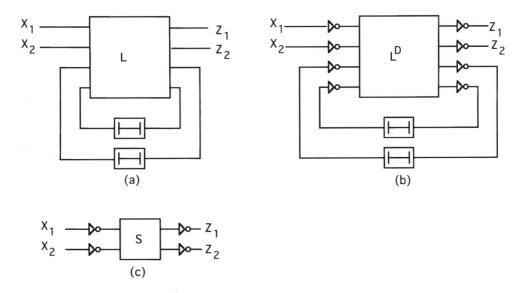

Fig. 6.36. Block diagrams of sequence circuits and their duals.

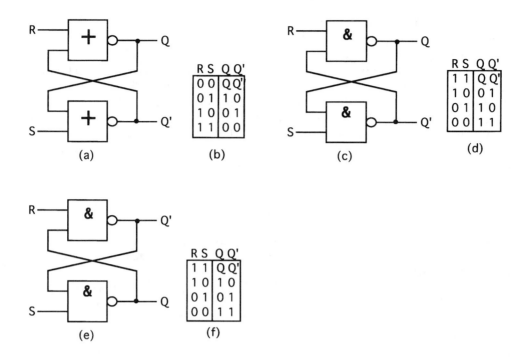

Fig. 6.37. Dual form of the SR-FF.

the dual function. The sequential function realized by the circuit of Figure 6.36c is the dual of that realized by the same circuit without the inverters. The principles involved are illustrated in the following example.

Example 6.15

A circuit for an SR-FF, developed in Figure 6.14, is reproduced in Figure 6.37a. The accompanying truth table, Figure 6.37b, summarizes the behavior of this FF. For the row corresponding to $S = R = 0$, the Q and Q' entries indicate that the Q and Q' entries remain stable at whatever values they had prior to this input state. (The reason for using Q' rather than $\overline{Q}$ for the second output is the entry for $R = S = 1$, where Q and Q' are *both* equal to 0; hence, $Q' \neq \overline{Q}$.)

Now suppose we replace the NOR-gates with NAND-gates, obtaining the dual shown in Figure 6.37c. According to the preceding discussion, the effect is as though inverters were placed in the R, S, Q, and Q' branches. Thus, the truth table is altered by replacing all of its 0s by 1s, and all of its 1s by 0s, yielding the result shown in Figure 6.37d. Readers are invited to verify the accuracy of this table by inspection of the circuit.

We can think of an $S = 0$ signal as setting the FF, if we consider $Q = 0$ as designating the set state. In order to avoid this contortion, the usual practice is to interchange the labels of the Q and Q' outputs, obtaining the circuit and accompanying table shown as Figure 6.37e and *f*, respectively. With this transformation we can retain the association of $Q = 1$ with the set state. But the S and R signals must be thought of as *active low*, in the sense that we set with $S = 0$ and reset with $R = 0$. The dormant input state is where $S = R = 1$. The FF is *set*-dominated in the sense that if *both* S and R are active (i.e., at

0), the FF goes to its set state (note though that Q' as well as Q is on when S = R = 0). The cross-coupled NAND-gate version of the SR-FF is commonly used where the technology favors NAND- rather than NOR-gates, as in transistor–transistor logic (TTL).

6.4. DESIGNING SEQUENTIAL CIRCUITS: AN INFORMAL APPROACH

In Section 6.1 a formal design procedure for SSCs is introduced, based on the use of flow tables. This procedure is developed further in Section 6.5, along with a flow chart approach to the problem. Formal methods for synthesizing ASCs are treated in Section 6.2. But many of the subsystems encountered in digital design, while involving too many variables to be treated formally as arbitrary sequential logic functions, are sufficiently simple to permit the use of more informal and direct methods. There are other situations where informal approaches suffice. Some important examples of such subsystems are treated here, beginning with a simple circuit demonstrating the value of the SR-FF.

6.4.1. The SR-FF and a Debouncing Circuit

The SR-FF is the simplest storage element to implement, requiring only two NOR-gates (See Figure 6.38a below for the flow table, Figure 6.37a for the logic circuit, and the discussions in Example 6.5, p. 171 and Example 6.12, p. 185.) A simple, but important, application of the SR-FF is presented in the following example.

Example 6.16

Mechanical switches, including those incorporated in keyboards, are widely used at the interface between people and computers. An annoying property of these switches is that they often "bounce" when operated. That is, instead of altering state cleanly, they may open and close several times before settling down to the new position. This typically occurs over an interval of a few milliseconds. If this is interpreted by the system as repeated operation of the switch, undesired results are likely. For this reason, it is customary to use *debouncing circuits* in conjunction with such interfacing switches. Such a circuit is shown in Figure 6.38b.

The SR-FF is represented by the box with inputs S and R, and outputs Q and Q'. Suppose the switch starts out in the position shown, in which the S-terminal of the FF is

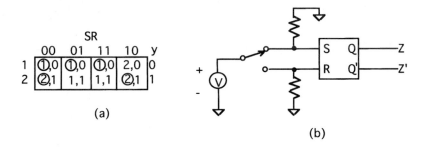

Fig. 6.38. A switch debouncing circuit.

connected to the positive-voltage terminal. Then, assuming positive logic, this is interpreted as a 1-signal, and the FF must be in the set state with Q = 1.

When the switch is operated, it disconnects the S-terminal from the voltage source so
that its voltage is pulled down to 0 by the grounded resistor. When the switch arm
reaches the other terminal (these mechanical movements take place at a snail's-pace
from the point of view of electronic circuits), the R-terminal receives a 1-signal, resetting
the FF, and hence turning off Q. If the switch bounces, then R is disconnected from the
source, returning it to 0. This has no effect on the FF, or on Q. We assume, and this
conforms with the behavior of real switches, that the switch arm will not rebound far
enough to make contact with the upper terminal again. Rather it will return to the lower
terminal, bounce off again, etc., for a number of cycles, with no further effect on the
output. A similar process occurs in the reverse direction when the switch is operated
again to turn on S and set the FF.

A third input, C, is sometimes added to an SR-FF to make it a clocked element (see
Problem 6.30). But this is less common than the ETDFF or the device considered next,
the latch.

6.4.2. The Latch: Some Designs and an Interesting Application

The latch, described by the flow table of Figure 6.39a, was introduced in Example 6.9,
p. 177, and it was subsequently shown that a hazard-free logic expression for the
associated combinational logic circuit is $Y = Q = CD + \bar{C}y + Dy$. Factoring this expression yields

$$Y = CD + (\bar{C} + D)y = CD + \left(\overline{\overline{CD}}\right)y,$$

from which the NAND-gate realization of Figure 6.39b follows. A latch implementation
of a T-FF is shown in p. 180. It is shown later in this chapter how latches are used to
realize arbitrary flow tables or to construct regular devices such as shift registers and
counters. In the next example, the capabilities of the latch are demonstrated by showing
how it can be used to construct a PETDFF (introduced on p. 167). But first let us
examine some alternative realizations of this very useful device.

A latch realized with a complex complementary MOS (CMOS) gate is shown in Figure
6.40a. It requires 12 transistors, including an inverter to generate $\bar{C}$. A more economical
design, Figure 6.40b, can be constructed with a 2–1 multiplexer (MUX) realized with

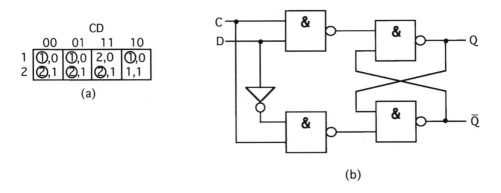

(a)

(b)

Fig. 6.39. NAND-gate implementation of a latch.

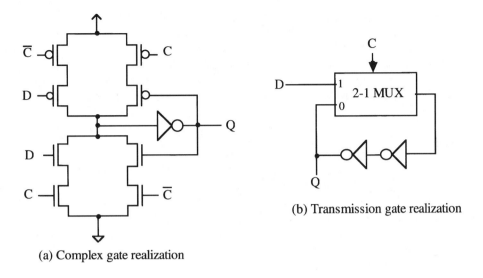

(a) Complex gate realization

(b) Transmission gate realization

Fig. 6.40. Latch.

transmission gates (see Figure 4.9*b* in Subsection 4.1.2). This version requires a total of 10 transistors.

Look more carefully now at the complex gate realization. Suppose $C = D = Q = 1$ and C changes to 0. The pulldown network originally provided a path to ground via the C and D transistors, while the path to V_{dd} through the pullup network was blocked by the transistors receiving the D and Q signals. When C goes to 0, the CD pulldown path is opened, and a new path, through the $\overline{C}$ and Q transistors, is enabled. It is quite possible that there will be some brief, but nonzero, interval during which the first path will be cut and the second path not yet established. This situation corresponds precisely to the combinational hazard introduced for another version of the latch in Example 6.9 on p. 177. Does this mean that the complex gate circuit is defective?

What happens to the Q-signal during the time when neither path exists? Unlike the corresponding situation for the gate realization of Example 6.9, no false signal is fed to Q. Rather, the Q-inverter input is left unconnected (the term used for this is *floating*) for that brief interval. In effect we have the stray capacitances attached to it charged to V_{dd} with nowhere to go except to ground through an extremely high leakage resistance. For any normal physical circuit of this kind, the time constant involved is very large compared with the length of time that node is left floating. Therefore, no significant change in the voltage at Q will occur. For a different initial state, a similar situation arises with respect to the pullup circuit (see Problem 6.34), also with no serious consequences. The given circuit is not defective. The same problem of a briefly floating node exists for the MUX-based realization of Figure 6.40*b*, where a similar analysis indicates that no unpleasant consequences are implied.

Note that, whereas, in general, transient *interruptions* of a switching circuit path do not cause serious problems, this is *not* the case for hazards that cause momentarily false paths to be established. Such paths can lead to the rapid charge or discharge of stray capacitance via relatively low-resistance paths through activated transistors. In the case of Figure 6.40*a*, the pulldown network corresponds directly to the logic expression $CD + \bar{C}Q$. If this network were replaced by the logically equivalent expression $(C + Q)(\bar{C} + D)$, then when $C = Q = D = 0$ and C changed to 1, there could be a brief interval during which a false path would exist through this network. This could lead to a malfunction, namely the manifestation of a 1-hazard at the input to the inverter generating Q. In general, the lesson to be drawn is that it is all right to use switching networks with 1-hazards in the expressions describing them, but 0-hazards can cause real trouble. This applies to pullup, to pulldown, and to pass networks.

Example 6.17

Refer now to Figure 6.41*a* and *b*, respectively, where the flow table for the PETDFF and a circuit realizing it with latches are shown. This particular circuit can be derived formally using advanced sequential logic methods (decomposition theory), but one might also have found it by "playing around" with latches. Let us see how it works. When $C = 0$, the output of the second latch, and, hence, of the overall circuit is frozen, while the outputs of the first latch follows the value of input D. When C goes on, the output of the first latch is frozen, and the output of the second latch copies the output of the first latch, which is the value of the D-input at the leading edge of the clock-pulse. The output of the first latch remains fixed while C is on, independent of any changes in the D-input, so that when C goes off, there is no change in the input to the second latch. Hence, the circuit responds to D only on the leading edge of the input clock-pulse.

As in all realizations of ETDFFs, it is important that the D-input be stabilized for a sufficient *setup* interval prior to the triggering edge of the clock-pulse, and that it remain constant for a sufficient *hold* time interval after the triggering edge. This circuit, is conceptually neat and easy to understand. Compare its cost with that of the realization of Figure 6.31 analyzed in Subsection 6.2.9.

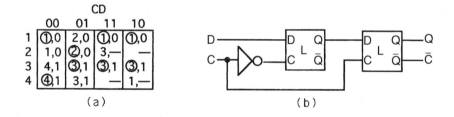

Fig. 6.41. ETDFF constructed with two latches.

6.4.3. Designing Registers

An important building block in every digital system is the *register*, a linear array of 1-bit storage devices, such as FFs or latches. These are used for the temporary storage of data, and when additional logic elements are built in, to manipulate data. A series of register designs is presented in this subsection using ETDFFs (it doesn't matter whether triggering occurs on the positive or negative edge of the clock-pulses). Additional designs are then presented, adding a counting function and introducing a different type of storage device, the JK-FF. The subsection concludes with an explanation of how to accomplish the same function with ETDFFs. We begin with a simple register capable of accepting data in parallel.

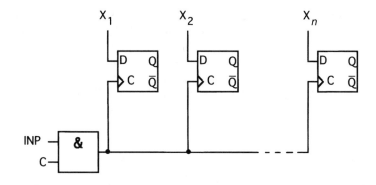

Fig. 6.42. Simple ETDFF register with parallel inputs and gated clock.

Example 6.18

The simplest possible arrangement for a register able to accept input data in parallel is shown in Figure 6.42. (Note the symbol used for the PETDFF, the triangle at the C-terminal, indicates edge triggering. If the triggering edge were negative, a negation bubble would have been shown at that terminal.) Whenever the signal INP is set up prior to clock-pulse signal C, the contents of the register FFs change to conform to the inputs $X_1, X_2, \ldots, X_n$. As long as INP remains 0, no changes occur in the stored data. Where logic elements appear in the circuit feeding the clock inputs to the FFs, we say that the clock is *gated*. While sometimes, as in this case, this may lead to simplified systems, it is detrimental with respect to timing precision in a way that is mentioned in Section 6.6. Next we look at a register that permits data to be shifted.

Example 6.19

A *shift register* is a register that allows the contents of each of its FFs to be moved to the FF immediately to its right. More generally, left shifts or shifts of more than one position may also be provided for. In Figure 4.11, p. 112, it is shown how to do shifting with combinational logic. In Figure 6.43 we present a shift register with the capability of executing right shifts of one position. As in the preceding example, control is exercised by gating the clock. When the control signal RSH is on at the onset of a clock-pulse, the signal on the X-input is read into the leftmost FF, and the contents of each FF is copied to the FF on its right. The bit in the rightmost FF is lost, unless provision is made for

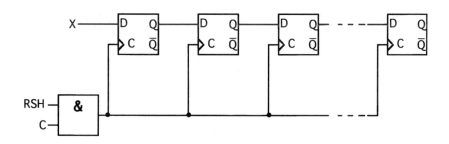

Fig. 6.43. ETDFF shift register with gated clock.

transferring it elsewhere during this clock cycle. It is assumed that by the time the output of a FF changes, the hold time interval for that clock-pulse has elapsed. Hence, each FF will sense only one D-signal in the neighborhood of the triggering edge of the clock-pulse, and so will behave properly. This is a fundamental assumption when ETDFFs are used. If it is not valid, then delay elements must be inserted at the output of each FF to delay the arrival of the signal at the next FF for a sufficient time. A register performing precisely the same function *without* gating the clock is the subject of the next example.

Example 6.20

Suppose the clock signal is fed directly to each of the FFs. Then, if no shift is to take place during a particular clock cycle, the D-input must be the same as the Q-output for each FF in order for the register contents to remain unchanged. In the circuit shown in Figure 6.44, the D-input to each FF is equal either to the output of the FF on its left or to its own output, depending on whether the shift signal is or is not activated. Where the FFs are implemented in such a way that the D-input goes to just one gate, and where that gate is an OR- or NOR-gate, the OR-gates in the Figure 6.44 circuit can be absorbed into the FF circuits by merging them with the gates receiving the D-inputs. Merging external logic into the FFs in this manner is useful in that it not only saves power and chip area, but also cuts down logic propagation delays. Next we see how several features can be combined in one register.

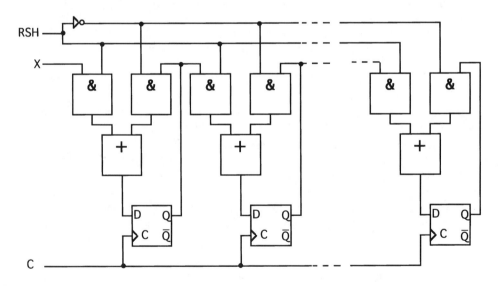

Fig. 6.44. An ETDFF shift register with *un*gated clock.

Example 6.21

The circuit of Figure 6.45 depicts the ith stage of a register that has both the parallel feed and right shift features. The INP and SHR signals cause each FF to receive data either from an X-terminal or from the FF to its left, respectively. If INP = SHR = 0,

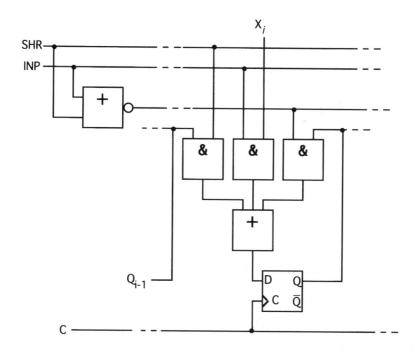

Fig. 6.45. The *i*th stage of an ETDFF register with right shift and parallel feed capabilities.

then the output of the NOR-gate goes on, causing the Q-output of each FF to be gated back to the corresponding D-input, so that all FFs remain stable. Note that a 3-1 MUX could be used here.

When the contents of a register represents a binary number, it is often important to be able to increment the value of that number by 1. A register with this capability is said to be a *counter*, or to have a *counting* or *incrementing* feature. The discussion of counters is simplified if we assume the availability of a different type of storage device, the clocked *toggle*-, or T-FF, which has two input terminals, C and T. If the T-input is on prior to the occurrence of the leading edge of a pulse applied to C, the state of the FF changes (*toggles*). Otherwise, it remains stable.

In order to increment a binary number by 1, we must, starting with the least significant bit, *change* the value of all bits up to and including the first 0 from the right. For example, to increment 01100111, we change the values of every member of the block of three rightmost 1s and we also change the value of the rightmost 0, obtaining 01101000. If the rightmost bit is a 0, then it is the only bit changed. Note that we *always* change the value of the rightmost bit.

Example 6.22

This process is implemented by the circuit shown in Figure 6.46. The INC-signal is fed directly to the T-input of the rightmost T-FF, holding the least significant bit. If that bit is a 1, then the AND-gate enables INC to get through to toggle the next FF. If the rightmost two FFs are both set, then the INC-signal gets through to the third stage. This three-stage register counts from 0 through 7 and then recycles to 0 on the 8th pulse. The combinational logic part is in the form of a linear iterative circuit (Subsection 5.2.3),

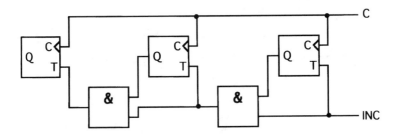

Fig. 6.46. Binary counter with linear logic.

each cell consisting of a single AND-gate. Propagation delay for the worst-case condition (*all* FFs changing state) is a linear function of the number of stages. The subject of the next example is a faster counter with the AND-gates structured in tree form.

Example 6.23

Propagation delay in a counter can be made to vary *logarithmically* with the number of stages if the logic is reorganized along the lines discussed in Section 5.3 (see Figure 5.19). A four-stage counter with tree-form logic is blocked out in Figure 6.47. Specifying the logic in the I-cells is left to the reader (see Problem 6.37).

Toggle-FFs are not a common device. A more powerful device with a toggling capability is the JK-FF (see Example 6.14, p. 190), which is widely available as an SSI circuit component. It has three inputs, the usual C, or clock input, as well as J- and K-inputs. When used individually, J and K are equivalent to the S and R inputs, respectively, of the SR-FF. That is, if J is on when C is on, then the FF is set, while if K is on, when C is on, the FF is reset. (There are clocked SR-FFs that behave precisely in this manner.) The special feature of the JK-FF, and the one of particular interest here, is that if *both* J and K are on when C is on, the FF toggles, i.e., it changes state. Hence, a JK-FF behaves as a T-FF if the T-signal is sent *both* to the J and the K terminals. Used

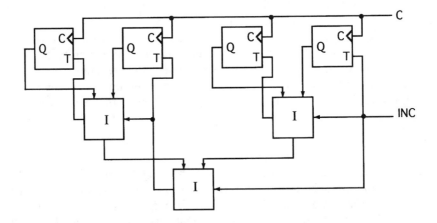

Fig. 6.47. Binary counter with tree-form logic.

in this way, JK-FFs can be substituted for the T-FFs in the circuits of Figure 6.46 and Figure 6.47.

An ETDFF can be made to behave as a T-FF by using an XOR-gate as shown in Figure 6.48a. But this is seldom desirable. An example of how an ETDFF can be used to perform several functions, *including* toggling, is the circuit of Figure 6.48b. Here we have $D = (INP)X + T\overline{Q} + (\overline{INP + T})Q$. Thus, if INP = 1, the FF accepts input X; if T = 1, the Q output of the FF is complemented; and if INP = T = 0, the state of the FF is unchanged. (Setting INP = T = 1 would make the next value of Q equal to the Boolean sum of the complement of the present value and the value of X.)

6.5. FORMAL DESIGN OF SYNCHRONOUS CIRCUITS

All formal design procedures must begin with formal statements specifying the system to be designed. We have seen that, for combinational logic functions, such statements can be in the form of truth tables, Karnaugh maps, Boolean algebra expressions, or even logic-circuit diagrams. For sequential functions, flow tables and state graphs have been used in previous sections. In the next subsection, several additional tools are introduced that are useful in this critical area, particularly for more complicated problems. The second subsection concerns the problem of generating logic circuits from the formal descriptions.

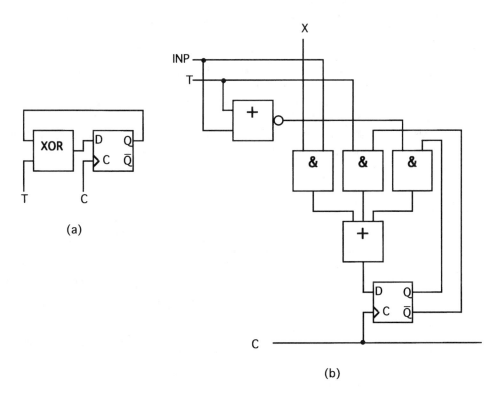

Fig. 6.48. Toggling with an ETDFF.

6.5.1. Problem Specification with Structured Statement Sequences, ASM Charts, and Flow Tables

It is seldom easy to formulate even a moderately complex algorithm with precision. Designers need all the help they can get in the form of good techniques for formalizing their mental images of how processes can best be executed. Some useful ideas can be borrowed from computer programmers, who have given the problem a great deal of thought. In addition to graphic aids in the form of flow charts, they often employ pseudoprograms, written in high-level structured languages. The latter are often of a semiformal nature, intended not for formal translation, but only as aids to clarify their thinking. Such a technique is illustrated in the following example. The algorithm is then recast more precisely in flow chart form, and, finally a flow table description is developed.

Example 6.24

The system to be designed is a controller for a multiplier. The multiplier, after receiving a start signal, uses the standard add-and-shift algorithm to find the product of two 32-bit unsigned binary numbers, and then generates a completion signal. The registers are shown in Figure 6.49a. The multiplier operand is initially in the MQ register, and the other operand, the multiplicand, is in the MPCND register. The E, AC, and MQ registers are linked together so that they can behave as one 65-bit shift register (for right shifts). A binary adder is provided that can place the sum of MPCND and AC in AC (with overflow in E). The result of the multiplication process is a 64-bit number in the combined AC–MQ register, with the most significant 32 bits in the AC. A 5-bit counter, CNTR, is used to keep track of the steps in the process that is described below.

Our initial description of the algorithm is in an informal language resembling programming languages such as Pascal. The meaning of most of the statements should be self-evident. The first statement, WAIT UNTIL start, requires some explanation, as it does not correspond to any common programming language construct. It means that the procedure is frozen at this point until the signal "start" becomes equal to 1.

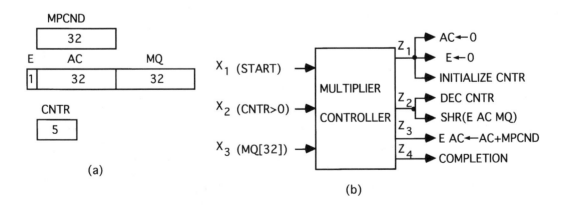

Fig. 6.49. Registers and controller block diagram for a multiplier.

BEGIN Multiplication
  WAIT UNTIL start
  AC := 0; E := 0;
  REPEAT 32 times
  BEGIN
    IF MQ[32] = 1 THEN E AC := AC + MPCND;
    RIGHTSHIFT(E AC MQ);
  END
  Completion := 1
END Multiplication

Note that the procedure executes the adding and shifting that we would carry out in a hand computation, except that instead of shifting the multiplicand to the left at each step, we shift the cumulative sum to the right *relative* to the multiplicand. Also, the least significant 32 bits of the cumulative sum gradually replace the multiplier in the MQ, as the multiplier bits are shifted out after use. A good way for readers to satisfy themselves that the algorithm works is to execute it for a pair of 2-bit operands, replacing the 32 in the algorithm with a 2. It is essential that the initial statement of the algorithm being implemented be very carefully checked before proceeding to the next stage. It is easier to correct any errors at this point. Particular attention should be paid to special cases of the inputs to ensure that the procedure is valid "at the margins." Now let us see exactly what it is that we are designing.

We assume that the necessary registers, data paths, adder, etc., needed to carry out the computation are in place. Our task is to design the control circuit that directs the operation. A block diagram of the controller is shown in Figure 6.49b. It is assumed that the system is clocked, but the clock signals are not shown explicitly at this point in the design process. The inputs are X_1 (the start signal), X_2 (equal to 1 if contents of CNTR $\neq$ 0), and X_3 (MQ[32], i.e., least significant bit of the MQ register). The controller generates four control signals that are used to make the right things happen at the right times. The Z_1-signal fans out to clear the accumulator (AC), to clear the E-bit, and to initialize the counter (which is used to keep track of the number of multiplier bits that have been used in the computation). The Z_2-signal decrements CNTR and causes a right shift (one position) of the contents of E, AC, and MQ to take place, with the three registers treated as one for this purpose. (A 0 is shifted into E.) The Z_3-signal causes the sum of AC and MPCND to be placed in AC, with any overflow in E. The Z_4-signal generates a completion signal that is fed back to whatever system initiated the action with the start signal.

The next step in our example is to convert the algorithm into a form that is closer to a hardware implementation. A decision must be made as to how to implement the REPEAT statement. This might be done by assigning at least one state of the controller for each step of the 32 cycles to be executed. This would be neither efficient nor elegant. A better idea is to use a counter external to the control unit. This is the purpose of the counter CNTR. States of the controller are assigned to represent the key points in the procedure. The initial state, in which waiting for the start signal occurs, is an obvious choice for one such point. We can assign another state to the condition following the initialization statements (in which E and AC are cleared). The IF statement and possibly the addition are initiated from this second state. A third state can then be used to initiate the shift operation, followed by either a return to the second state for a test of the new rightmost bit of MQ, or by the termination of the algorithm. One way to proceed from here is to cast the procedure, enhanced by the concepts involving the states, into a form resembling both a flow chart and a state diagram. It is called *algorithm state machine* (ASM) diagram or chart.

In this type of diagram (see Figure 6.50), the rectangular boxes represent states, numbered in the lower left corners. Any outputs that are *always* associated with a state

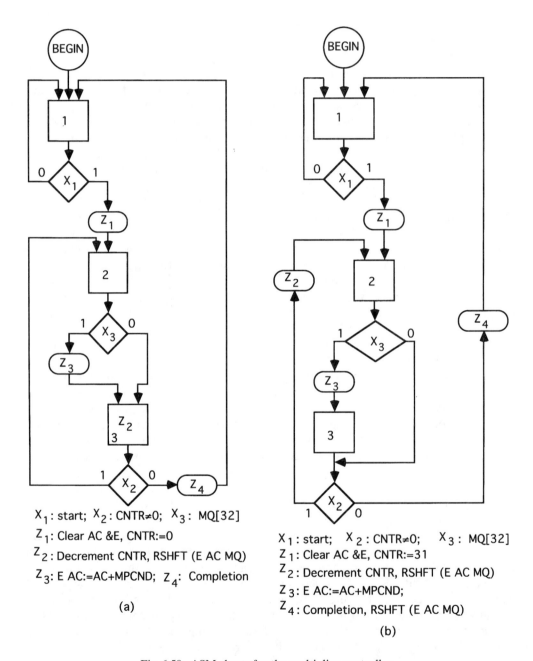

X_1: start; X_2: CNTR$\neq$0; X_3: MQ[32]

Z_1: Clear AC &E, CNTR:=0

Z_2: Decrement CNTR, RSHFT (E AC MQ)

Z_3: E AC:=AC+MPCND; Z_4: Completion

(a)

X_1: start; X_2: CNTR$\neq$0; X_3: MQ[32]

Z_1: Clear AC &E, CNTR:=31

Z_2: Decrement CNTR, RSHFT (E AC MQ)

Z_3: E AC:=AC+MPCND;

Z_4: Completion, RSHFT (E AC MQ)

(b)

Fig. 6.50. ASM charts for the multiplier controller.

are specified inside the box. Diamond-shaped boxes represent decisions made on the logic variables or functions shown inside them. Boxes with rounded corners contain outputs that are generated only if the process state takes a path through them. The passage of the system state from one box to another, including all decision made en route and the generation of any outputs specified takes place in one clock cycle. Every closed path must include at least one state box. Let us see how Figure 6.50a represents a straightforward implementation of the algorithm.

The BEGIN bubble points to the initial state, labeled 1. In this state, at each clock cycle, the X_1-input is inspected (as indicated by the decision box below the state-1 box). As long as there is no start signal, i.e., while X_1 remains 0, the system keeps looping around state-1, implementing the WAIT operation. If X_1 is found to be 1, then the 1 exit of the decision box is taken and Z_1 is set to 1 as specified in the "output" box through which the system passes en route to state box 2. This output is on for one clock cycle, which suffices for the execution of the micro-operations initiated by X_1, i.e., the clearing of E and AC, and the initialization to 0 of the counter CNTR. (If the 5-bit counter is initialized to 0, then the result of subtracting 1 during the first pass is to set it to 31 *before* it is tested. The effect is the same as if it had been initialized at 32, but a 5-bit counter suffices.) In state-2 the value of X_3 (the rightmost bit of MQ) is tested, and if it is equal to 1, the control signal Z_3 is generated to carry out the addition operation. In any event, the next state is state-3. In this state, control signal Z_2 is always generated, decrementing the counter and causing the right shift to take place. If $X_2 = 1$, then the count in CNTR has not yet been reduced to 0 and so the system returns to state-2 to process the next bit of the multiplier. Otherwise, the completion signal Z_4 is generated and the system returns to state-1 to await another start signal. While this is a valid and reasonable implementation of the algorithm, it is not the only possibility. The alternative shown in Figure 6.50b is considered next.

From state-2 in the Figure 6.50a chart, the addition is carried out if $X_3 = 1$, or else nothing at all is done during that clock cycle. In all cases, the shift operation is performed during the next cycle from state-3. Thus, regardless of the value of the multiplier, two cycles are consumed for each of the 32 passages around the inner loop; each involves going through both state-1 and state-2. But this is not necessary. Where $X_3 = MQ[32] = 0$, no addition is necessary from state-2, and so the *shift* can be performed during that cycle and state-3 can be skipped, saving one cycle. This is what is done in the ASM chart of Figure 6.50b (where, among other changes, CNTR is initialized at 31 to take into account the fact that it is tested *before* it is decremented). The result is that on average about 16 cycles can be saved for each multiplication. This illustrates the value of carefully examining the algorithm *before* getting down to the more detailed levels of the design.

It is possible, and sometimes convenient, at this point to assign y-states to the states of the ASM chart and proceed to generate the logic expressions directly. An alternative is to produce a flow table from the ASM chart and to complete the synthesis by using an elaboration of the techniques introduced in Section 6.1. The flow table corresponding to the Figure 6.50b chart is shown in Figure 6.51. The don't care entries result from the assumption that the start signal, X_1, is on for only one cycle, so that it cannot be on for any state other than the initial state. The other entries follow directly from the chart.

The next subsection deals with implementation of the logic, which will enable readers to complete the multiplier design by working problems 6.42 or 6.46. But first another example illustrating the specification of a somewhat more complicated system is presented.

$$X_1 X_2 X_3$$

	000	001	011	010	110	111	101	100
1	1,0000	1,0000	1,0000	1,0000	2,1000	2,1000	2,1000	2,1000
2	1,0001	3,0010	3,0010	2,0100	—	—	—	—
3	1,0001	1,0001	2,0100	2,0100	—	—	—	—

Fig. 6.51. Flow table corresponding to the ASM chart of Figure 6.50b.

Example 6.25

A bit of background is necessary for this example, which involves a method for reducing access times for the main memories of computers. For virtually every technology, reading from or writing to main memory is, compared with logic operations, a slow process. But it is not difficult to build small (e.g., 5000 words) random-access memories (RAMs) that are an order of magnitude faster than a large main memory (e.g., 10 million words). If we could organize matters so that at any give time most of the words being referred to in a program are in such a small memory (called a *cache*), then considerable speed advantages would result. Observe now that most programs do in fact have a *locality* property in that they tend to access repeatedly the same information from main memory. We need only think of a tight program loop that is executed perhaps hundreds of times consecutively. The same instructions are repeatedly fetched from memory, and in most cases certain constants are used over and over again. It is also the case that memory read instructions are executed perhaps five times as often as memory write instructions. With this in mind, we can now outline how a simple cache might work.

At any time, the cache is filled with copies of words from various parts of main memory. Whenever a read command is issued to main memory, the cache is checked to see if the word is also there. Methods exist that we need not discuss here for making such checks very quickly. If the word is indeed found in the cache (this is called a *hit*), then the memory read instruction is aborted and the copy of the word in the cache is used. If the word is not in the cache (a *miss*), then the main memory read operation is allowed to go to completion and that word is not only returned in response to the read command but is also copied into the cache. Hence, the next time that word is the subject of a read command, there will be a cache hit. Since, in the steady state, the cache is always full, it is necessary to eject an old cache word any time a new one is stored in it. There are several strategies for selecting that word. For example, the least recently used word might be chosen. Readers are referred elsewhere (see Sources at the end of the chapter) for information on this issue as well as for other information on caches. On write commands, a common strategy, which will be followed here, is called *write through*. All write commands result in main memory being updated in the usual way. The cache is involved only if the word at the address being written into is in it, in which case the cache contents are appropriate updated.

The subject of our current example is the design of a *cache interface controller* (CIC). The CIC issues control signals to the cache and main memory (MM) to execute read or write commands received from an outside unit, such as the central processor of the computer using the memory system. It is assumed that the cache unit has its own controller to handle such problems as determining if a given word is stored in the cache or deciding what word to evict to make room for a new word. A block diagram showing the relations among CIC, MM, and cache is shown in Figure 6.52. The register labeled MAR is the memory address register. It receives from outside the memory system the address of the word to be accessed, and its contents are available both to the cache and to MM. The memory buffer register (MBR) can both send and receive data from the outside, and is also accessible to MM and to the cache. Both the cache and MM send signals to the controller to indicate when they have completed assigned tasks. The controller similarly reports back to the unit commanding it when a read or write command has been executed. Note that the entire system is synchronous, including the CIC itself, although clock signals are not shown.

When a read command is received by the CIC, it is assumed that the address of the location to be read has already been transmitted to the MAR. The CIC sends read signals to the cache and to MM. If a hit signal is returned from the cache, meaning that the desired data is in the cache, then the CIC sends an abort signal to MM, canceling the read order, and, upon receipt of a finish signal from the cache, the CIC relays a finish signal to the "client" who issued the read command, indicating that the data is now in the MBR. If the response from the cache to the read command is a miss signal, then the CIC allows the MM read to go to completion, whereupon it issues a write command to the cache. This causes the word that was just read to be stored in the cache; note that data is in the MBR and the associated address is in the MAR.

A write command causes the CIC to issue a write command to MM. It also signals the cache to "check" to see if it is storing data from the location involved. If the answer is "yes," in the form of a hit signal, then the cache is signaled to acquire the new contents of that location. The following is a more formal description of the function performed by the CIC.

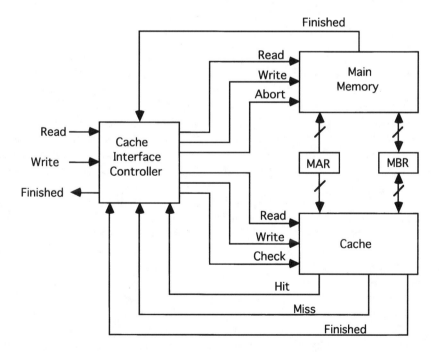

Fig. 6.52. The cache interface controller, main memory, and cache.

```
BEGIN Cache interface controller function
   WAIT UNTIL read or write command occurs
   IF read THEN
   BEGIN
      MM read: = 1; cache read: = 1;
      WAIT UNTIL hit or miss
      IF hit THEN
      BEGIN
         Abort: = 1
         WAIT UNTIL cache finished
      END
      ELSE
      BEGIN
         WAIT UNIT MM finished
         cache write: = 1;
         WAIT UNTIL cache finished
      END
   END {end of read}
   ELSE {write command}
   BEGIN
      write MM: = 1; check cache: = 1
      WAIT UNTIL hit or miss
      IF hit THEN write cache: = 1;
      WAIT UNTIL MM finished
   END {end of write}
   CIC finished: = 1;
END
```

In several places in the preceding process it is assumed that cache operations are much faster than MM operations (if this were not so, then why would we want to use a cache?). But otherwise, due to the "wait until" operations utilizing the finish signals, the algorithm is not sensitive to the speed of the cache and the MM. An ASM chart for the CIC is presented in Figure 6.53. There are six states, six input variables, and seven output signals. The use of meaningful names for the variables on this chart makes it easier to verify that the algorithm described is correct. At a later stage in the design the more concise symbols listed on the chart are substituted to facilitate the necessary algebraic manipulations.

The state graph of Figure 6.54 is easily obtained from Figure 6.53. The states are the same and are numbered the same. As explained in Subsection 6.1.1, the label X/Z on an arc indicates that the arc is selected if X is true, and that the corresponding output is Z. The state graph could also have been derived directly from the structured sequence expression of the algorithm. This is a good example of a case where the state graph represents the function more effectively than a flow table would. Since there are six input variables, the full flow table would have 64 columns, one for each input state. But because at each internal state only one or two of the input variables affect the next state or output, the state graph is not particularly complicated. We turn now to the problem of generating logic circuits from formal representations of synchronous sequential functions.

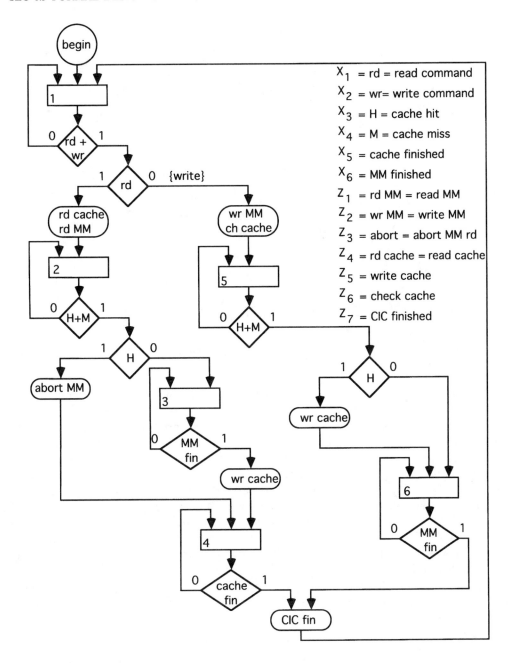

X_1 = rd = read command
X_2 = wr = write command
X_3 = H = cache hit
X_4 = M = cache miss
X_5 = cache finished
X_6 = MM finished
Z_1 = rd MM = read MM
Z_2 = wr MM = write MM
Z_3 = abort = abort MM rd
Z_4 = rd cache = read cache
Z_5 = write cache
Z_6 = check cache
Z_7 = CIC finished

Fig. 6.53. ASM chart for the cache interface controller.

6.5.2. *Generating the Logic*

Clocked sequential circuits can be generated from flow tables by following the procedure introduced in Subsection 6.1.3. The ideal delays are replaced by clocked storage devices such as latches or ETDFFs, each of which receives, at its C-input, pulses from the same clock source. Then, whenever a clock-pulse arrives, each device transmits

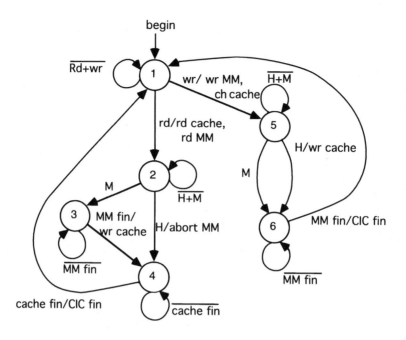

Fig. 6.54. State diagram for the cache interface controller.

the signal at its D-input to its Q-output. These signals, in combination with external input signals, are processed by the combinational logic block to produce Z-output signals and signals (usually labeled as Ys) that are fed to the D-inputs of the latches or FFs to affect their values during the *next* clock-pulse cycle. Thus, the storage devices, in conjunction with the clock signals behave as the ideal unit delay elements postulated in Subsection 6.1.3. The conditions relative to clock-pulse periods, stray delay values, etc., under which such systems will actually work properly are examined in Section 6.6.

The next example, in which an important device is introduced, illustrates how we can design circuits effectively, starting with a reduced flow table. Implementations are shown with latches, ETDFFs, and JK-FFs. We will also see two different ways of dealing with output signals. Then, in Example 6.27 we return to the cache interface controller, using this problem to illustrate several approaches, including one that may be useful for more complex problems. This is a tabular method leading to the development of variable-entered Karnaugh maps (Subsection 3.3.5).

Example 6.26

An arbiter is a device that coordinates the use of a facility needed by two or more processes. The flow table of Figure 6.55 (ignore the state assignment for the moment) describes one variation that might be used to arbitrate between two processes that use the same clock. (The more troublesome situation in which the competing processes do *not* share a common clock is treated in Subsection 6.7.4.) The device described here is "fair" in the sense that if the two processes simultaneously request use of the facility, it will be granted first to the one that was *not* the one that used it last. (Other definitions of fairness are often used.)

AB

	00	01	11	10	y_1 y_2
1	1,00	3,01	2,10	2,10	0 0
2	4,00	3,01	2,10	2,10	0 1
3	1,00	3,01	3,01	2,10	1 0
4	4,00	3,01	3,01	2,10	1 1

$Z_1 Z_2$

Fig. 6.55. Flow matrix for a synchronous arbiter.

In synchronous mode, input signals are examined for each clock-pulse. (It is assumed here that changes in A or B, controlled by the same clock that controls the arbiter, will occur only at times that will cause no timing problems. That is, when the clock goes on, the values of A and B seen by the arbiter are stable. We can see from the flow table how the arbiter works. Suppose that with the arbiter initially in state-1, Process A requests the facility by turning on input A. The arbiter turns on Z_1, which assigns the facility to A, and goes to state-2. As long as A remains on, Z_1 remains on, even if a request from B arrives (i.e., if B goes on). Once Process A releases the facility by turning off A, the arbiter turns off Z_1. If B = 0 at that time, then the arbiter goes to state-4. Otherwise, if B = 1, it turns on Z_2 and goes to state-3. If both A and B go on when the arbiter is in state-1, then it gives the facility to process A. Note that if process B has the facility and then gives it up, the arbiter goes to state-1, where A has the priority in case of a conflict. If process A has the facility, then after it finishes using it the arbiter goes to state-4. From this state, in the event of simultaneous requests, Z_2 is turned on, granting the facility to process B. Thus, the process that most recently used the facility will always see the other process given priority in case of simultaneous requests. Now let us see how this flow table can be realized with various types of storage elements.

The logic design is precisely the same for the latches or D-FFs. After arriving at a minimal-row flow table, the next step is to encode the states. The basic requirement here is that each state be assigned a unique code. For a four-row table, this means we need at least two internal variables (we call these y-variables). Depending on the particular encoding used, complexity of the resulting circuit may vary considerably. Unfortunately, finding the best coding is a very difficult problem, and will not be treated in this book. However, in this case, we can take advantage of one simple heuristic: if two rows have many next-state and output entries that are the same, then it is a good idea to assign adjacent y-states to them. (This tends to increase the number of adjacent cells in the resulting K-maps that have the same values, which generally leads to simpler logic expressions.) Examining the given table very quickly reveals that rows 1 and 3 are quite similar, as are rows 2 and 4. We can thus exploit this heuristic by using the state assignment shown in Figure 6.55.

In order to facilitate the use of K-maps for this problem, it is convenient now to transform the flow table so that it will correspond directly to a K-map with rows specified by the y-variables. (The columns are already ordered as in a K-map where they are specified by the inputs.) Interchanging rows 3 and 4 will accomplish this. We first interchange rows 3 and 4, and then replace all next-state entries of 3s with 4s and vice versa. The result is the table of Figure 6.56*a*. Now we can easily generate the Y-matrix, Figure 6.56*b*, which leads directly to K-maps for Y_1 and Y_2. The K-maps for Z_1 and Z_2 are obtained directly from the flow matrix. These too are shown in Figure 6.56*c*. The next step is to find minimal logic expressions for the set of four functions. The steps involved in choosing subcubes are shown on the maps, within the ovals. No branching is necessary. This leads to the logic expressions below. (Note that all of the product terms are shared.)

$$Y_1 = \overline{A}B + By_1 + \overline{A}\,\overline{B}y_2, \qquad Y_2 = A\overline{y}_1 + A\overline{B} + \underline{\overline{A}\,\overline{B}y_2}$$

$$Z_1 = \underline{A\overline{y}_1} + \underline{A\overline{B}}, \qquad Z_2 = \underline{\overline{A}B} + \underline{By_1}.$$

There are five AND-gates and four OR-gates. The circuit in which latches are used is shown in Figure 6.57. It is not hard to show that, by factoring, a simpler circuit can be found with a total of three AND-gates and four OR-gates (Problem 6.43). Edge-triggered FFs could be substituted directly for the latches.

	AB = 00	01	11	10	$y_1 y_2$
1	1,00	4,01	2,10	2,10	0 0
2	3,00	4,01	2,10	2,10	0 1
3	3,00	4,01	4,01	2,10	1 1
4	1,00	4,01	4,01	2,10	1 0
					Z_1Z_2

(a) Transformed flow matrix

	AB = 00	01	11	10	$y_1 y_2$
1	00	10	01	01	0 0
2	11	10	01	01	0 1
3	11	10	10	01	1 1
4	00	10	10	01	1 0
					Y_1Y_2

(b) Y-matrix

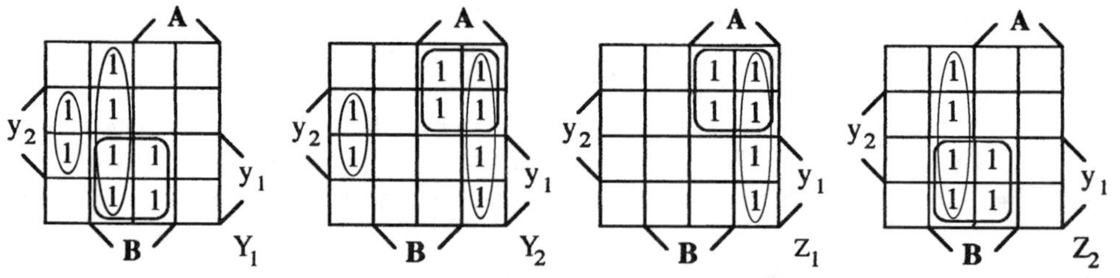

(c) K-maps

Fig. 6.56. Matrices and maps for arbiter.

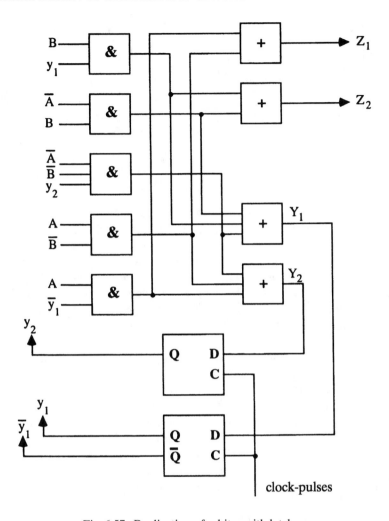

Fig. 6.57. Realization of arbiter with latches.

Now let us see how the same function can be realized with JK-FFs. The output is treated in exactly the same way; no changes are needed in the Z-functions. Each latch is replaced by a JK-FF, so that the Y_1-function must be replaced by J_1- and K_1-functions and the Y_2-function must be replaced by J_2- and K_2-functions. We can generate Karnaugh maps for these functions by inspection of the Karnaugh maps for the Y-functions (Figure 6.58).

Consider first the J-function. The J-input should be set at 1 when we wish the state of the FF to change from 0 to 1. Then in any position in the Y_1-map for which $y_1 = 0$ and $Y_1 = 1$, we must set $J_1 = 1$. If $y_1 = 0$ and $Y_1 = 0$, then we do not wish the y_1-value to change, and therefore J_1 must be specified to be 0. For all states in which $y_1 = 1$, it does not matter what value J_1 is given, since a change to the 0-state will occur if and only if $K = 1$, regardless of the value of J. Recall that if J and K are *both* 1, then the FF toggles (changes state). Thus, for all map positions in which $y_1 = 1$, there should be don't care

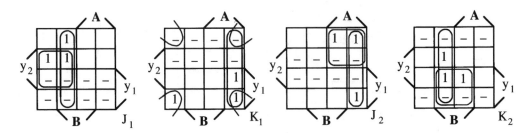

Fig. 6.58. Karnaugh maps for JK-FF realization of arbiter.

entries for J_1. These rules have been followed in the J_1 map of Figure 6.58. A similar set of rules applies for the K-functions. The K-input must be set to 1 wherever the value of the corresponding FF-output is to be changed from 1 to 0. Hence, if $y_1 = 1$ and $Y_1 = 0$, we must have $K_1 = 1$. Wherever $y_1 = 0$, the value of K_1 is irrelevant, and may therefore be left as a don't care. This is reflected in the K_1-map of Figure 6.58. Of course, the same rules account for the values of J_2 and K_2 in the same figure. The JK-FF requires that logic be generated for two inputs, J and K, rather than for a single D-input. This drawback is eased by the fact that half of the map entries for these functions are don't cares.

There are several ways of dealing with outputs in synchronous circuits. One question is whether they should be generated as pulses or levels (steady-state signals). In the designs developed here, the Z-signals are considered as levels. If it is desired to make them pulses, then they can be gated by the clock-pulse. That is, each Z-signal can be fed to an AND-gate along with the clock signal. The output of that gate can then be considered as the pulsed Z-signal. Another distinction is how the outputs are to be associated with inputs. In the flow tables used here, an output is associated with each total state, in the sense that, for a given internal state, a different output may be specified for each input. This is known as *Mealy mode operation* (after George Mealy, a pioneer in this field).

An alternative approach is referred to as *Moore mode operation* (after Edward Moore, another pioneer in the development of the theory of sequential circuits). Here, outputs are associated only with the internal states (or rows of the flow table). If the outputs in Moore mode operation are to be level signals, then they can be made functions only of the y-variables. Alternatively, if we wish to treat the outputs as pulses, we can assume that if a transition is made to a particular state, then this transition is accompanied by output pulses specified by the output state associated with the destination row.

Minimal SOP expressions for the JK-FF realization are:

$$J_1 = \overline{A}B + \overline{A}y_2, \quad K_1 = A\overline{B} + \overline{B}\overline{y}_2, \quad J_2 = A\overline{y}_1 + A\overline{B}, \quad K_2 = \overline{A}B + By_1,$$

$$Z_1 = J_2, \quad Z_2 = K_2.$$

There are six AND-gates and four OR-gates, so this circuit is somewhat more costly than the SOP logic for the latch realization (in addition to the fact that the JK-FF itself is more costly than a latch). As in the previous case, factoring can be used to simplify the circuit. In this case, four AND-gates and four OR-gates are sufficient.

Example 6.27

We return now to the design of the cache interface controller, initiated in Example 6.25 on p. 208. Our initial approach is one that can be applied to fairly complicated problems. It is seldom optimal with respect to the gate counts of the resulting circuits, but it is very direct and is likely to lead to results that are easy to understand. We shall work directly from the ASM chart (the state diagram could also be used with minor modifications of the method).

As is the case for all methods, we must first satisfy ourselves that the ASM chart is correct and that it cannot be simplified in any significant way. Assuming this is the case, then the next step is to encode the states in terms of a set of binary variables. Here is where we make a major decision to trade off circuit elements (in effect chip area and possibly power) for ease of design. This is done by using a 1-hot code; a single y-variable is assigned to each state. Let us assign y_i to state-i. Since there are six states in the Figure 6.53 chart, we require six y-variables. Under this assignment, state-1 has the code 100000, state-2 is encoded as 010000, etc. The logic expressions are generated directly from Figure 6.53 (refer to the listing there identifying the X- and Z-variables).

We begin with the Z-variables. An examination of the ASM chart indicates that the rd MM operation, corresponding to the output variable Z_1, occurs at only one point. This follows state-1 and the 1-exits from the next two decision boxes. The first box, "rd + wr," corresponds to $X_1 + X_2$, and the second to X_1. Hence Z_1 is to be on in state-1 when $(X_1 + X_2)X_1 = 1$. Since state-1 is uniquely specified by $y_1 = 1$, this means that $Z_1 = (X_1 + X_2)X_1y_1$. This can be simplified to $Z_1 = X_1y_1$.

A more interesting case is Z_5, representing the "write cache" operation. This appears in two places on the chart, after state-3 and after state-5. The first appearance follows the 1 branch from the "MM fin" decision box, corresponding to X_6. The second appearance follows the 1-branches of two decisions corresponding to $(X_3 + X_4)X_3$. Putting this together, we obtain $Z_5 = X_6y_3 + (X_3 + X_4)X_3y_5 = X_6y_3 + X_3y_5$. The other Z expressions can be obtained in a similar manner as Boolean sums of terms, each consisting of a product of a y-variable corresponding to a state and an expression describing a path through one or more decision boxes. In cases where an output is *always* associated with a state (indicated by the appearance of the output within the state box—see, for example, the Z_2-entry in the state-3 box of the ASM chart for the multiplier controller, Figure 6.50a), the term consists *only* of the y-variable.

Y-variable expressions (the Y-signals are fed to the D-inputs of the latches or FFs) are obtained in a similar manner. Consider Y_4. Three branches terminate with arrows on the state-4 box. Each generates a component of a sum expression for Y_4. Thus the rightmost arrow impinging on the box produces the term X_6y_3, since it comes from state-3 via the 1-branch from the X_6 decision box. The other two terms (from right to left) are $X_3(X_3 + X_4)y_2 = X_3y_2$ and $\overline{X}_5y_4$. So we have $Y_4 = X_6y_3 + X_3y_2 + \overline{X}_5y_4$. Note that finding the Y-expressions corresponding to a 1-hot code for an SSC is quite different from the analogous process for an ASC (see Example 6.8). Finding the remainder of the Y- and Z-expressions is the topic of Problem 6.44.

A practical problem not yet addressed is that of ensuring that a system is in the correct initial state when the power is first turned on. In this example, the "begin" bubble that points to state-1 might be thought of as specifying what that initial state is. It is often necessary to provide some special *start-up* circuitry to initialize all of the storage devices correctly. In this case, we might specify extra logic to be associated with these devices to permit a special start-up signal to reset all of them to 0, except for the y_1 device, which must be set to 1. This problem is often overlooked by inexperienced designers.

In the next example, another approach to synthesizing the CIC is explored. This involves the use of state assignments with fewer y-variables. It is shown how logic expressions can be generated for this case in several different ways.

Example 6.28

The ASM chart for the CIC is redrawn in Figure 6.59 with X- and Z-symbols substituted to facilitate the discussion. Since there are six states, three y-variables are sufficient to encode them. Suppose that the states are assigned the values of $y_1y_2y_3$ shown within the state boxes. State-1 is assigned the code 000, state-3 is encoded as 110, etc. This encoding was made in an arbitrary manner. Different encodings might result in solutions of significantly different complexity. There are methods for making the choice so as to make economical realizations more likely, but these are not discussed in this book.

Logic expressions can be found directly from the chart in essentially the same manner used in the previous example. As pointed out earlier, Z_1 is on only in state-1 when X_1 is on. This leads immediately to $Z_1 = X_1\bar{y}_1\bar{y}_2\bar{y}_3$. There are two terms in the expression for Z_5, corresponding to states 3 and 5. Again it takes a product of three y's to identify a state, as opposed to the single y that sufficed when we used the 1-hot code. The resulting expression is $Z_5 = X_6y_1y_2\bar{y}_3 + X_3\bar{y}_1y_2y_3$.

The first step in finding an expression for Y_1 is to identify the states for which $y_1 = 1$, namely 3 and 4. A Y_1-expression can then be constructed as the Boolean sum of terms generating these states. State-3 follows state-2 if $(X_3 + X_4)\bar{X}_3 = \bar{X}_3X_4 = 1$. (Since a hit and a miss cannot *both* occur, X_3X_4 can be considered as a don't care term. This can be used to reduce $\bar{X}_3X_4$ to X_4.) State-2 must be identified in the expression by specifying the full y-state 010 (unlike the analogous case when the 1-hot code was used making a single y-variable sufficient). Thus, the first term is $X_4\bar{y}_1y_2\bar{y}_3$. A second way to reach state-3 is the path from state-3 itself via the 0 branch from the X_6 decision box. This corresponds to $\bar{X}_6y_1y_2\bar{y}_3$. In a similar manner, the paths to state-4 contribute three terms, so that the overall expression is

$$Y_1 = X_4\bar{y}_1y_2\bar{y}_3 + \bar{X}_6y_1y_2\bar{y}_3 + \bar{X}_5y_1\bar{y}_2\bar{y}_3 + X_3\bar{y}_1y_2\bar{y}_3 + X_6y_1y_2\bar{y}_3.$$

These expressions are not necessarily minimal; they may be used as starting points for combinational logic minimization procedures. An obvious simplification of the Y_1-expression is obtained by applying the uniting theorem to the second and last terms to obtain $Y_1 = X_4\bar{y}_1y_2\bar{y}_3 + y_1y_2\bar{y}_3 + \bar{X}_5y_1\bar{y}_2\bar{y}_3 + X_3\bar{y}_1y_2\bar{y}_3$. This procedure, while often

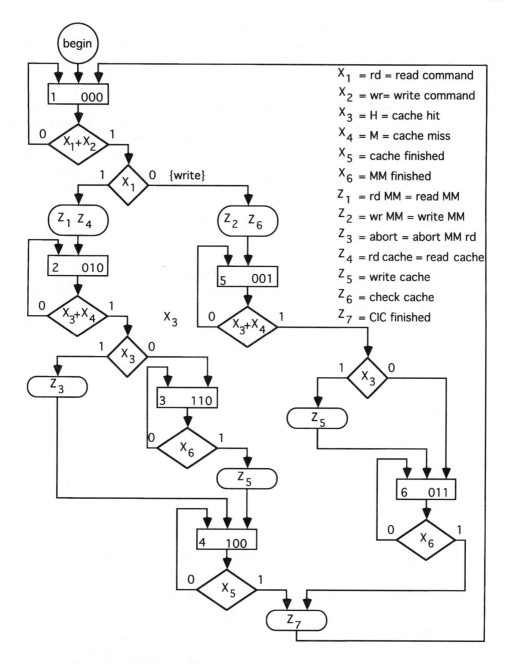

Fig. 6.59. ASM chart with state assignment for cache interface controller.

feasible, has the drawback that it requires additional work to optimize the logic expressions, e.g., to plot them on K-maps. A more basic problem with it is that it does not exploit don't care conditions resulting from unused y-states (in this case, two of the eight y-states are unused). The technique described next addresses both of these problems.

After a state assignment has been made, the information in an ASM chart (Figure 6.59 will be used as an example) can be usefully organized in the form of a K-map, as shown in Figure 6.60. Each cell of the map, which we shall call a *state map*, represents a y-state. Those that are not used are marked with dashes as don't care states. Each of the others is assigned to a state of the system as specified in the lower left corners. Other entries in the state cells indicate the conditions under which various Y- and Z-variables are activated. Thus, in the state-1 cell, there is a line, $X_1: Z_1 Z_4 Y_2$, indicating (see Figure 6.59) that, in state-1, when $X_1 = 1$, Z_1 and Z_2 are turned on and a transition is initiated to a state-2 for which $y_2 = 1$. The second line in this cell is $X_2: Z_2 Z_6 Y_3$. It indicates that, when the present state is 1 and X_2 is on, the Z_2- and Z_6-outputs will be on and the next state will be one for which $Y_3 = 1$ (state-5). Where a Y or Z is 1 for some state, regardless of the input, the entry in the cell lists these values after a colon that is not preceded by any X. An example of this is the state-5 cell, where Y_3 is listed in this manner. (An examination of the ASM chart indicates that from state-5, the next state is either 5 or 6, both of which are encoded with $y_3 = 1$.)

The state map can also be generated, even more easily, from the state graph (Figure 6.54). It is fully equivalent to the state graph and to the ASM chart in that either can be derived from the state map. Variable entered K-maps for the Y- and Z-functions are easily obtained from the state map. These are shown in Figure 6.61 for Y_1, Y_2, Z_1, and Z_5. Using the methods introduced in Subsection 3.3.5, we can derive minimal SOP expressions

$$Y_1 = y_1 y_2 + \overline{X}_5 y_1 + X_3 y_2 \overline{y}_3 + X_4 y_2 \overline{y}_3,$$

$$Y_2 = X_1 \overline{y}_1 \overline{y}_2 \overline{y}_3 + X_3 \overline{y}_2 y_3 + X_4 \overline{y}_2 y_3 + \overline{X}_3 \overline{y}_1 y_2 \overline{y}_3 + \overline{X}_6 y_2 y_3 + \overline{X}_6 y_1 y_2,$$

$$Z_1 = X_1 \overline{y}_1 \overline{y}_2 \overline{y}_3, \qquad Z_5 = X_3 \overline{y}_2 y_3 + X_6 y_1 y_2.$$

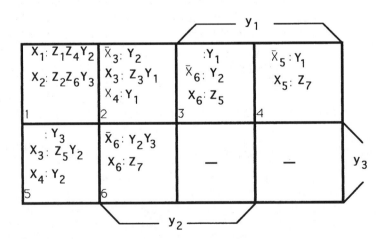

Fig. 6.60. State-map representation of cache interface controller.

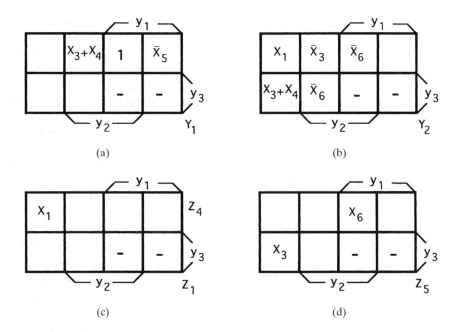

Fig. 6.61. Variable entered K-maps for the cache interface controller.

Note that the Y_1-expression has 14 gate inputs, as compared to 19 for the Y_1 expression derived directly from the ASM chart using the same state assignment. The savings result from the fact that the state map permitted us to exploit the don't care conditions. The state-map approach is well suited to problems where there are many input signals, only a few of which are significant for any particular internal state, a situation often encountered in the design of controllers. Completing the design for this example is the subject of Problem 6.45.

6.6. CLOCKING SCHEMES

In an almost literal sense, the clocking system of a computer is its heart. All operations take place in synchronism with the clock-pulses, so that clock malfunctions have catastrophic effects. The clock-pulse period directly determines the rate at which the machine operates, so that decreasing it yields immediate, across-the-board speed increases.

Clocked, or synchronous, systems have the advantage of freeing the logic designer from having to consider time-related problems such as hazards and critical races. The price for this is the need to devise an overall clocking scheme that guarantees correct operation if certain constraints are satisfied. These constraints are in the form of simple design rules restricting minimum and maximum path delays. A common starting point for the design of a clocking scheme is the maximum delay path through the logic (D_{LM}, the long-path delay), specifications for the storage devices to be used, and tolerances on the arrival times of clock-pulse edges (discussed below). Given this information, and the general method to be used (e.g., a 1-phase clock), the goal is to find the best possible trade-off between the clock-pulse period (P) and the lower bound on the minimum delay through the logic (D_{Lm}, the short-path delay). Since there is always a practical limit on how close D_{Lm} can be to D_{LM}, it may be necessary to increase D_{LM} in order to make it feasible to satisfy the D_{LmB} limit.

Clock-pulses for any digital system are generated by a single oscillator, usually crystal controlled for accuracy, and then put through special "shaping" circuitry to adjust the pulse width. Other circuitry may be used to produce additional phases of the signal, i.e., other versions displaced in time by a fixed amount. The clock-pulses are then distributed via a tree-like network to the various storage devices, which may be located at widely varying points, often on different chips, or even on different circuit boards. In the course of this overall process, it is inevitable that inaccuracies occur and accumulate. It is impossible to have a particular edge of a clock-pulse arrive simultaneously at the C-input of every latch or FF. Discrepancies among arrival times of corresponding edges are often referred to as *clock-pulse skew*, or just *skew*. We shall use a slightly different terminology. Assume that the *nominal* arrival time of a clock-pulse edge is the time when it would have arrived if the generation and distribution system were perfectly accurate and uniform. Then the *clock-pulse edge tolerance*, T, is defined as the absolute value of the maximum deviation (early or late) of the arrival time at any device. Subscripts on T are used to specify the edge involved, i.e., T_L refers to the leading edge and T_T to the trailing edge.

The approach taken here is to design systems that will work under worst-case conditions. Thus, it is always assumed that all parameters may simultaneously equal the extreme values that are most likely to cause trouble. Three different clocking systems are considered: one-phase systems with latches, two-phase systems with latches, and the simplest system, which we consider first, one-phase systems with ETDFFs.

6.6.1. One-Phase Clocking with Edge-Triggered D-FFs

A block diagram for a one-phase clocked system is depicted in Figure 6.62. The storage elements in the feedback paths are assumed in this subsection to be PETDFFs. (It would make no difference in our treatment if negative edge triggering were used instead.) The relevant FF parameters are D_{CQ} (the propagation delay from clock input

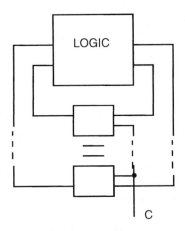

Fig. 6.62. Block diagram of a system with a one-phase clock.

to Q-output), C_{Wm} (the minimum clock-pulse width), S (the setup time), and H (the hold time). Setup and hold times were mentioned in Subsection 6.4.2. If the D-input changes during the interval beginning S before the triggering edge of the clock-pulse and ending H after that edge, the FF may behave in an unpredictable and most unpleasant manner (this is discussed further in Section 6.7). For some FFs, H may be nonpositive, in which case the triggering edge is outside the interval, but there is always such a nonzero interval. It is assumed that S and H are maximum values (always the worst case), and that as usual the subscripts M and m, when appended to another subscript such as CQ in D_{CQ} designate maximum and minimum values, respectively.

Our first objective is to find a constraint that ensures that for each cycle data appears at the D-inputs of the FFs in time to allow the FFs to respond properly. This means that inputs must be stable at least S (the setup time) prior to the leading edges of the clock-pulses, even under worst-case conditions. Refer now to Figure 6.63. Here we see that if the nominal arrival time of a clock-pulse leading edge at a FF is t = 0, then, at the latest, a resulting change in the output of that FF will appear at another (or the same) FF input at $t = D_{CQM} + D_{LM} + T_L$. This is the case where the delays through the FF and the logic are both maximal and where the clock-pulse edge is late at this FF by the maximum time, T_L. In the worst case, the triggering clock-pulse edge at the FF receiving this change will arrive at the *earliest* possible time, $P - T_L$. Therefore, in order for the data from the transmitting FF to be on time in the worst case, it must arrive at $P - T_L - S$. Thus, we have the inequality

$$D_{CQM} + D_{LM} + T_L < P - T_L - S \qquad \text{or}$$
$$P > D_{CQM} + D_{LM} + 2T_L + S. \qquad (3)$$

Next, we must find a constraint that ensures that signals do not arrive too *early* at the FFs; this is the short-path constraint. Signals starting out at the input to a FF on the leading edge of a clock-pulse must not be able to get through the FF and the logic fast enough to arrive at some FF input before the hold time for that clock cycle has elapsed. Refer now to Figure 6.64. If the nominal arrival time of a clock-pulse leading edge is t = 0, then that edge may occur as soon as $t = -T_L$. A signal starting through a transmitting FF at that time could get through the logic and arrive at a FF input $D_{CQm} + D_{Lm}$ later, arriving as early as $D_{CQm} + D_{Lm} - T_L$. The same clock-pulse edge might arrive at a receiving FF as late as $+T_L$. Since trouble could occur if an input change occurred before the hold time H has elapsed after the triggering edge, it is necessary to ensure that the arrival time of the edge at the receiving FF is no earlier

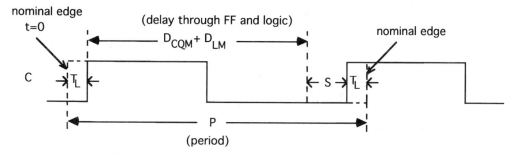

Fig. 6.63. Calculating the long-path constraint for the one-phase case with ETDFFs.

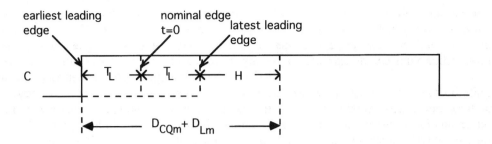

Fig. 6.64. Short-path constraint for one-phase clocking with ETDFFs.

than $T_L + H$. Thus, we have $D_{CQm} + D_{Lm} - T_L > T_L + H$. This gives us the bound on the short-path delay:

$$D_{Lm} > D_{LmB} = 2T_L + H - D_{CQm}. \tag{4}$$

One of the FF parameters is C_{Wm}, the minimum allowable width that a clock-pulse must have for proper operation. Since it is possible for the leading edge of a pulse to arrive early and the trailing edge to arrive late at the same FF, the width of the clock-pulse may be reduced from its nominal value by as much as $T_L + T_T$. We therefore have a third constraint for the clocking system, namely one on the minimum allowable value of the clock-pulse width W (the distance between the *nominal* leading and trailing edges of the clock-pulse)

$$W > C_{Wm} + T_L + T_T. \tag{5}$$

(In addition to the minimum clock-pulse width constraint for the FF, there is also a lower bound on the space *between* consecutive clock-pulses. Since this is rarely a problem, no more will be said about it.) Constraint (5) is seldom a problem for the ETDFF case. Unless the skew is very large, constraint (4) on the short-path delay is usually not very burdensome. The long-path constraint (3) usually determines the allowable clock-pulse period. The application of these constraints is illustrated in the next example.

Example 6.29

Suppose the ETDFF has parameters as follows (the time units can be thought of as being in nanoseconds): $D_{CQM} = 3$, $D_{CQm} = 1.5$, $S = 2.5$, $H = 2$, $C_{Wm} = 2$. Assume that we are informed that the clock-pulse edge tolerances are $T_L = 1.5$ and $T_T = 2$. If our estimate of the maximum number of levels of logic we require, along with estimates of gate and wiring delays, yields $D_{LM} = 14$, then we can proceed as follows to determine the clocking parameters and the lower bound on the short-path delay. From (3) we obtain $P = D_{CQM} + D_{LM} + 2T_L + S = 3 + 14 + 2(1.5) + 2.5 = 22.5$. (We use equality, since obviously we want the period to be as small as possible.)

Constraint (4) gives us $D_{LmB} = 2T_L + H - D_{CQm} = 2(1.5) + 2 - 1.5 = 3.5$. Since this number is only a fraction of the long-path delay, satisfying this bound should raise no serious problems. It is only necessary to make sure that if there are any logic paths consisting of one or two gates, that some extra delays, perhaps in the form of pairs of inverters, are added.

Constraint (5) on the pulse width gives us $W > C_{Wm} + T_L + T_T = 2 + 1.5 + 2 = 5.5$. This is less than half of the period, which allows us to use a pulse width equal to $P/2$, a value often favored for electrical reasons.

The ease with which satisfactory clocking parameters can be chosen without incurring penalties in the form of severe constraints on short-path delays makes this type of system attractive in situations where maximum performance is not the objective. The presence of S and $2T_L$ in the expression for P is the major drawback. The other two systems that we shall consider allow us, at a price, to attain smaller values of P by eliminating such terms from the long-path constraints.

6.6.2. One-Phase Clocking with Latches

The block diagram of Figure 6.62 also applies to this case, with the storage devices now assumed to be latches. Latches also have setup and hold times analogous to those for ETDFFs, but the edge involved is the *trailing* edge of the C-pulse. In addition to the parameters defined for the ETDFF, another pertinent latch parameter is D_{DQ}, the propagation delay from the D-input to the Q-output, assuming that C has been on for at least D_{CQ} prior to the D-change. More precisely, we assume that D_{CQ} and D_{DQ} are related as follows: If the D-input changes at t_D, and the C-pulse goes on at t_C, then the Q-output changes at $t_Q = MAX(t_D + D_{DQ}, t_C + D_{CQ})$. While not always exactly true, this is usually a valid approximation. Next, the principal constraints for this type of system are developed, beginning with two constraints involving the long-path delay.

First, we observe that in one clock period, it must be possible, even in the worst case, for a signal to propagate from the D-input of a latch, through the latch and the logic block to reach the D-input of a receiving latch. If getting around the loop takes longer than a clock period by some amount e, then, in n such cycles, the arrival time of a signal relative to the trailing edge of C would be later by ne. Hence, for sufficiently large n, the setup time constraint would be violated, leading to a malfunction. In the worst case, the loop delay is at least as long as $D_{DQ} + D_{LM}$. Therefore, the first long-path constraint, a *necessary* condition for the period to be sufficiently long relative to the loop delay, is

$$P > D_{DQ} + D_{LM}. \tag{6}$$

The second long-path constraint involves propagation delay from the C-input. Refer now to Figure 6.65, which illustrates what happens to a signal change present at the D-input of a latch (the transmitting latch) well before the C-signal goes on. Starting at the leading edge of the C-pulse, this signal propagates through the latch, then through the logic block to the D-input of a receiving latch. The time for this transmission is at most $D_{CQM} + D_{LM}$. It must reach the receiving D-input no later than S prior to the *trailing* edge of the *next* clock-pulse, nominally in time $P + W - S$. As can be seen in Figure 6.65, if the clock-pulse leading edge at the transmitting latch is late by T_L, and the clock-pulse trailing edge at the receiving latch is early by T_T, then the time for transmission is further reduced. From the diagram we can see that it is necessary to satisfy the inequality $P + W > T_L + D_{CQM} + D_{LM} + S + T_T$. Transposing terms yields the second long-path contraint:

$$P > D_{CQM} + D_{LM} + S + T_L + T_T - W. \tag{7}$$

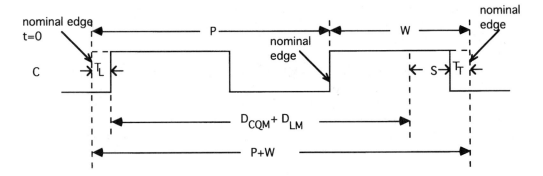

Fig. 6.65. Calculating the second long-path constraint for the one-phase case with latches.

Of course *both* (6) and (7) must be satisfied. Since (7) includes W as a negative term, it appears that the pulse width might be made sufficiently large as to make the right side of (7), equal to the right side of (6), giving us a value for P that is independent of both S and the clock-pulse edge tolerances. This possibility is indeed a major reason for using this type of clocking system. But now let us consider the cost of making W sufficiently large. This is incurred in terms of the short-path constraint developed next.

As in the ETDFF case, it is necessary to avoid the possibility of a signal racing through a latch and the logic block fast enough to arrive at a receiving latch before the hold time for the current cycle has expired. Referring to Figure 6.66, the worst case is when the leading edge of the clock-pulse is early at the transmitting FF and late at the receiving FF. Since the signal can start with the leading edge of the pulse, while the hold time starts at the trailing edge, the clock-pulse width is added to the time available for the signal to complete its journey. The fastest time for the trip is $D_{CQm} + D_{Lm}$. It may start out as early as $-T_L$, and must not arrive sooner than $W + T_T + H$. As is evident from an inspection of the diagram, the inequality to be satisfied for proper operation is $D_{CQm} + D_{Lm} > T_L + W + T_T + H$. Transposing leads to

$$D_{Lm} > W + T_L + T_T + H - D_{CQm}.$$

Let us call the right side of this inequality D_{LmB}, the lower bound on D_{Lm}. Then we have the following, which is called the short-path constraint:

$$D_{Lm} > D_{LmB} = W + T_L + T_T + H - D_{CQm}. \tag{8}$$

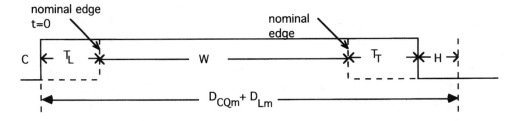

Fig. 6.66. Short-path constraint for one-phase clocking with latches.

The short-path constraint is the place where the price is paid for reducing the period by increasing the clock-pulse width. Since W appears as an additive factor for the short-path bound, the larger we make W, the harder it will be to satisfy the short-path constraint. That is, if W is increased, then we must make the delay along the shortest path through the logic closer to the delay along the longest path. This can become a very difficult problem if D_{LmB} is very close to D_{LM}. Of course, it becomes physically impossible if $D_{LmB} > D_{LM}$.

The fourth constraint involves the minimum clock-pulse width. It is the same as the corresponding one for the ETDFF case:

$$W > C_{Wm} + T_L + T_T. \tag{9}$$

The application of these four constraints is illustrated next.

Example 6.30

Assume we are using the same logic block and clock distribution network as in Example 6.29, and that the latch parameters are similar to the ETDFF parameters, which means we have $D_{LM} = 14$, $T_L = 1.5$, and $T_T = 2$, $D_{CQM} = 3$, $D_{CQm} = 1.5$, $S = 2.5$, $H = 2$, $C_{Wm} = 2$, and we add $D_{DQM} = 2$.

The first step is to find the lower bound on P due to (6). This is

$$D_{DQM} + D_{LM} = 2 + 14 = 16.$$

Applying this value of P in (7) and solving for W, yields

$$W > D_{CQM} + D_{LM} + S + T_L + T_T - P = 3 + 14 + 2.5 + 1.5 + 2 - 16 = 7.$$

What does this entail with respect to the short-path bound?

Substituting $W = 7$ into relation (8) gives us

$$D_{Lm} > D_{LmB} = W + T_L + T_T + H - D_{CQm} = 7 + 1.5 + 2 + 2 - 1.5 = 11.$$

Even if we added dummy logic elements as necessary to place the same number of gates in every path, the delay variations from gate to gate would, in many technologies, make it very difficult to ensure that the delays in all paths are between 11 and 14.

Where it is impossible, or simply undesirable to meet the bound imposed by (8) for the desired value of W, we must compromise by setting D_{LmB} at the largest value that we feel is reasonable and solving (8) for W. Substituting this value in (7), we can then find the lowest achievable value of P (which will, of course, exceed the bound imposed by (6)). In this case, let us choose 10 as the highest value reasonably attainable for D_{LmB}. From (8) it is evident that decreasing D_{LmB} requires W to be decreased by the same amount, in this case by 1, so W becomes 6. From (7) we can see that decreasing W causes P to increase by the same amount, again by 1, so that we now have $P = 16 + 1 = 17$. We must also check to see that W meets the lower bound imposed by (9). In this case, that is $W > C_{Wm} + T_L + T_T = 2 + 1.5 + 2 = 5.5$. Since it does, we have a solution, namely $P = 17$ and $W = 6$. Note that the period is significantly shorter than the 22.5 achieved in Example 6.29 with similar parameters for the ETDFF case.

Suppose that the value of W found in the preceding example had failed to satisfy the minimum clock-pulse width constraint (9)? This would have happened, for example, if C_{Wm} had been 3 instead of 2, increasing the lower bound on W to 6.5. This, in turn, means that D_{LmB} increases to 10.5. In such cases, assuming we cannot reduce the edge tolerances or usefully alter some other parameters, it would be necessary to increase D_{LM}. This might be done by allowing the maximum number of stages of logic to be increased, which might be beneficial in other respects, or we could simply add fixed delays at the outputs of all of the latches. Let us see how this would work in the present case if the value of D_{LmB} had to be increased by 0.5 over what was presumably the maximum attainable limit of 10. Our strategy would be to design the system to meet the lower bound of 10 and then add delays to all paths (the latch outputs would be convenient locations) with *minimum* values of 0.5. The *maximum* values of these delays would then have to be estimated or given to us in the same way that the ranges of latch propagation delays was supplied. Suppose, for example, that we are told that the ratio of maximum to minimum values for the available delay elements is 1.4 (i.e., equal to D_{LM}/DL_{mB}). Then D_{LM} would increase by 1.4 × 0.5 = 0.7. Since the P increases by the same increment as D_{LM}, this means that in our example we now have P = 17 + 0.7 = 17.7, while W becomes 6.5.

Since estimates of delay values and edge tolerances are just that—estimates—it may be discovered after a machine has actually been built that bounds (8) and (9) cannot be reconciled without the use of extra delay as earlier. Adding them at this stage is a very troublesome process, both costly and time-consuming. An advantage of the two-phase clocking system discussed next is that late stage corrections of timing problems can usually be confined to the clock-pulse distribution system itself. The insertion of extra delay elements is never necessary. A wider range of trade-offs is also possible between the clock period and the lower bound on the short-path delays.

6.6.3. Two-Phase Clocking with Latches

In a two-phase system (see Figure 6.67), pairs of latches (L_1, L_2) replace each of the latches of a one-phase system. The L_1-latches are fed by C_1-pulses, and the L_2-latches receive C_2-pulses. The latter are generated by the same oscillator as the C_1-pulses but are out of phase and may have different pulse widths. During the C_1-phase of the clock cycle, the L_1-latches receive inputs from the logic block dependent on the contents of the L_2-latches. During the C_2-phase, the L_2-latches copy the contents of the corresponding L_1-latches. The phase relation between the C_1- and C_2-pulses is specified by the *overlap* V. The clock-pulse widths are not very important; V assumes the role played by W in the one-phase case. As in the case of W, V is defined in terms of *nominal* edges: the length of the interval beginning with the leading edge of C_2 and ending with the trailing edge of the C_1-pulse for the *same* cycle. It is negative if the C_2-pulse begins after the C_1-pulse ends.

As in the previous case, the process of determining a satisfactory set of parameters for the clocking system begins with the derivation of two long-path constraints and one short-path constraint. These closely parallel the corresponding constraints for the one-phase case. (Subscripts are prefixed by 1 or 2 to distinguish between the latches.) We begin by observing that the minimum time around a loop, starting and ending at a D-input to a latch, must not exceed P. This immediately gives us the first lower bound on

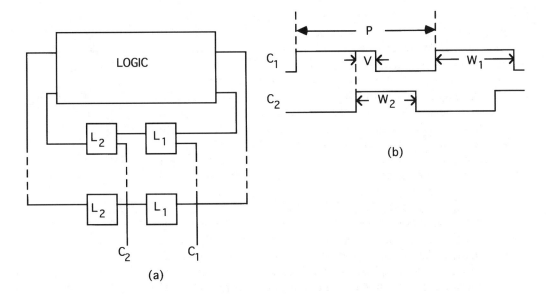

Fig. 6.67. Two-phase clocking with latches.

P (analogous to (6) for the one-phase case)

$$P > D_{1DQM} + D_{2DQM} + D_{LM}. \tag{10}$$

The second long-path constraint is found by considering a signal established at the D-input to an L_2-latch. It must be able to propagate through that latch and the logic block in time to be set up at an L_1-latch input at least S_1 units of time prior to the trailing edge of the *next* C_1-pulse. The situation, including the effect of edge tolerances, is depicted in Figure 6.68. Starting at $t = 0$, defined as the arrival time of the nominal leading edge of C_2 at the transmitting latch, $P + V$ is the distance to the nominal trailing edge of the next C_1-pulse. It is evident from the diagram that the same interval can be broken up into the sum of components, $T_{2L} + D_{2CQM} + D_{LM} + S_1 + T_{1T}$. From this we can deduce the need to satisfy $P + V > T_{2L} + D_{2CQM} + D_{LM} + S_1 + T_{1T}$. Transposing V yields the second lower bound on P (analogous to (7) for the one-phase case)

$$P > T_{2L} + D_{2CQM} + D_{LM} + S_1 + T_{1T} - V. \tag{11}$$

Since (10) and (11) must both be satisfied, setting V at a value that equalizes the two bounds would yield the minimum value for P. The next step indicates how the price for a large V must be paid in terms of a larger value of D_{LmB}. This entails the derivation of a constraint to prevent a signal starting at an L_2-latch from getting through that latch and the logic block quickly enough to arrive at an L_1-latch input before the expiration of the hold time for that latch.

Referring to Figure 6.69, we note that starting at the actual arrival time of a C_2 leading edge, the interval until the expiration of the hold time for a receiving L_1-latch is the sum of T_{2L} (the leading edge of C_2 may be early), V, T_{1T} (the C_1 trailing edge may

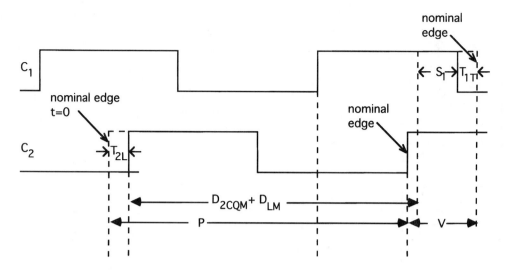

Fig. 6.68. Calculating the second long-path constraint for the two-phase case with latches.

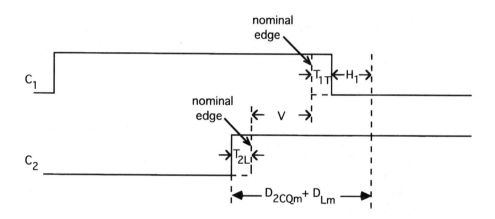

Fig. 6.69. Short-path constraint for two-phase clocking with latches.

be late), and H_1. If it arrived sufficiently early at the D-input to the transmitting latch, then beginning at that same C_2 edge, the minimum time for a signal to get through an L_2-latch and the logic is $D_{2CQm} + D_{Lm}$. Hence, the following constraint ensures that no signal can get through in time to cause trouble: $D_{2CQm} + D_{Lm} > T_{2L} + V + T_{1T} + H_1$. This yields the short-path bound (analogous to (8) for the one-phase case):

$$D_{Lm} > D_{LmB} = T_{2L} + V + T_{1T} + H_1 - D_{2CQm}. \tag{12}$$

Constraints (10), (11), and (12), supplemented by lower bounds on clock-pulse widths W_1 and W_2 in the form of (9) are usually necessary and sufficient for specifying clocking parameters for two-phase clocking systems with latches. Their application is illustrated in the next example.

Example 6.31

The parameters for this example are similar to, but not the same as, those used in Example 6.30 for the one-phase system: $D_{LM} = 14$, $T_{1L} = 2$, $T_{1T} = 1.5$, $T_{2L} = 1.5$, $T_{2T} = 2$, $D_{1DQM} = 2$, $D_{1DQm} = 1$, $D_{2DQM} = 2$, $D_{2DQm} = 1$, $D_{1CQM} = 3.5$, $D_{1CQm} = 1.5$, $D_{2CQM} = 3$, $D_{2CQm} = 1.5$, $S_1 = 2.5$, $H_1 = 2$, $S_2 = 3$, $H_2 = 2.5$, $C_{1Wm} = 3$, $C_{2Wm} = 3$. Note that the edge tolerances are not all the same. Since T_{1T} and T_{2L} play a key role in determining the performance limits of the system, it is desirable (and possible in practice) to design the clock generation and distribution system so as to minimize these at the expense of increasing the other two edge tolerances. Similar differences in the roles of the L_1- and L_2-latch parameters make it desirable to vary the latch designs to optimize performance. Let us now proceed to apply the constraints.

From (10) we find $P > D_{1DQM} + D_{2DQM} + D_{LM} = 2 + 2 + 14 = 18$. If we are to achieve this value, then we must choose V so that constraint (11) will be satisfied with $P = 18$. Thus, we require that

$$V > T_{2L} + D_{2CQM} + D_{LM} + S_1 + T_{1T} - P = 1.5 + 3 + 14 + 2.5 + 1.5 - 18 = 4.5.$$

With $V = 4.5$, (12) is now invoked to determine D_{LmB}. This yields

$$D_{LmB} = T_{2L} + V + T_{1T} + H_1 - D_{2CQm} = 1.5 + 4.5 + 1.5 + 2 - 1.5 = 8.$$

In many cases, this would be a reasonable lower bound as compared with a long-path delay of 14. If it is satisfactory, then we would indeed be able to operate the clock with minimum period of 18.

Suppose that a still lower value of D_{LmB} is desired. We need only reduce V by the same amount that we wish to reduce D_{LmB}. Thus, to reduce D_{LmB} to 5, V would be reduced by 3 to 1.5. As indicated by (11), the period would then have to be *increased* by the same amount, i.e., to 21. The lower bound can be eliminated altogether, i.e., reduced to 0, by reducing V by 8 to -3.5, requiring P to rise to 26. (In many treatments of two-phase systems, it is assumed that C_1 and C_2 have such a negative overlap. We can see here that the price in terms of reduced performance can be substantial.)

Nothing has been said so far about the other two parameters of the clocking system, pulse widths W_1 and W_2. These play a distinctly secondary role. They must each satisfy lower bounds in the form (9), (6.5 for both W_1 and W_2). There are additional constraints in the form of both upper and lower bounds on $W_1 + W_2$, as well as another pair of lower bounds on W_1 and W_2. These are not derived here, but are listed below for reference:

$$D_{1CQm} + V + P - H_2 - T_{1L} - T_{2T} > W_1 + W_2 > V + S_2 + D_{1CQM} + T_{1L} + T_{2T} \quad (13)$$

$$W_1 > -P + D_{1CQM} + D_{2DQM} + S_1 + D_{LM} + T_{1L} + T_{1T} \quad (14)$$

$$W_2 > V + S_2 - S_1 + D_{1DQM} + T_{2T} - T_{1T}. \quad (15)$$

For the system of Example 6.31, the bounds on the Ws, assuming $P = 18$ and $V = 4.5$, are $17.5 > W_1 + W_2 > 15$, $W_1 > 7.5$, and $W_2 > 7.5$. Setting $W_1 = W_2 = 8.2$ satisfies all of these bounds comfortably.

Of tertiary importance is the fact that (12) can be replaced by the usually more restrictive constraint

$$D_{Lm} > D_{LmB} = W_1 + T_{1T} + T_{1L} + H_1 - D_{1CQm} - D_{2DQm}, \tag{16}$$

and the upper bound on $W_1 + W_2$ of (13) can be replaced by the usually more restrictive constraint

$$H_1 - H_2 + D_{1DQm} + V + T_{1T} - T_{2T} > W_2. \tag{17}$$

Neither replacement would be helpful in this case.

Each of the clocking systems described here is useful in various situations. Where high performance is the top priority, as in certain supercomputers, single-phase clocking with latches yields maximum speed, provided that stringent efforts are made to reduce skew and to make delay paths uniform. The two-phase system is very flexible, allowing the widest range of trade-offs between easing the bound on the short-path delays and minimizing P. It is often chosen for mainframe computers. Using ETDFFs is a good option for reasonably safe systems at reasonable cost.

6.7. THE SYNCHRONIZATION PROBLEM

As is evident from the material of the previous section, it is possible to design clocking schemes for synchronous systems that can ensure freedom from timing problems. But this is true only where *all* of the signals encountered by the system are synchronized to the same clock. If signals must be exchanged among two or more systems, each with an independent clock, there are serious problems of a fundamental nature. These are related to several other problems that strain the simple model we have been using in which all signals can be represented as 1s or 0s. Much of the discussion and arguments in this section are of an intuitive nature. More rigorous developments exist, but are beyond the scope of this book.

6.7.1. Runt Pulses

Thus far in this book, nothing has been lost by approximating signal waveforms as having straight, perpendicular edges. By so doing, we have avoided unnecessary complexity. But now it is necessary to recognize that real signal waveforms are never exactly of this form. The rising and falling edges can be more accurately approximated as sloped straight lines. Upon still closer examination they are revealed to be more irregular in shape.

Suppose that, as shown in Figure 6.70a, signals A and B feed an AND-gate. Let these signals be uncorrelated in the sense that the relative times at which they can change are completely uncontrolled. That is, if A changes at $t = t_A$, B may change at $t = t_A + \Delta$, where Δ is any real number, positive or negative. Assume now that $A = 0$ and $B = 1$; then $Z = 0$. If A changes to 1 at $t = 0$, and B changes to 0 at some later time Δ (now assume $\Delta > 0$), then a 1-pulse of duration Δ appears at the Z-output. For small values of Δ, the pulse is, of course, narrow. What happens if we keep reducing Δ? In the ideal case where no delays are involved, the width of the output pulse simply shrinks along with Δ, as shown in Figure 6.70b.

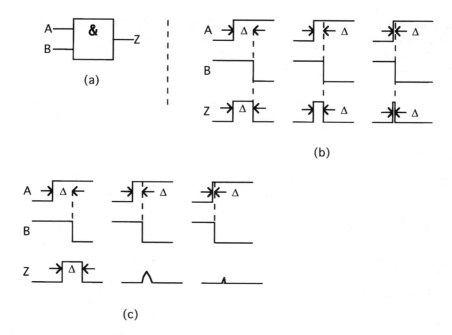

Fig. 6.70. Ideal and real effects of changing AND-gate inputs in opposite directions.

Now consider more carefully the question of output waveforms. (It is not necessary, however, to consider less regular *input* waveforms.) A better approximation to reality is shown in Figure 6.70c. Here it is assumed that there is a delay associated with the gate resulting from various physical effects, including resistance and capacitance. The output is therefore shifted in time. But this is not the important point here. More significant is that once the pulse width is of the order of the delay magnitude, the wave shape of the output can no longer be reasonably approximated as rectangular. In the first case, where Δ is relatively large, the output is still essentially a rectangular pulse. In the third case, where Δ is clearly less than the delay value, the output is reduced by inertial effects to a barely noticeable blip that would be treated as a 0 by any logic element.

The second case is the most interesting one. Here Δ is of the same order as the delays. The output is attenuated in magnitude and is no longer even approximately rectangular in shape. It is what is known as a *runt* pulse: a signal of a marginal nature that, as an input to any given logic element, might be treated as a 1 *or* as a 0. In certain cases, it might be so marginal as to cause behavior that falls entirely outside our usual model of digital system operation. The nature of this type of irregularity is examined below. But first it must be emphasized that in situations such as shown in Figure 6.70, there is no way to eliminate the possibility of runt pulses occurring. Changing the physical design of the gate, for example, can only *alter* the range of values of Δ for which runt pulses are produced. Nor, as is shown later in this section, is it possible to filter out runt pulses with inertial delay elements.

The fundamental problem is related to the concept of physical continuity. If input signals I_1 and I_2 to any system differ by ever decreasing amounts, we cannot expect the difference between the outputs $f(I_1)$ and $f(I_2)$ to be bounded below by some clearly measurable amount. Eventually, the outputs begin to converge—to approach one another in a continuous manner. In our example, as Δ approaches 0, the waveform of Z shrinks in magnitude as well as in duration, assuming a continuum of forms between the rectangular pulse shown for the first value of Δ and no deviation at all from 0.

6.7.2. The Unrealizability of Ideal Inertial Delays

The preceding argument (which is of an intuitive nature rather than being a rigorous proof), applies very specifically to the possibility of constructing an ideal inertial delay element. A strict interpretation of the definition of an inertial delay (ID) would require that for any $e > 0$, however small, a pulse of length $D + e$ applied to the input of an ID of magnitude D, would get through unchanged except for a time shift of $D + e$. On the other hand, a pulse of length $D - e$ would be completely filtered out. Such behavior is unrealizable due to the continuity argument presented earlier.

Any real ID of magnitude D presented with pulses of duration varying from $D + e$ to $D - e$ for some small value of e, will produce outputs consisting of pulses with waveforms ranging from a near rectangular pulse of width $D + e$ to no pulse at all, with a continuum of waveforms in between. A sampling of what such waveforms might look like in this range is shown in Figure 6.71. The waveforms labeled "a" through "f" represent responses to progressively narrower input pulses in the range from $D + e$ to $D - e$. Thus, an inertial delay can be expected to approximate ideal behavior only when the durations of input changes are not too close to the delay magnitude.

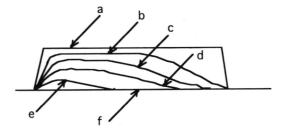

Fig. 6.71. Responses of an inertial delay to input pulses of duration in the neighborhood of D.

6.7.3. Metastable States

Next, to get some insight into the effects of runt pulses, let us consider what happens when a runt pulse is fed to the S-input of an SR-FF initially in the reset state. (See Subsection 6.2.4 to review the behavior and construction of this device.) An examination of Figure 6.72a (reproduction of the Figure 6.14a FF circuit) indicates that when $S = R = 0$, the circuit is reduced to a pair of inverters connected in a loop as shown in Figure 6.72c. This of course is the condition following the application of a pulse to either input. If the runt pulse succeeds in making $Q = 1$ and $v = 0$, then the circuit will be stable in the set state. If the runt pulse fails to change Q sufficiently, then Q falls back to 0 and v remains at 1, leaving the FF in the reset state.

But what if Q is brought to a voltage value midway between the 0 and 1 values? It can be shown that, for every inverter, there is an input voltage that produces an output voltage of the same value. At this value, let us call it $\frac{1}{2}$ (although it is not necessarily

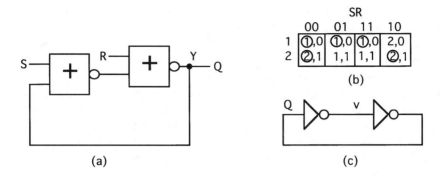

Fig. 6.72. SR-FF and equivalent circuit when S = R = 0.

exactly halfway between the 0 and 1 voltage values), both inverters have input and output values of $\frac{1}{2}$, and so the system is stable. But suppose Q should, perhaps as a result of random noise, increase slightly. The output of the inverter would then decrease, causing v to decrease. This in turn would cause the output of the other inverter to *in*crease, further increasing Q. The result of this process would be that the FF would rapidly go to the set state. A similar scenario involving a decrease in Q (or an increase in v) would drive the system to the reset state. Because of the precarious nature of the $\frac{1}{2}$ stable point, it is referred to as a point of *unstable equilibrium* (or as a *metastable* state). The set and reset states are both strongly stable in that small perturbations do *not* build up in this manner to cause state changes.

Every SR-FF can be brought to a metastable state, generally of very short duration. The probability of a device remaining in the metastable state for a period exceeding t is approximately e^{-at}, where a is a parameter determined by its physical characteristics. Thus, for t a sufficiently large multiple of $1/a$, we can say that it is very unlikely that the system will remain in the metastable state for longer than t, but we can never be *certain* that this will not happen. Due to complex dynamic factors, systems in this condition do not usually have constant voltage outputs. Rather, they tend to oscillate in some region between the 0 and 1 voltage levels. Formal proofs of all of this exist, but they are more in the realm of analog than of digital systems and so are not reproduced here. The potential for metastability exists for virtually all devices with multiple inputs, and can be detected by examination of flow tables.

Figure 6.72*b* is a flow table description of an SR-FF. The situation just explored is where the system starts in 1-00 and then goes to or toward 1-10. If it remains in 1-10 long enough to *start*, but not to complete the transition to row 2, and then the input reverts to 00, it is not clear as to whether it will land in 1-00, 2-00, or in between, in the metastable state. No such problem exists for a *single* input change starting from any other input, because in every other input column there is only one stable state. If, for example, we start in 2-10 and apply a runt pulse to the R-input, there is no problem because when the input state reverts to 10 following the runt pulse, the internal state can only return to 2. At worst there might be a transient 0 at the Q-output.

Metastability can also arise following a multiple input change from the state 1-11. If the relative timing of S and R changes in combination with the stray delays is such that S appears to change first, then the system ends up in 1-00. But if it appears that R changes first, the final state may be 2-00. Here again we have a situation in which an arbitrarily small change in the input, in this case the relative times at which S and R change, is ideally supposed to cause results that differ in a discrete manner. If the relative time changes of the variables are almost exactly on the border between the case where the final state would be 1 and the case where the final state would be 2, the system may enter the metastable state. In general, for any flow table, wherever a multiple input change could lead to any of several different stable states, we have a danger of entering the metastable state. For the SR-FF, there are no other ways in which this can happen. For example, starting in 1-00 and turning on S and R simultaneously cannot lead to metastability, because, regardless of the order in which they change, the final state must be 1-11. (We are not concerned here with possible output glitches of bounded duration.) A sequential circuit realizing a function with more than two internal states can also become metastable if the state assignment has a critical race. We consider next another important situation in which metastability is of concern.

6.7.4. The Arbiter

Certain digital facilities, for example most random access memories, can be used by only one processor at a time. Where a system includes more than one processor that may simultaneously request access to such a facility, an orderly allocation procedure is needed. The usual way of handling this problem is to use a device called an *arbiter*. Such a device for coping with two processors is described by the Figure 6.73 flow table. (The arbiter discussed here is specified to work for processors that do not share a common clock, so that a clocked circuit cannot be used. This is in contrast to the arbiter treated in Example 6.26, p. 212, which arbitrates between two processors controlled by the same clock.)

The arbiter works as follows. Inputs R_1 and R_2 are request signals from processors P_1 and P_2. If R_1 is turned on, signifying a request by P_1 for the facility, while the facility is idle and $R_2 = 0$, the arbiter grants the request by turning on output Z_1. As long as R_1 remains on, P_1 retains control of the facility, regardless of any requests made by P_2. Thus, if R_2 goes on while the facility is held by P_1, no change is made in the output values. Only after P_1 relinquishes the facility by turning off R_1, will the arbiter grant the facility to P_2. Readers should be able to verify that the arbiter described in the flow table behaves this way. Note that a crucial requirement is that Z_1 and Z_2 never both be on.

Suppose that, starting in state 1-00, R_1 and R_2 go on simultaneously (i.e., that both processors are requesting the facility at the same time). We can see that if the R_1-signal had been first, then the state path would have been from 1-00 to 1-10 to 1-11, and P_1 would have been granted the facility (i.e., Z_1 would have been turned on). If R_2 had been the first to change, then the path would have been 1-00 to 1-01 to 2-01 to 2-11. In that case, Z_2 would have been turned on, granting the facility to P_2. Thus, the result is

$$R_1 R_2$$

	00	01	11	10
1	①,00	2,00	①,10	①,10
2	1,00	②,01	②,01	1,00

Fig. 6.73. Flow table for an arbiter.

dependent on the order of change, and so we have the situation described in the previous subsection where the system might enter a metastable state. Thus, it is impossible to build an arbiter for systems not sharing a common clock in which the possibility of a metastable state being encountered is eliminated. The *probability* of such an occurrence *can* be made small enough to be negligible through careful design.

6.7.5. The Synchronizer

The problem mentioned at the start of this section was that of passing data between systems operating under independent clocks. Let us examine this problem more closely. Suppose that a signal X, generated by a peripheral unit such as a disk drive with its own clock, must be passed to a processor. (It might, for example, represent an interrupt signal.) Starting at some time that is completely arbitrary with respect to the processor's clock, X changes from 0 to 1. What is required is an interface device, called a *synchronizer*, that receives X and at some time after this change, produces an output signal Z that also goes from 0 to 1, but at a time that is correlated to the processor's clock signals. The exact phase relationship need not concern us here. As long as we know what it is, we can always shift it appropriately within some reasonable degree of precision by means of delay elements. Figure 6.74 illustrates one satisfactory mode of operation for a synchronizer, namely where, following a change of X from 0 to 1, Z is turned on by the next leading edge of a clock-pulse.

The mode of behavior illustrated here conforms precisely to the specifications of the positive edge-triggered D-FF discussed in several places elsewhere in this chapter. But in order to use an ETFDD properly, it is necessary to adhere to setup and hold time constraints, i.e., to avoid changing the D-input signal in some neighborhood of the triggering edge of the clock-pulse. An inspection of the flow table for this device, shown in Figure 6.9, p. 167, indicates that these constraints reflect the problems discussed before. They amount to prohibiting multiple input changes from states 1-00 and 4-00 that could lead to metastable behavior. Since, by definition of the synchronizing problem, we have no control over the relative times of X- and C-changes, it is possible for X to change at precisely the right (or wrong!) time in the neighborhood of the leading edge of the C-pulse to put the synchronizer into the metastable state. It would then produce some garbled output that might, in turn, result in a malfunction of the system receiving the signal. Thus, it is evident that the problem of building a synchronizer falls into the category of problems introduced in this section. In fact, as suggested by this section's title, this class of problems is sometimes referred to as "the synchronization problem."

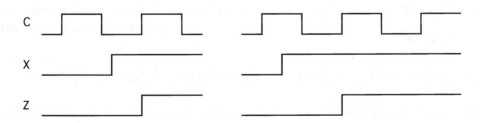

Fig. 6.74. What a synchronizer has to do.

All of these problems are equivalent in the sense that, if any could be solved, then all of them are solvable. That is, if we could build an ideal ID, we could also build an ideal synchronizer, an ideal arbiter, etc., all of which would be free of the menace of metastability. Similarly, it has been shown that, given an ideal synchronizer, we could build an ideal ID, etc. Unfortunately, there exist proofs that it is impossible to build such devices. The next best thing to producing devices that are immune to this type of failure, is to build devices that under the conditions of their normal use, are extremely unlikely to fall into the metastable state. A solution of this type is given next for the synchronization problem.

Example 6.32

Even if no special design precautions are taken, the likelihood of metastable behavior is almost always very small. One might expect, for example, that a synchronizer consisting simply of a PETDFF might correctly handle many thousands of signals before failing in this manner. Furthermore, when a metastable state is entered, the probability of it lasting as long as a full clock cycle is usually extremely small. (There are ways of estimating these factors that are not addressed here.) We can exploit this low initial probability of failure to produce a device that has a vastly lower failure probability.

Refer now to the circuit shown in Figure 6.75a, composed of two PETDFFs. Both receive the clock signal to which input X is to be synchronized. The output Q_1 of the first FF feeds the D-input of the second FF. Q_1 will go on cleanly to synchronize almost all input signals. Assuming that the propagation delay is sufficient so that the Q_1-change occurs after the hold time has elapsed for that clock-pulse, then at the next clock-pulse leading edge, the second FF responds to the change by turning on Z.

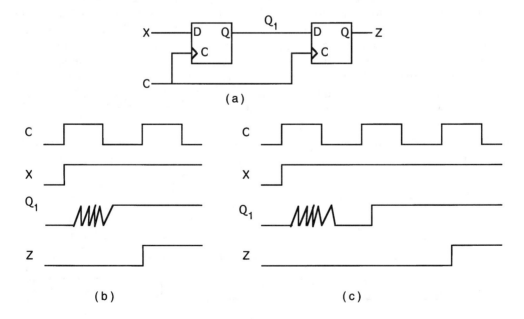

Fig. 6.75. A practical solution to the synchronization problem.

Suppose now that the unlikely occurs and an X-change brings the first FF into a metastable state. We assume then that the Q_1-signal oscillates in some erratic manner before settling down either to a 1-value, as shown in Figure 6.74b, or to a 0-value, as shown in Figure 6.74c. Our assumption is that the metastable state will almost certainly be over in time for the D-signal to the second FF to be set up properly in time for the next clock-pulse. Then the Z-signal will either change at that time, or remain at 0 and change in response to the next clock-pulse. Thus, there is no possibility of a synchronization failure unless (1) an X-change causes the first FF to enter the metastable state, *and* (2) that state lasts a full clock cycle (less the setup time). Even then, there is a real possibility that the second FF will still produce a proper output. Further reductions in the failure probability can be made by such means as adding a third FF to the chain, or by transmitting only every other clock-pulse to the second FF, so as to more than double the available time for the first FF to recover from the metastable state.

6.8. PULSE-MODE OPERATION

It is sometimes desirable to operate sequential circuits in a mode with no clocks, but such that each input signal consists of a pulse. Only one input at a time is permitted. This is called pulse-mode operation. It is generally confined to special purpose circuits that are not parts of larger systems.

Figure 6.76 is a flow table that can be interpreted as describing a pulse-mode function. As in the case of synchronous systems, the device at any time is assumed to be in an internal state corresponding to one of the rows. Nothing can happen until an input appears, i.e., a pulse on input terminal A or B. A pulse is then generated on the output terminal if there is a 1 in the output position of the cell in the flow table corresponding to the row and input column. The next state is as specified in that cell. Thus, if with the system in state-2, a pulse appears on input terminal A, an output pulse is generated, and the next state will be 1. If a B-pulse appears with the system in state-1, then no output is produced, but the state changes to 2.

With the given state assignment, it is clear that an expression for the output Z is

$$Z = Ay.$$

This means that an input pulse on A is gated to the output terminal Z whenever the device is in state-2, as indicated by state-variable y having the value 1. It is convenient to use toggle FFs (Subsection 6.2.7, p. 183) as state devices. State changes for this table

	A	B	y
1	1,0	2,0	0
2	1,1	1,0	1

Z

Fig. 6.76. Flow matrix for a pulse-mode function.

occur when a B-pulse occurs or when an A-pulse occurs with y = 1. Thus, an expression for the T-input to the T-FF is

$$T = B + Ay.$$

The resulting circuit is shown in Figure 6.77.

It might appear from the flow matrix that we could use a simpler expression for T, namely T = B + y. But this would not work, because no pulse would be generated at T for the case when A = 1, B = 0, and y = 1. It is necessary that every product term include an input variable, since these are the only sources of pulses in the circuit (recall that there is no clock). For any particular physical implementation of this circuit, correct operation should result as long as the input pulses exceed some minimum width, and as long as the spacing between inputs exceeds some minimum interval. The concepts introduced in Section 6.6 apply here.

Next we consider a more complex design example. The example does not just illustrate pulse-mode operation. It also gives some insight into the nature of design in the real world.

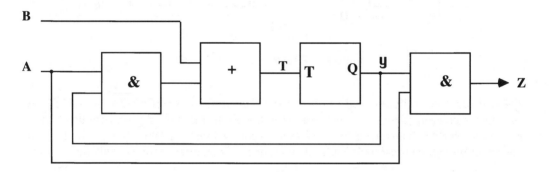

Fig. 6.77. Realization of a simple pulse-mode circuit.

Example 6.33

Suppose we are to design the controller for a simple candy machine vendor. This particular machine accepts nickels, dimes, or quarters and releases a (mini) candy bar priced at 15¢. The machine gives change. It also includes a refund button (r), which returns whatever is due the customer. Thus, if the customer deposits two dimes, the machine ejects a candy bar and a nickel. Assume that, for each input, no more than one candy bar and/or one coin can be ejected. So, if a quarter is deposited after a nickel, the machine returns a candy bar and a dime, and also remembers that the customer has a balance of 5¢. If a dime is then deposited, the machine releases another candy bar.

Assume that only one coin at a time can be deposited, and that there is a mechanism that generates a pulse for each such action. Thus, depositing a nickel results in a pulse on input terminal n, a dime generates a pulse on d, and a quarter generates a pulse on q. Pushing the refund button generates a pulse on r. These signals, n, d, q, and r, serve as the inputs to the control circuit being designed. Assume that the outputs of the circuit are pulses C, D, and N. These signals control the ejection mechanism of the machine. The result of a C output is the release of a candy bar. The D and N signals, respectively, cause the machine to emit a dime or a nickel. The mechanism can eject one coin and one candy bar simultaneously. That is, it cannot accept simultaneous D and N signals, but it can accept a C-signal along with a D or N.

These specifications constitute constraints on what the logic circuit designer can do. We shall see that, as we go along, there will be various options in the design process. How these are utilized will, in general, affect the cost or quality of the resulting product. The first step is to construct a flow table that precisely describes the desired behavior of the controller.

For each input, a column of the table is necessary. In this case, we need four columns. In many cases, the number of rows corresponding to the number of internal states becomes known only after the specification process is complete. But here we can determine in advance how many rows are needed. The controller need only remember how much money the customer is owed by the machine. The possible amounts (call this the balance) are 0, 5¢, and 10¢. Let rows 1, 2, and 3, respectively, represent these amounts. Now we are in position to specify the next-state entries and outputs. For each cell of the table, the first entry will be the next state, and the second entry will be the output, if any. If there are two outputs, there will be three entries. As in many cases of pulse-mode design, it is a good idea to develop the flow table one column at a time. Consider first the n-column, i.e., the column representing the deposit of a nickel.

For row 1, where the balance is 0, the next state is 2, which indicates that there is now a balance of 5¢. There is no output, since the machine does not return anything. Similarly, the next-state entry for row 2 in this column is 3, indicating a new balance of 10¢, and again, there is no output. If a nickel is inserted with the controller in state-3 (10¢ balance), then a candy bar is ejected and the balance is set to 0, so the entry is 1, C. This part of the specification is shown in Figure 6.78a. Before reading on to see how

Balance		n	d	q	r
0	1	2			
5¢	2	3			
10¢	3	1,C			

(a)

Balance		n	d	q	r	$y_1 y_2$
0	1	2	3	1,C,D	1	0 0
5¢	2	3	1,C	2,C,D	1,N	0 1
10¢	3	1,C	2,C	3,C,D	1,D	1 0

(b)

Fig. 6.78. (a) Partial flow table for candy machine controller. (b) Complete flow table (with state assignment) for candy machine controller.

column d is handled, the reader might wish to pause and consider the problem independently.

Clearly the effect of depositing a dime from state-1, where the balance is 0, is to increase the balance to 10¢, so that the entry in row 1 of column d should be 3. If the balance is 5¢, then inserting a dime should cause a candy bar to be ejected, and the new balance becomes 0. So the entry in row 2 of column d should be 1, C. The situation for row 3 of column d is interesting. The customer has inserted a total of 20¢ and is entitled to one candy bar and 5¢ in change, or a balance in the machine of 5¢. Hence, we could specify the entry as 1, C, N (ejecting a candy bar and a nickel, and returning the balance to 0). Or, we could make the entry 2, C (ejecting a candy bar and setting the balance to 5¢). The second option has been chosen since it simplifies the logic for the N-function.

Next, we consider what happens when a quarter is deposited (column q). The row 1 entry could be 1, C, D which would return a candy bar and a dime, retaining the balance at 0. It would also be reasonable to make the entry 3, C, which would mean that a candy bar is ejected and the balance set to 10¢. A third alternative would be to return a candy bar and a nickel, and to set the balance to 5¢, which would be accomplished by the entry 2, C, N. For row 2, the result of depositing a quarter is an immediate balance of 30¢. Since only one candy bar can be ejected, this would leave the machine owing 15¢. But the machine cannot eject two coins at once. Therefore, we could have the machine eject a candy bar and a dime and set the balance at 5¢ (the corresponding entry is 2, C, D). Alternatively, we could specify the ejection of a candy bar and a nickel and set the balance to 10¢ (entry 3, C, N). Starting in state-3 (balance 10¢), there is no alternative to entry 3, C, D, i.e., ejecting a candy bar and a dime and keeping the balance at 10¢. Considering all of the various alternatives, it becomes clear that we have the option of ejecting a candy bar and a dime for each of the three rows of column q. This would certainly simplify the logic for the functions C, D, and N. If this is done, then the next-state entries in column q would be 1, 2, and 3, respectively. That is, all of the states in that column would be stable, which means that neither T-FF changes state in column q. This also simplifies the logic for the T-FF functions.

The column r entries (refund button) are simple. In each row, the balance is refunded and the new balance is set to 0 (row 01). The complete flow table is shown in Figure 6.78*b* (for the moment disregard the state assignment). At this point, we should carefully check it out to see that it represents a valid set of choices. Note that the table is clearly not reducible.

Once we are satisfied with the flow table, the next step is to encode the rows. The complexity of the resulting logic circuit may be significantly affected by this state assignment, but, as mentioned earlier, we will not discuss general methods for dealing with this problem. Let us choose the assignment shown in Figure 6.78*b*.

We can now construct a matrix indicating, for each position in the flow table, the values of T_1, T_2, C, D, and N. For example, in row 2 of column d, the next state is 1, so the y-state must change from 01 (for row 10) to 00 (for row 1). This means that the y_1-FF must remain stable, while the state of the y_2-FF must change. So the values of T_1 and T_2 must be 0 and 1, respectively. This accounts for the 01 in the first two positions of the row 2, column d entry of the matrix shown in Figure 6.79. Since a candy bar is to be ejected in this situation, there is a 1 in the next position for the same cell of the matrix. No coins are released, so there are 0s in the positions for D and N. In column q, all three states are stable, so the values of T_1 and T_2 are both 0 in every row of that column. Since C and D, but not N, signals are to be generated in each row of column q, the last

	n	d	q	r	$y_1 \ y_2$
1	01000	10000	00110	00000	0 0
2	11000	01100	00110	01001	0 1
3	10100	11100	00110	10010	1 0

Fig. 6.79. $T_1 T_2 CDN$ matrix for candy machine controller.

three symbols are 110. Completing the matrix is a straightforward process. Next, we generate the logic expressions for T_1, T_2, C, D, and N.

We are dealing here with five functions of six variables. This may sound intimidating, but most of the input states are don't cares for all of the outputs. A good approach is to treat each input column separately for each function. Thus, we reduce the problem to finding sums of simple functions, each of three variables, y_1 and y_2, and an input (n, d, q, or r). This yields

$$T_1 = (ny_1 + ny_2) + (d\bar{y}_2) + (ry_1)$$

$$T_2 = (n\bar{y}_1) + (dy_1 + dy_2) + (ry_2)$$

$$C = (ny_1) + (dy_1 + dy_2) + (q)$$

$$D = (q) + (ry_1)$$

$$N = ry_2.$$

Let us start simplifying these expressions by factoring common terms. This gives us

$$T_1 = n(y_1 + y_2) + d\bar{y}_2 + ry_1$$

$$T_2 = n\bar{y}_1 + d(y_1 + y_2) + ry_2$$

$$C = ny_1 + d(y_1 + y_2) + q$$

$$D = q + ry_1$$

$$N = ry_2.$$

Now we underscore repeated subexpressions (to indicate repeated subexpressions that can be exploited by sharing logic), producing

$$T_1 = n(y_1 + y_2) + d\bar{y}_2 + ry_1$$

$$T_2 = n\bar{y}_1 + d\underline{(y_1 + y_2)} + ry_2$$

$$C = ny_1 + \underline{d(y_1 + y_2)} + q$$

$$D = q + \underline{ry_1}$$

$$N = \underline{ry_2}.$$

The resulting circuit is shown in Figure 6.80. Note the fanout from gates producing shared subexpressions $(y_1 + y_2)$, $d(y_1 + y_2)$, ry_1, and ry_2.

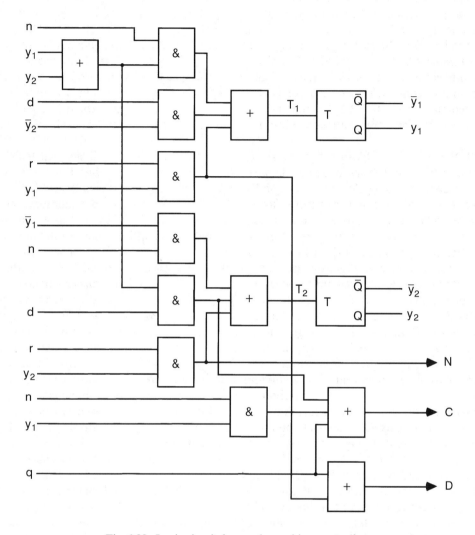

Fig. 6.80. Logic circuit for candy machine controller.

We have now produced a logic circuit that conforms to the specification of the Figure 6.78b flow table. Does this mean that the design process has been successfully concluded? Even if, after careful checking, we verify the correctness of the circuit, the design is complete only in an academic sense. That is, if this were a real-world problem, the designer would not really be finished. The reason is that, even for a relatively simple system, it is very difficult to specify the circuit requirements in such a way as to take into account all of the possible situations that may arise when the system is put into operation. These requirements are never fully known until after extensive use of the system in the field.

One obvious feature that is necessary in every system, but easy to overlook during the design process, is that of system initialization. When the power is turned on, the initial state (sometimes called the *power-up state*) of the system will be determined by factors not generally under our control. But it is usually important that before being put into operation, the system be in some particular starting state. For the candy machine, we want this to be the "0 balance" state, i.e., row 1 of the flow table. Hence, we need some mechanism for bringing the system to row 1 regardless of the power-up state. In this case, the problem appears to be very simple. All that is necessary is to push the r-button. Then, since all next-state entries in the r column are 1s, the system will go to state-1. (A coin will be ejected if the power-up state is 2 or 3, but this is not a cause for concern.)

Now we must consider another trap for the unwary. We used two state variables to realize this 3-row flow table. This means that one of the four possible states of these variables, namely 11, has not been assigned to any row. What happens if the power-up state is 11? An inspection of the logic expressions for T_1 and T_2 (or of the circuit) indicates that fortune is with us in this case. If $y_1 = y_2 = 1$, then r-pulses will be routed both to T_1 and to T_2, sending the system to the desired state. (What, if any, coins are released?) Sad to say, the initialization problem is seldom this simple. It is often necessary to modify designs, usually by adding gates, to take care of this problem.

It would be very natural for the mechanical design of the candy machine to be such that only one coin at a time could be entered. This would be the case if there were a single coin slot, from which nickels, dimes, and quarters would be channeled appropriately. It is therefore reasonable to assume that n, d, and q pulses would never coincide. But the refund button, r, would clearly be independent, and the possibility of pushing it while at the same time inserting a coin cannot be ignored. Thus, depending on the exact timing, the possibility exists that both an r-pulse and an n-, d-, or q-pulse might occur simultaneously. This situation was not considered in the original problem specification.

Looking at the flow table (or the logic expressions, or the circuits) we can see that, with the system initially in state-3, simultaneous n- and r-signals will produce a candy bar and a dime, while resetting the state to row 1. Thus, it is possible to get a candy bar for only a nickel. This is a good example of a system failure due to an unanticipated mode of operation.

There are at least two ways to deal with this particular problem. One is to gate the r-signal through an AND-gate controlled by the complement of the sum of r, n, and q. The other is to have some sort of mechanical interlock that prevents the r-button from being pressed while a coin is being inserted.

Consider now another contingency. What should happen when the machine runs out of candy bars? Or nickels? Or dimes? Clearly, this set of problems necessitates more than just changes in the controller specifications. Various sensing devices are necessary as well as displays to notify customers of the situation. We will not explore this class of problems. Note, however, that choices made earlier with respect to the column-d and column-q entries affect the likelihood of running out of nickels or dimes. Our decisions were based on simplifying the control logic. But, depending on how these machines are used, i.e., the proportions of nickels, dimes, and quarters deposited by customers, we might have made other choices that would maximize the likely number of transactions before the machine runs out of nickels or dimes. Optimization in this respect would require accumulating data on inputs to such machines. Probably, usage differs significantly according to geographical or sociological factors.

Finally, we might note that the decisions about giving change versus building up balances in the machine may affect the behavior of customers. For example, in our design, when two dimes are deposited, the result is a candy bar and a balance of 5¢ left in the machine. This might make it slightly more likely that a customer would deposit a third dime to buy a second candy bar, than would be the case if the machine returned a candy bar and a nickel.

Thus, we can see that, even in the simple case of the candy machine, interesting issues arise that are by no means evident at first sight. It should not be surprising, then, that where more complex systems are to be designed, it is easy to overlook significant aspects. Where system failures may have consequences far more serious than a candy machine occasionally being swindled out of a dime, it is clear that designers must exercise the utmost care and skill.

6.9. Self-Timed Systems

As shown in Section 6.6, the speed with which synchronous circuits can operate is quite sensitive to the range of variations in the stray delays along paths both within the combinational logic block and in the clock distribution network. As integrated-circuit (IC) technology progresses, and logic elements are scaled down further in size, the *relative* size of on-chip wiring delays grows. This causes skew to become relatively larger. For this reason, clock periods are not decreasing as fast as the long-path delays. Designers must resort to increasingly complex schemes in order to cope with the problem. Clock distribution networks, which must be carefully designed to minimize skew, take up significant chip area and consume significant amounts of power.

As a result, more consideration is being given to the possibility of designing systems, or parts of systems, that operate without clocks at speeds determined by their own internal parameters. These are called *self-timed systems*. (Related terms are *speed-independent systems*, *delay-insensitive systems*, and simply, *asynchronous systems*.)

The idea is to have systems that provide feedback to the input source by providing *completion signals*, to indicate when they are ready for new data. Consider, for example, the binary adder (see Example 5.4, p. 139). If k is the time required to add two 1-bit numbers, then the worst-case time for an n-bit ripple carry adder is kn. The best-case time is k, and the average is about $k \log_2 n$. An adder has been designed that reports back when the computation is complete. In a system that can exploit this information, the speedup, on average, is comparable to that obtainable with a full-carry lookahead adder (Subsection 5.3.6).

There is a substantial body of theory applicable to the implementation of both sequential and combinational functions. The key concept of handshaking is briefly sketched below. Next, the C-element, an important device, is introduced. Finally, a detailed design of a self-timed buffer is presented.

6.9.1. Handshaking

It is often necessary for two subcircuits within a larger circuit to communicate with one another. For example, one might have to pass data to the other. Or one might have to signal to the other that it has completed some task. Ideally, we would prefer that such signaling be independent of delays both in the circuits and in the communications channels linking them. One common solution is referred to as four-phase handshaking (level signaling and return to zero signaling are alternative names for this).

Suppose that a subsystem, p1, must tell another subsystem, p2, to initiate some action. Further suppose that p2 must inform p1 when that action has been completed. Refer now to Figure 6.81. Subsystem p1 transmits the request by switching to 1 the signal on a *request* line R that runs from p1 to p2. When p2 has completed the action, it informs p1 by switching to 1 the signal on an *acknowledge* line A, running from p2 and p1. Upon receipt of the acknowledge signal, p1 immediately turns off R, and the response of p2 to this is to turn off A. This completes the exchange, and both signal lines are back to normal, ready for another round. Note that the last two of the four signal changes convey no real information; they simply restore the signal levels to 0. In the figure, a second handshake occurs after the completion of the first.

An alternative approach that eliminates the need for restoring the signal levels is the two-phase handshake (also referred to as transition signaling). Again, two signal lines, one in each direction, connect the two subcircuits. But now attention is focused, not on the *values* of the signals (0 or 1), but on *changes* in signal values. Let us designate the two signals as S (for start) and F (for finish). As shown in Figure 6.82, action is initiated when p1 changes the value of S, and completion is signaled by p2 changing the value of

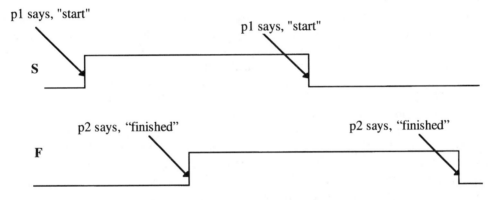

Fig. 6.82. Two-phase handshake.

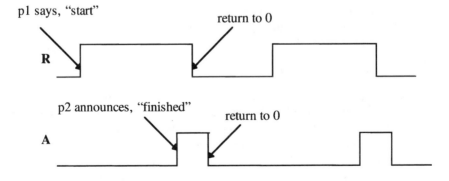

Fig. 6.81. Four-phase handshake.

F. Two consecutive actions are initiated by p1: the first by turning on S, and the second by turning off S.

The two-phase handshake leads to somewhat faster operation by eliminating the return-to-zero steps. However, the amount of time involved is not usually large, and it is often possible to overlap the return-to-zero with other operations. In many cases, the logic associated with the two-phase handshake is more complex than that for the four-phase case. Both kinds of signaling are used. It is even possible to use one type of handshaking in one part of a system, and the other type elsewhere in the same system. This is facilitated by the fact that circuits translating two-phase to four-phase are relatively simple, as are circuits for the reverse translation.

An example of a situation in which handshaking might be used is where a central processing unit (CPU) must transmit a data word to a remote memory unit (MU). The CPU might place the word on a bus accessible to the MU, and then turn on a ready signal R on some line connected to the MU. When the MU sees R go on, it takes the data off the bus, and then turns on an acknowledge signal A, visible to the CPU. The CPU completes its part of the handshake by turning off R. When the MU sees R go off, it turns off A, terminating the process. A similar process could be carried out using a two-phase handshake.

The advantage of using a handshake procedure, is that it is not sensitive to the relative speeds of the CPU and the MU. In fact, if the MU were replaced by a faster unit, no changes would be needed in the interface circuitry. The process would simply take place in less time. Thus, the use of handshaking makes it feasible to design subsystems independently of one another, a major advantage.

There is an assumption about timing implicit in the handshaking scheme just sketched. It is essential that there are no unusually large delays in any of the paths carrying data from the CPU to the MU. When the R signal is received by the MU, there must be no doubt that all bits of the data word on the bus have arrived. In order to ensure that this is true, a worst-case assumption about path delays must be made. The R-signal must not be turned on until enough time has elapsed so that by the time it reaches the MU, the data bit on the longest path will also have arrived. This may be taken care of by building in a sufficiently large delay in the R-signal path.

One way to organize this is to consider the R-signal as another data bit, but with a delay in its path from the CPU to the MU that is guaranteed to exceed the delay in any of the information bit paths. Then, at the same time that the CPU puts the data on the bus, it also sets R to 1. This approach is called *data bundling*. A different approach to the problem of ensuring that the receiving subsystem knows when the data it must process has arrived is introduced in the discussion of the self-timed buffer.

6.9.2. The Concurrency- or C-Element

A device called a *C-element* (where the C stands for *concurrency*) is used in many systems, particularly where parallel processing is involved. In particular, it is used as a component in the self-timed buffer to be discussed shortly.

A flow table description is shown in Figure 6.83. The C-element is a two-state circuit with inputs A and B and output Z. When A = B, Z assumes the same value, otherwise Z remains constant. Thus, if A = B = 0, Z will be 0. Changing only A or only B leaves Z unchanged. Only when *both* A and B have changed to 1, will Z also change to 1. The

AB

	00	01	11	10
1	①,0	①,0	2,0	①,0
2	1,1	②,1	②,1	②,1

Fig. 6.83. Flow table for a C-element.

implementation of the C-element is left to the reader (Problem 6.56). Now let us see what a self-timed buffer is.

6.9.3. A Self-Timed Buffer

Consider a processor P that operates on n-bit words, taking varying amounts of time for each word. Suppose that the input words are available at varying intervals from a source, S (possibly another processor), without regard to whether or not P is ready for them. Then it would be useful to have a *buffer* that (1) accepts words from S as soon as they become available, and (2) presents a new word to P, if possible, as soon as it finishes with the current word. It may be, for example, that the inputs arrive at a relatively uniform rate, while P operates somewhat faster than S on average, handling some inputs rather quickly, while taking much longer for others. By using the buffer, both S and P can be used more efficiently, since there will be fewer occasions when S must wait for P to be free, or when P must await an input from S. The buffer to be discussed here is *speed independent*; it should work properly regardless of the relative or absolute values of the delays associated with its logic elements. However, we shall see later that it is not quite immune to variations in wiring delays.

The form of the buffer is shown in Figure 6.84. It consists of a chain of elements, each capable of holding one n-bit data word. The first element, B_1, receives data in parallel from S, and can pass it along to B_2. Each internal element receives data from the element on its left and eventually passes it to the element on its right. The rightmost element sends its output to P. A 1-bit feedback signal, F_j, is passed back from each element B_j to the element on its left (or in the case of B_1, to S), and the rightmost element receives an F-signal from P. These feedback signals play a key role in controlling the flow of data. They play a role similar to that of the acknowledge signal in the four-phase handshake. When F_j changes from 1 to 0, it indicates that B_j is ready to receive a new data word. A change of F_j from 0 to 1 indicates that B_j has copied the word stored in B_{j-1}, so that B_{j-1} should now be cleared in preparation for the receipt of new data.

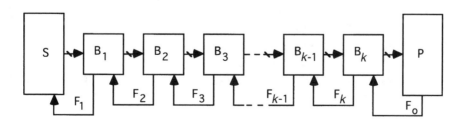

Fig. 6.84. Block diagram of a self-timed buffer.

How are the F-signals generated? How is it determined that a word has been received? Since the system must be able to work regardless of the values of the stray delays, it cannot be assumed that the transmission time for a word is bounded. The determination that a word has been received must be made on some basis other than the mere lapse of time. Thus, we cannot rely on an R-signal as described in connection with the four-phase handshake. No matter how much the launching of the R-signal might be delayed, or how large a delay element is inserted in the path it must follow to the next stage, a sufficiently large delay in one of the data-bit paths would cause the R-signal to be received before all of the received data is correct.

The method used is a variation of double-rail encoding, called *dual-rail* encoding (a special case of what is termed *autosynchronous coding*). It works as follows. Each bit X_i of a word is encoded in terms of two binary signals, X_i^0 and X_i^1. When $X_i = 0$, the encoding is $X_i^0 = 1$ and $X_i^1 = 0$. For $X_i = 1$, the encoding is $X_i^0 = 0$ and $X_i^1 = 1$. Under no circumstances is $X_i^0 = X_i^1 = 1$ permitted. But $X_i^0 = X_i^1 = 0$ *is* permitted; it is interpreted as X_i being *undefined*. If, for *every* i, X_i is undefined, then we have what is called a *spacer*, the significance of which will soon be made clear.

Every buffer element has a register consisting of n pairs of reset-dominated SR-FFs (see Subsection 6.2.4), each pair corresponding to one bit of a data word. The pair of FFs for X_i stores X_i^0 and X_i^1. If every FF of a register is in the 0- or reset state, the register is said to contain a *spacer*. If every pair of FFs corresponds to a defined bit, the register is said to contain a *data* word. The buffer operates in such a manner that each register alternates between holding a spacer and a data word. In the course of changing from a spacer to a data word, one FF of each pair is turned on. At any intermediate stage of this transition, the contents of the register is neither a spacer nor a data word; some bits are defined and some are undefined. At no point does the register contain any valid data word other than the one it is changing to. The same is true of transitions from data to spacer, where one FF of each pair is turned *off*. Now the method for detecting when a register has received data can be explained.

Assume that buffer element B_i is initially clear, i.e., that its register holds a spacer. Then, if B_{i-1} sends it a data word, reception of that word is complete when the B_i-register contains a data word. This can be detected by checking to see if one FF of each bit pair is set. Operation of the overall buffer can now be described in more detail.

Assume that the system starts with all buffers empty (i.e., holding spacers), with no data available from S and with P idle. (Some means must always be provided for initializing the system when it is first turned on by clearing all FFs, etc.) When a buffer element, B_i, is in an idle state, it is ready to accept data, and it indicates this to B_{i-1} by setting the F_i-signal to 0. This is the situation initially for B_1, and so when S starts putting the first data word, D_1, in its output register, D_1 flows immediately into the B_1-register. In fact, since *all* the buffer elements are in this idle state, D_1 passes right through from one buffer register to the next, and on into P's input register as fast as the wiring delays and FF response times allow. Due to variations in these delays, even if all the bits of D_1 are simultaneously placed in the S output register, different bits of the word may reach various registers at different times. When B_1 has received the complete word, it turns F_1 on to command S to clear its output register, since B_1 no longer needs D_1. In general, each buffer element does the same with respect to the element on its left. Thus, when B_2 has received D_1, it turns on F_2, thereby commanding B_1 to clear its register to a spacer. When B_i *and* B_{i+1} are clear, B_i can receive a new data word from B_{i-1}.

Data words supplied by S flow through the buffer as far as they can go before either reaching P or a buffer element that is already holding the previous data word. Actually, a new data word cannot get quite that far, since an element with a spacer must separate each new data word from the element holding the previous word. As words are removed from the buffer by P, the remaining contents can move forward. The design of the B_i, discussed next, is governed by two requirements. One is that consecutive words must not overlap. The second is that no word be transmitted more than once, even if the relative delays in the components have arbitrary values. Refer to Figure 6.85, a logic circuit diagram of stage i of the buffer. (In order to keep the size of the figure small, only a 2-bit register is shown, but the extension to larger word sizes is obvious.)

The coded X-inputs from B_{i-1} are fed directly to the set inputs of the SR-FFs. An OR-gate assigned to each bit of the word determines if the corresponding bit of the register is defined. The AND-gate output goes to 1 if there is a data word in the register. This will be the case if each OR-gate feeding the AND-gate reports that one of the pair of FFs it is monitoring is set. The OR-gate that serves as the other input to C-element 1

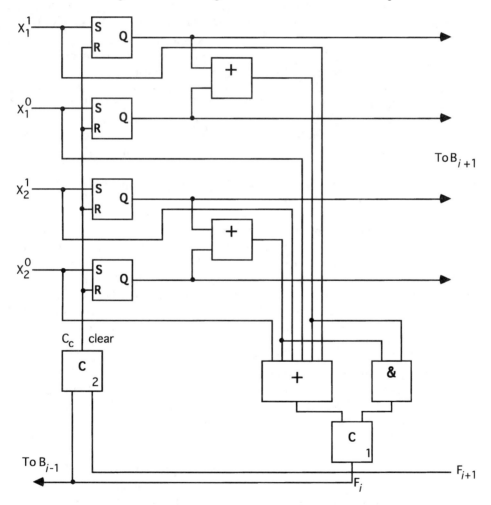

Fig. 6.85. Logic for buffer stage B_i.

(referred to henceforth as C_1) has the same inputs as the AND-gate as well as all the X-inputs to B_i. Hence, when the AND-gate goes on, the OR-gate will also go on. After both of these gates are on, the C-element goes to its 1 state, thereby turning on F_i. (The role of the OR-gate may seem unclear at this point, as it adds no information pertinent to the turning *on* of F_i. Its function is to control the turning *off* of F_i.)

Once B_i has received a data word, it notifies B_{i-1} by turning on F_i. It then remains quiescent until the *next* stage, B_{i+1}, informs it, by turning on F_{i+1}, that it now has a copy of that word in its register. This is the signal for B_i to clear its register in preparation for the next data word. The decision to clear is made by the second C-element, C_2, upon receiving 1-signals from both C_1 and F_{i+1}. When its output, C_c, goes to 1, all the FFs of B_i receive reset signals. How long should the reset signals be on? Having this depend on the delay parameters of the FFs would be against the rules for speed-independent circuits. Therefore, the reset signals must remain on until there is a positive indication that all the FFs are indeed in the reset state. Such an indication would be when F_i returns to 0, which could only happen after *both* of the C_1 inputs are at 0. But the clearing of all the FFs is not sufficient to turn off C_1, since the OR-gate also has inputs from the Xs. It is necessary that the B_{i-1} FFs *also* be clear before F_i can go off. Note that as long as C_c remains on, the FFs cannot be set again, because they are *reset dominated*. Thus, by keeping C_c on until the previous stage has been cleared of the old data word, we avoid having pieces of that word put back into the B_i-FFs. Even turning off F_i is not enough to extinguish C_c, since *both* inputs to C_2 must go to 0 before this can occur. Suppose that new data were allowed into B_i while $F_{i+1} = 1$. Then, since data in the B_i-FFs is connected to the set inputs of the B_{i+1}-FFs, the data word currently in the B_{i+1}-FFs could be corrupted by bits from the next word coming into B_i. Therefore, before allowing new data into B_i, it is necessary that F_{i+1} also return to 0, indicating that B_{i+1} is ready for new data.

The circuit described here is indeed speed independent. It is totally insensitive to the delays associated with the outputs of all the logic elements (except of course that its *speed* is affected). But, along with all other speed-independent circuits, it may be vulnerable to wiring delays following fanout points, which cannot be lumped in with gate output delays. For example, consider the X_1^1-input to the Figure 6.85 circuit for B_i. It fans out to the set input of the top FF and to the OR-gate feeding C_1. Suppose there is a very large delay in the branch leading to the FF, and a negligible delay in the other branch. Assume (1) B_i receives a data word; (2) F_i goes on to tell B_{i-1} to clear its FFs; (3) B_{i+1} receives the word and turns on F_{i+1}; (4) as a result of the previous two items, C_c goes on to clear the B_i-FFs; (5) the B_{i-1}-FFs are all reset, including the FF feeding X_1^1, which *was* on; (6) all the inputs to the C_1 OR-gate (as well as the AND-gate) go to 0, so F_i is turned off; (7) the B_{i+1}-FFs are all reset, so that F_{i+1} also goes off; (8) both inputs to C_2 are 0, so it turns off C_c, allowing the B_i-FFs to accept new data. Now the time bomb represented by the big delay in the FF-input wire goes off. Due to this delay, the set input to the top FF is *still* active! With the reset signal removed, this signal can now set the top FF again. Given Murphy's Law, we may be confident that if this sequence of events occurs, X_1 will be 0 in the next data word, so that this will indeed constitute a malfunction.

Situations of this type must be noted and appropriate precautions taken. For example, by ensuring that the kinds of unusual delays that would cause the trouble are not present, or are compensated for by other delays. Buffers of the type treated here are sometimes called *first-in, first-out* (*FIFO*) buffers or *fall-through* shift registers. They are sometimes designed to be only partly speed independent (less so than the one just discussed), in the sense that they will work despite reasonable, but not extreme, ranges of delay values.

Speed-independent buffer elements such as the one described here can be used to build speed-independent processors of a very general nature. To start with, combinational logic blocks can be inserted between stages, with the F-signals bypassing the logic, which would have to be fully dual-railed, generating both polarities of every function. The effect of these blocks on the buffers would be the same as wiring delays. A more elaborate arrangement would entail having networks of buffers, with logic circuits in between. It would then be necessary to make provisions for buffers to receive F-signals from several buffers, and to send their own F-signals back to several buffers. Finally, it is possible to allow feedback in these networks, so as to be able to handle sequential functions. Readers may enjoy thinking about solutions to the problems involved.

6.10. FAULT DETECTION IN SEQUENTIAL LOGIC CIRCUITS

Testing a digital system, such as a microprocessor, is equivalent to testing a sequential logic circuit. The nature of the faults to be tested for and the fault model used are as described in Section 3.9 with respect to the testing of combinational logic. In the combinational logic case the problem can always be solved in a straightforward way by simply applying all possible input states. The challenge is to find more efficient solutions. There is no counterpart to this straightforward solution for sequential logic. What would constitute applying "all possible input states" to a sequential logic circuit? We would have to be able to drive the circuit into each of its possible internal states, and, for each of these states, apply each possible input. Then, following each application of an input, we would have to verify that the resulting output and internal state is correct. How, or even *if* this can be done is by no means obvious. There are indeed general solutions to this problem in the form of *checking sequences* for sequential machines. But for even relatively simple machines, the computations involved in finding checking sequences are so complex, and the lengths of the sequences are so long, that this approach is impractical in most situations. Thus, the discussion here is centered on two other general techniques. The first is *scanning*, a technique for converting the problem to that of testing combinational logic. The second approach (compatible with the first) is to incorporate testing circuitry on the same chip with the logic being tested, and to utilize pseudorandom sets of tests.

6.10.1. The Scan Method

One way to ease the problem of testing sequential logic circuits is to incorporate a switch in each feedback path. These switches can be kept closed during normal operation and can be opened during testing. With the feedback paths broken, the circuit is combinational and, if alternate paths are made available for feeding test signals into the points normally served by the opened switches, procedures for testing combinational logic can then be applied. Apart from the added cost of the special circuitry, the addition of circuit elements in the feedback paths entails added delay that would degrade normal operation. A better technique, described below, is to provide means for directly controlling and directly observing the internal state.

Assume two-phase clocking with latches (Subsection 6.6.3). The idea is that in test mode the latch pairs are all linked to form a single shift register. A known y-state is shifted in from a single input point and the external inputs are set up. Then, the C_1- and C_2-clocks are allowed to operate for *one* cycle so that the combinational logic generates the next y-state, which appears in the L_1-latches. The system is then put back in test mode, with the latches again organized as a shift register, and the y-state is shifted out for comparison with the expected value.

This scheme can be implemented by proving the L_1-latch with an added pair of inputs, as shown in Figure 6.86a. It is assumed that C_A and C_B are never both on simultaneously. When C_A is on, Q copies the value of D_A, and when C_B is on, Q is set equal to D_B. If *both* clock inputs are off, the value of Q is frozen. The L_2-latches are ordinary latches, receiving as inputs C_2 and the outputs of the L_1-latches (see Figure 6.86b, which shows a system with three latch pairs). The A-inputs are the usual operating inputs to the L_1-latches, receiving the Y- and C_1-signals. A new clock signal, C_T, dedicated to the test mode, is fed to the C_B-input, and the D_B-inputs are y's; the first latch pair in the chain receives the scan input X_S instead. The last latch pair in the chain feeds its y-output to the scan output, Z_S. In Figure 6.86b, the lowest L_1-latch receives X_S at its D_B-terminal, and y_1, the output of the corresponding L_2-latch, goes to the scan input of the next highest latch. The y_2-signal goes to the upper L_1-latch, and y_3 goes to Z_S. Thus, the three latch pairs behave as a shift register with respect to C_T-signals. Let us see how this system works.

In normal mode operation, C_T is fixed at 0, so that the scan connections have no effect. Suppose that the circuit is to be tested with input $X_1 = 1, y_1y_2y_3 = 011$. Then the state 011 must be shifted into the latches. This can be done as follows: Turn off C_1 and C_2 and set $X_S = 1$. Then, put a pulse on C_T to read a 1 into the L_1-latch for y_1. Move

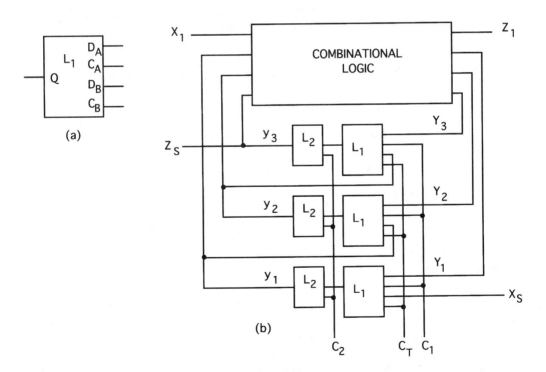

Fig. 6.86. Latch and block diagrams for scan method.

this signal to the L_2-latch by pulsing C_2. Leaving X_S at 1, repeat this process, which will move the first 1-signal to the L_2-latch of y_2 while leaving y_1 at 1. Now set $X_S = 0$ and again apply pulses to C_T and then to C_2. This completes the process of setting the y-state to 011. We can now leave C_T at 0 and pulse first C_1 and then C_2, allowing the circuit to generate the output Z_1 and the next state $y_1y_2y_3$. The output Z_1 can be inspected directly (we assume the outputs are observable). In order to inspect the y-state, we return to the test mode, and with alternate C_T- and C_2-pulses, read out the y's to Z_S, also assumed to be observable. In this manner any specified tests can be applied to the combinational logic. It is a simple matter to test the latches (at least for stuck faults) by simply shifting in strings of 0s and 1s from X_S and observing the results at Z_S.

In addition to converting the test problem for sequential circuits to a test problem for combinational circuits, another important test problem is addressed by the scan method. In order for an external unit to be able to test a logic circuit on a chip, data specifying the tests to be performed must be transmitted to the chip, and the resulting outputs must be passed back from the chip to the tester. But the number of terminals (or *pins*) on a chip is generally small compared to the number of gates. Especially for VLSI chips, this is a limiting factor for many applications. Any test method requiring the use of many pins would thereby be reducing the usefulness of the chip involved by devouring a scarce commodity. The scan method is very undemanding in this respect. It requires only three pins: for X_S, C_T, and Z_S.

A side benefit of the scan method is that it solves the power-up problem mentioned on p. 218 and on p. 245. Initial states of all latches can be scanned in via the shift register chain.

There are many variations possible within the world of the scan method. A principal variation is to use multiplexers and demultiplexers to control and observe the state devices rather than to link them up to form a shift register. The scheme outlined here is the most widely used. Under the name *level-sensitive scan design* (LSSD), it is incorporated in all mainframe IBM computers. (The phrase *level sensitive* refers to the fact that the signals are levels rather than pulses, so that the system is less sensitive to delay variations and the clock may be slowed down if desired without causing trouble.) There are many interesting and important issues concerning the best ways to exploit the scan method, but these are not treated here. An approach to testing large systems without the need for expensive apparatus is discussed next.

6.10.2. Built-in Test

Consider what is involved in testing a complex system implemented on an IC chip, assuming that external testing apparatus is available. A set of tests must be devised and stored in the tester, along with the correct results for each test. Then, one by one, the tests must be communicated to the circuit via pins on the chip, and the responses returned to the tester for comparison with the correct answers. The scan method described earlier deals effectively with the communication problem, and with the general problem of dealing with sequential logic. But serious difficulties remain.

It is highly desirable that the test equipment be capable of operating at the same rate as the circuit being tested. This makes the testing more likely to uncover timing-related problems. But what if the circuit being tested is at the leading edge with respect to speed? Then the test equipment must be upgraded to meet the challenge. The continuing need to upgrade testers adds greatly to their cost, and causes production delays while this is taking place. Furthermore, while it is possible to generate adequate test sets for any system, the process of doing so *and* the process of applying the tests have become excessively time-consuming and costly for systems of the sizes being built today. And future systems are not getting any simpler.

Suppose we try to solve the problem by exploiting the very factor that is exacerbating it: the ever increasing number of logic elements that can be embedded on a chip. That is, consider the possibility of including the tester on the chip containing the circuit to be tested. This would essentially eliminate the problem of communications with off-chip apparatus. The apparent difficulty is that the test circuitry, particularly the memory needed to store the tests and results, might occupy more space than the circuit being tested! Here is where several new ideas come into play to eliminate the need for storage.

One idea is to use exhaustive testing. If there are n inputs to a circuit, set up an n-bit binary counter and run it through all 2^n states, using each as a test. This indeed might be feasible with respect to modest values of n, *or* if it is possible, to partition the inputs into disjoint sets corresponding to the outputs each affects. Research is being done on this approach, but we shall follow a more traveled road. Recall that the commonly used fault models only approximate reality, so that, even if a set of tests is complete, there is a small, but nonzero, probability of a fault escaping detection. In that case, why not accept the fact that no test procedure can be guaranteed to find every possible fault, and use a probabilistic approach? That is, where the number of input states is very large, let us choose at random a subset of the inputs. If this subset is itself very large, say hundreds of thousands or a few million, the likelihood of catching any particular fault is very high. (This point has been established by such means as computer simulations.) The advantage of this approach is that there are simple circuits for generating pseudorandom sequences of 1s and 0s as fast as they can be utilized. This effectively eliminates the need for storing test sets. But what about checking the responses to the tests?

Once in the realm of probabilistic testing, we might as well go all the way and use a similar philosophy for response checking. Suppose we proceed as follows to test a system with n inputs and k outputs: (1) generate an n-bit pseudorandom sequence that is then used as a test input; (2) store the response R_1 as a k-bit vector R; (3) generate and use another n-bit sequence; (4) call the k-bit response R_2. Set $R = f(R, R_2)$, where f is a deterministic function to be described shortly. Repeat steps 3 and 4 until a predetermined number of tests have been applied. The result of the tests is the final value of the k-bit vector R. Suppose that this process is carried out several times on a circuit known to be free of all faults using precisely the same pseudorandom sequence. Then, since the entire process is deterministic, the final value of R should be exactly the same every time. The value of R for a fault-free circuit is called R_S, the *signature* of the test. If there is a fault in the circuit, *and* if that fault is detected by a member of the test set, then R *might* differ from R_S. Here is where the nature of the function f becomes an issue.

An ideal f would have the following property: If a single bit of any response vector at any point in the test process is incorrect, then, for every bit r_i of the result vector R, there is a probability of 0.5 that r_i differs from the corresponding bit of R_S. This implies that every bit of $f(R_1, R_2)$ is sensitive to every bit of both arguments. If f has this property, then the probability that an error detected by a test will fail to affect R (this is called *aliasing*) is 2^{-k}. This is less than one chance in a million for $k = 20$. In addition to being sensitive, a good f function should be realizable by a fast-acting simple circuit. Fortunately, there exists a function satisfying all of these requirements to a remarkable degree. It is based on principles that also apply to the problem of generating pseudoran-dom test sequences. The generation problem is treated first.

The basic device involved here is the *linear feedback shift register* (LFSR); a simple example is shown in Figure 6.87. An LFSR is a shift register receiving, at the left end, the output of an XOR-gate whose inputs come from a subset of the stages of the shift register. (The X-input is used only for initializing the register contents. It is normally held fixed at 0.) *If* the connections to the XOR-gate are properly chosen, then, starting in any state other than the all-zero state, an n-stage LFSR will cycle through all 2^n-1 other states. The sequence of XOR-gate outputs is periodic, with period 2^n-1. If any subsequence of a period were analyzed, it would have the principal characteristics of a random sequence of 1s and 0s. For the Figure 6.87 circuit if $y_1 y_2 y_3$ is initially 001, then the sequence of states is 100, 110, 111, 011, 101, 010, 001, 100, and so on.

Such circuits can be built with any reasonable number of stages. For a properly designed 30-bit LFSR, the period is well over a billion. Since any subsequence of length less than the period is pseudorandom, we have here an excellent source of pseudoran-dom test inputs. Optimal LFSRs with no more than four input connections to the XOR-gates are known for all lengths less than 64. Thus, the circuits are not very complicated. Speed is also no problem, since a shift can take place with each clock-pulse. Next it is shown that, with only a minor elaboration, the same device can serve to generate the test signatures.

Suppose that an n-stage LFSR is in initial state S_0 and that m shifts are executed, where m is much greater than n. Let the state reached be designated as S_m. Suppose next that S_0 were altered in exactly one bit position and that once again m shifts were executed. If the altered bit and the value of m were randomly chosen then, on average, each final state would differ from S_m in $n/2$ positions. Roughly speaking, the result of changing any one bit of S_0 is to make each position of S_m equally likely to be a 0 or a 1. Thus, the probability of a change in one or more bit positions of S_0 *not* altering S_m is about 2^{-n}. It is this characteristic of LFSRs that gives them the sensitivity property necessary for a good signature generator.

Perhaps the simplest way to use a LFSR to implement the f function discussed earlier, is as follows. Let the k-bit result vector of each test be serially fed into the X-input of a k-stage LFSR structured as in Figure 6.87, each input concurrent with a C-pulse that executes a shift. Then R is taken as the state of the LFSR at the end of the process. If failures are detected at any outputs for one or more tests, then the likelihood of aliasing, i.e., that $R = R_S$ despite a detected failure, is very small, of the order of 2^{-k}. Rather than have the test results fed in serially, the LFSR can be enhanced as shown in Figure

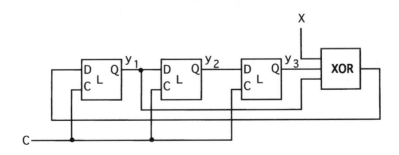

Fig. 6.87. A 3-bit linear feedback shift register.

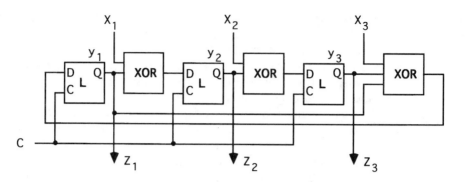

Fig. 6.88. A LFSR with parallel inputs.

6.88. The added XOR-gates make it possible to input an X-vector (representing test results) in parallel. The R-output is the vector composed of the Z-signals. Note that the same circuit can be used as the test generator. After initializing the LFSR with some fixed X-input, we could allow k shifts and then use the Z-vector as a test input, shift k more times, use the Z-vector again, and so on.

A built-in test facility on a chip could consist of a LFSR to generate test vectors, and another LFSR to process the results. Control circuitry, triggered by signals from off the chip, could direct the operation, counting the number of tests applied and checking the result vector against the test signature to determine if the circuitry is faulty. The test signature would have been determined in advance either by executing the tests on circuitry known to be fault free, or by means of a simulation program (Section 7.2). Scanning might be used to facilitate the application of the tests to complex sequential logic. A nice feature of this built-in-test approach is that it can be used not only for the initial screening of chips just manufactured, but also to aid in troubleshooting equipment already in the field.

The preceding discussion has been somewhat sketchy, as a more complete treatment would entail a presentation of the theory of linear sequential circuits, a very interesting topic, but one beyond the scope of this work. (Those familiar with conventional linear circuits may be interested to learn that superposition applies to LFSRs, where ordinary addition is replaced by modulo-2 addition.) Just to get a feel for the kind of things that can happen, readers might explore the ways in which aliasing can occur. For example, using the Figure 6.88 circuit, suppose a failure is detected via X_3. If, on the next test, failures are detected on *both* X_1 and X_3, the effects would be canceled, leaving the result vector still matching the test signature (assuming no other failures are detected). For LFSRs with say 20 or more stages, such aliasing problems seem not to be a serious factor in the light of the other weaknesses of our fault-detection processes (e.g., the fault model and the multiple-fault problem). There are many other problems and techniques that must be left for more specialized treatises.

OVERVIEW AND SUMMARY

Sequential circuits, or finite-state machines, fall into several categories, depending on what restrictions are placed on the inputs. If changes may occur (or are recognized) only at discrete intervals of time, we have a synchronous or clocked circuit (SSC). Where the only restrictions on inputs concern minimal spacing between consecutive changes, the circuit is asynchronous (an ASC). In the case of a speed independent circuit, input changes are made in response to feedback signals from the circuit that indicate when it is ready for new data.

Informal ideas as to what a sequential machine is supposed to do are made precise by constructing flow tables or state diagrams. Flow tables as originally constructed may have more states than is necessary. Formal procedures exist for finding minimal-row equivalents of any flow table. The next step after reducing the flow table is to assign y-states to the rows. For SSCs, the only issue is the complexity of the resulting logic circuits. For ASCs, the prime requirement is to avoid critical races: situations in which several y-variables start changing simultaneously and the final state depends on which variables finish first. There are a number of procedures for solving this problem, with trade-offs involving factors such as the number of y-variables, transition time, and designer effort.

A 1-*hazard* exists in a combinational logic circuit if there are both complementing and noncomplementing paths from a common point to an OR-gate, such that all other inputs to that gate are fixed at 0 at a time when the signal at the common point is changing. In terms of logic expressions, this corresponds to the existence of an input condition where all variables except X are fixed at such values that $Z = X + \overline{X}$. (The dual situation results in a 0-*hazard*.) The result is the possible appearance at the output of spurious pulses, or "glitches," depending on the relative values of the stray delays in the circuit. Such hazards can be eliminated for single-input-change (SIC) operation by appropriate logic design. Or they can be coped with by the judicious use of delay elements, either to ensure that the path delays are not such as to produce spurious pulses, or to filter out such pulses if they do occur. Filtering may be the only solution in the case of multiple-input change hazards. A related problem in ASCs is the *essential* hazard, which also involves races between complementary versions of a signal, but with paths including state branches. They are inherent in the function itself and constitute the principal reason for placing delay elements in feedback paths. When delay elements are used to combat the effects of essential hazards, we can compute the required minimum values, given long- and short-path delays. Extending this reasoning enables us to calculate how close together consecutive input changes may be spaced. *Maximizing* short-path delays helps increase allowable rates of input changes. Important examples of ASCs are storage elements, such as SR-FFs, latches, and edge-triggered D-FFs (ETDFFs). Many of these are used as components in synchronous systems.

A great deal of the logic design for digital systems is based on the use of standard "macros," some of which are combinational logic circuits such as binary adders, decoders, and MUXs. Others are sequential circuits, such as registers with facilities for counting, shifting, or complementing. Simple logic circuits are often designed informally, perhaps with some use of Boolean algebra. At the other extreme, it is sometimes necessary to implement rather complex sequential functions, for example, to direct the operations of a memory cache. In these cases, it may be useful to begin with specifications written in a high-level register transfer language, translate them into an ASM chart (a form of flow chart), and then generate either a flow table or a state map that can be used to generate the logic. Synchronous sequential circuits are usually implemented with latches or edge-triggered FFs as storage elements. JK-FFs are often used in small-scale integrated circuits. Duality applies to sequential as well as to combinational logic circuits.

The proper and efficient operation of synchronous systems is dependent on the selection of a clocking scheme and associated parameters such as period and pulse width. The simplest system is one-phase clocking with ETDFFs. Using latches with one- or two-phase clocks is more conducive to high-speed operation. In analyzing clocking schemes it is useful to think of the system as though it were one large sequential circuit. Sets of constraints in the form of linear inequalities are developed on the basis of worst-case analyses. These are then satisfied in such a way as to minimize the period while keeping the lower bound on the short-path delay at an achievable value. Wherever a circuit must process multiple input changes from sources that are not governed by a common clock, designers must be wary of additional failure modes. One is the generation of *runt* pulses, signals somewhere between 1s and 0s. The other is entry into a *metastable* state, whereby a system may, for an indefinite time, wobble between two states. The best that can be done is to reduce the probability of such failures to acceptable levels.

Pulse-mode circuits operate without clocks, but under the assumption that each input variable state can be implemented as a pulse signal. Toggle FFs are natural basic elements for such systems.

In order to avoid the growing problems associated with distributing clock signals on VLSI chips, self-timed systems are being developed that operate without clocks and furnish completion signals to regulate the flow of data to their inputs. An important technique for interfacing subsystems is two- or four-phase handshaking. Speed independent buffers are an important component of such systems. The potential advantages of self-timed systems are better support for modular design, reduced power consumption, and higher average computational speed. The drawbacks are added costs in logic circuitry (more chip area) and increased worst-case delays.

Fault detection in sequential circuits is a fundamentally difficult problem. By using specially designed latches with two sets of inputs, it is possible to operate a system in test mode with all of the latches arranged to form one big shift register. Data can then be shifted (*scanned*) in and out to set the initial state of the system and then to check the resulting state after one clock cycle has elapsed. This reduces the problem to that of testing combinational logic. By using linear feedback shift registers (LFSRs), random sets of tests can be generated quickly with simple circuitry built into the same chip as the circuit being tested. Another LFSR can then be used to compress the results of all of the tests, for comparison with a test *signature*, which is a vector obtained by executing the same process on a fault-free copy of the circuit.

SOURCES

The foundations of sequential logic-circuit theory, including treatments of the reduction of fully specified machines, were laid in [Huffman, 1954], [Mealy, 1955], and [Moore, 1956]. (The last two treat only SSCs.) The flow table reduction problem and general material on SSCs can be found in various texts on logic circuits and digital design such as [Fletcher, 1980], [Friedman, 1986], [Hayes, 1993], [Hill, 1981], [Kohavi, 1978], [Katz, 1994], [Mano, 1991], [McCluskey, 1986], [Wakerly, 1994], and [Winkel, 1980]. Incompletely specified flow tables are dealt with in [Paull, 1959] and in [Unger, 1969]. A thorough treatment of ASCs can be found in [Unger, 1969]. This source includes material on dynamic hazards and on MIC combinational hazards. For more recent results see [Unger, 1995]. See also [Friedman, 1975]. Jingshuang Yang contributed to the material on inertial delays connected in cascade (see problem 6.15 and its solution.) The original sources for the Liu and Tracey USTT state assignment methods are in [Liu, 1963] and in [Tracey, 1966]. The 1-hot code for ASCs was first described in [Huffman, 1954]. The ETDFF and the particular circuit shown in Figure 6.31, which is widely used, for example, in SSI chips, is due to Maley [Maley, 1971]. ASM diagrams were introduced in [Clare, 1973]. The material on clocking systems came from [Unger, 1986]. Other material on this topic and also on the synchronization problem can be found in [Mead, 1980]. Basic work on the synchronization problem is in [Chaney, 1973], [Hurtado, 1975], [Marino, 1977], [Kacprzak, 1988], and [Reyneri, 1990]. The concept of the speed-independent circuit was first developed by Muller, e.g., see [Muller, 1967]. The idea of the autosynchronous code originated in [Sims, 1958], and the buffer circuit described here works on principles derived from [Armstrong, 1969]. Other material on this subject can be found in [Friedman, 1975] and in [Unger, 1969]. More recent work on self-timed systems can be found in [Seitz, 1980], [Rosenberger, 1988], [Martin, 1989], [Sutherland, 1989], and [Unger, 1993]. A good survey is in [Birtwistle, 1995]. Fault detection is the subject of numerous published papers and the proceedings of many conferences, such as the annual IEEE Test Conference. General treatments can be found in [McCluskey, 1986], in the somewhat dated [Breuer, 1976], in [Fujiwara, 1985], in [Miczo, 1988], and in [Abramovici, 1994]. Basic papers on the scan method are [Williams, 1973], and [Eichelberger, 1977]. An early description of signature testing as used by the Hewlett-Packard Corporation is in [Chan, 1977]. Linear sequential circuits were introduced in [Huffman, 1956]. Further work, including a table of simple LFSRs is in [Gill, 1966], and there is an introduction to the subject in [Kohavi, 1978].

PROBLEMS

6.1. Considering the flow table of Figure 5.16 (p. 143) as describing a sequential function, specify the sequence of outputs and the sequence of states resulting when, starting in state-1, the following input sequence is applied: X : 0 0 1 1 1 0 1 0 0.

6.2. Repeat Problem 6.1, using the table of Figure 5.26, p. 153, with the input sequence:

$$AB: 00, 11, 11, 00, 01, 11, 10, 00, 01.$$

6.3. Construct a flow table with a minimal number of rows, with output Z and inputs A, B, and C, such that Z = 1 whenever a C appears immediately preceded by an odd-length string of Bs, which, in turn, is immediately preceded by an A, as illustrated below:

A A B B A C B A A B B C C B A B B B C C B A B C C B C B B B C B
0 0 0 0 0 0 0 0 0 0 0 0 0 0 0 0 0 0 1 0 0 0 0 1 0 0 0 0 0 0 0 0.

*6.4. Construct a minimal-row flow table describing a serial subtracter. Inputs are two streams of bits A and B representing two binary numbers, least significant bits coming first. The output, S, is to be the binary representation of the number A-B.

6.5. Construct state diagrams corresponding to the flow tables of Problems 6.7 and 6.8 below.

6.6. For a 10-state system $(0, 1, 2, 3, 4, 5, 6, 7, 8, 9)$ the following is a *partial* list of equivalent pairs: $\{14, 29, 38, 15, 20, 70\}$. On the basis of this information, specify the set of equivalence classes that this implies. If additional pairs are found, what *might* happen to these classes?

6.7. Find a minimal-row equivalent of the flow table of Figure 6.89.

$$X_1X_2$$

	00	01	11
1	3,0	4,0	2,0
2	1,0	3,0	4,1
3	1,0	2,0	2,0
4	1,0	1,0	2,1

Fig. 6.89.

*6.8. Find a minimal-row equivalent of the flow table of Figure 6.90.

	A	B	C	D
1	2,0	3,0	5,1	1,1
2	2,0	6,0	4,1	1,0
3	2,0	5,0	6,1	6,0
4	2,0	3,0	5,1	1,0
5	4,0	6,0	2,1	1,0
6	5,0	5,0	3,1	6,0

Fig. 6.90.

6.9. Find a minimal-row equivalent of the flow table of Figure 6.91.

	A	B	C	D
1	8,0	7,0	4,0	2,0
2	6,1	3,0	2,0	1,0
3	5,1	5,0	3,0	1,0
4	8,0	7,0	7,0	3,0
5	3,1	6,0	5,0	1,0
6	2,1	2,0	6,0	1,0
7	8,0	8,0	4,0	7,0
8	2,0	7,0	4,0	6,0

Fig. 6.91.

*6.10. Implement the flow table of Figure 5.13a (p. 140) as a clocked sequential circuit with ideal delay elements. Derive an economical realization using two-stage logic with NOR-gates.

6.11. Implement the flow table of Figure 5.16, p. 143, as a clocked sequential circuit with ideal delay elements. Derive economical SOP expressions for the logic.

6.12. Consider the following SOC asynchronous sequential function with inputs X_1 and X_2, and outputs Z_1 and Z_2. Operation is restricted to SIC. Following each input change, the output Z_1Z_2 assumes the value of the previous input state X_1X_2. How many stable states are necessary to realize this function? Construct a primitive flow table describing the function.

6.13. A *toggle-*, or T-FF, was discussed in Subsection 6.2.7. Specify a minimal-row primitive flow table for a more elaborate TR-FF that has inputs T and R. The FF is to toggle (i.e., change state) on the trailing edge of each T-pulse. On the trailing edge of each R-pulse, the FF goes to its reset state, where its Q-output is 0. Assume SIC operation, and allow for the possibility of R and T being simultaneously on.

6.14. Extend the design of the T-FF of Figure 6.28, p. 184, to add a reset input R, such that, regardless of the value of input T, whenever R is turned on, the FF is reset to the state with Q = 0. This reset action should occur immediately upon R going on, *not* on the trailing edge of R as in the preceding problem.

*6.15. In Figure 6.92 there are three waveforms (*a*, *b*, and *c*) and three circuits (1, 2, and 3) composed of inertial delay elements, with signals assumed to flow from left to right. Delay magnitudes are shown inside the delay symbols. Sketch the output waveforms for each of the given input waveforms when applied to each of the three circuits. Label your results $a1, a2, \ldots, c3$. Does it ever make any difference if the order of the delays in the circuit is reversed? If the input waveform is reversed, in the sense of reversing the flow of time, does this always result in the output waveform being reversed? Suppose that in circuits (2) and (3) the 2-unit inertial delay is replaced by a pure delay. What, if any, difference does this make? Can you make a general statement about such a transformation?

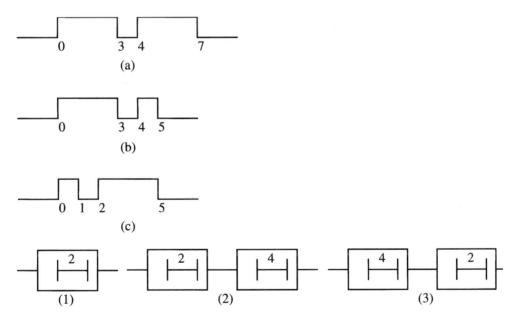

Fig. 6.92.

6.16. Complete the logic design of the edge-triggered D-FF by generating two-stage NAND-gate logic for the Ys and Q corresponding to the flow matrix of Figure 6.18*b*, p. 174. Assume single-rail inputs.

6.17. Consider the flow matrix in Figure 6.20, p. 175. For each of the transitions in columns 00 and 11 specify the initial and final y-states and the possible intermediate y-states.

*6.18. Generate two-stage NOR-gate logic for the flow matrix of Figure 6.19d, p. 174.

6.19. Generate two-stage NOR-gate logic for the flow matrix of Figure 6.20, p. 175, realizing the same function as in the previous problem, but with a USTT state assignment. Assume single-rail inputs.

6.20. Find minimal SOP logic expressions for 1-hot code realizations of the following flow tables:

(a) Figure 6.19a, p. 174   (b) Figure 6.16, p. 173

*6.21. Draw the circuit corresponding to the expression $Z = (A + BC)(\overline{AD} + E)$, discussed in Subsection 6.2.6 (p. 177). Complete the details of the analysis for combinational hazards.

6.22. Draw the circuit corresponding to the expression $Z = \overline{(A + B)C\overline{D}} + (A + B)\overline{C}D + ABC$, discussed in Subsection 6.2.6, p. 183. Show the details of the analysis for hazards, and show how best to modify the expression to make it hazard free.

6.23. Find minimal, hazard-free, SOP realizations of each of the functions mapped in Figure 3.46 (p. 97).

6.24. Identify a logic hazard in the 2–1 MUX design of Figure 4.9a (p. 111).

*6.25. Identify the essential hazards in the flow table of Figure 6.9, p. 167 defining the PETDFF.

6.26. Show where in the circuit of Figure 6.41b (p. 198) a large stray delay would cause a malfunction due to an essential hazard.

*6.27. Assume the flow table of Figure 6.93 below is realized with NOR-gates. Construct a flow table describing the function that would be realized if all the gates of this circuit were replaced by NAND-gates.

$$X_1 X_2$$

	00	01	11	10
1	1,0	1,0	2,1	3,1
2	2,1	2,0	2,1	3,1
3	1,0	2,0	3,0	3,1

Fig. 6.93.

6.28. Find an expression for Z_2 in Figure 6.31, p. 189. Compare it to $\overline{Z}_1$ in the stable states.

6.29. Analyze the NETDFF circuit of Figure 6.31 and show, for each of the two essential hazards, where rather large stray delays would be necessary to cause malfunctions. For which of these hazards would be the likelihood of having a sufficiently large delay be greater? Where in the circuit would the placement of a delay element counteract the presence of the more dangerous stray delay?

6.30. Design a FF with output Q and inputs C, S, and R that works as follows: S and R are not *both* permitted to be on while, or just before, or just after C = 1. Neither S nor R is permitted to change in the vicinity of a C change. The FF is set by the condition C = S = 1, and is reset by the condition C = R = 1. When C = 0, the state cannot change.

6.31. (a) Generate a four-row flow table for a NETDFF.
(b) Using the same components used in Figure 6.41b, p. 198 construct a NETDFF.

6.32. Redesign the Figure 6.44, p. 200 shift register stage using: *(a) NAND-gates; (b) NOR-gates.

6.33. Identify a static hazard in the logic circuits feeding the D-inputs of the ETDFFs in Figure 6.44, p. 200. Explain why it is tolerable in this case.

6.34. Consider the latch shown in Figure 6.40a (p. 197). Assume that, initially, C = 1, D = Q = 0. Is there a hazard for a change in C? Explain.

6.35. Add to the register of Figure 6.45 (p. 201) capabilities for *left* shifts and for complementing the contents of the register. Show the modified logic for one stage.

6.36. Generate a primitive flow table for a clocked T-FF that toggles on the leading edge of the C-pulse when T = 1.

6.37. Complete the design of the tree-form counter of Figure 6.47, p. 202, by specifying the contents of the I-cells. (Two AND-gates suffice.)

6.38. How does the worst-case propagation delay for a 64-stage counter built with linear logic compare with the corresponding delay for a 64-stage counter built with tree-structured logic?

6.39. Modify the logic for a four-stage counter using T-FFs and linear logic, so that it counts from 0 to 9 and then recycles to 0.

6.40. Construct a flow table corresponding to the ASM chart (for a multiplier) of Figure 6.50a, p. 206.

6.41. (a) Write a structured register transfer sequence description of the standard "comparison" method for division, described in texts on computer organization such as [Ercegovac, 1985], [Mano, 1993], and [Hennessy, 1994].
(b) Produce an ASM chart for a controller corresponding to this algorithm.
(c) Generate a flow table from the chart produced in part (b).
(d) Encode the states of either the ASM chart or the flow table in terms of y-variables and generate logic expressions for the Ys and the outputs of the controller.

6.42. Apply the methods of Subsection 6.1.3, p. 164, to the flow table derived in Figure 6.51, p. 208, for the multiplier controller. Encode the rows of the table as $y_1 y_2 = 00$, 01, and 11, respectively. Then produce K-maps for Y_1, Y_2, Z_1, Z_2, Z_3, and Z_4. Find minimal SOP expressions for these functions.

*6.43. Use factoring to simplify the logic for the arbiter of Figure 6.57 on p. 215.

6.44. Complete the synthesis of the cache interface controller started in Example 6.27 by finding logic expressions for the remaining Zs and Ys directly from the ASM chart of Figure 6.59, p. 219.

6.45. Find variable entered maps for the remaining Ys and Zs of the cache interface controller from the state map of Figure 6.60, p. 220. Then find minimal SOP expressions for each.

6.46. Generate a state map from the ASM chart (Figure 6.50b, p. 206) for the multiplier controller using the state assignment of Problem 6.42.

6.47. Find the minimum P value, a satisfactory value for W, and the associated value of D_{LmB} for a one-phase clocked system with PETDFFs:
(a) Given the parameters $D_{CQM} = 2$, $D_{CQm} = 0.5$, $H = -1$, $S = 3$, $C_{Wm} = 2$, $T_L = 2$, $T_T = 3$, $D_{LM} = 16$.
(b) What design changes, if any, are necessary if the maximum acceptable value of $D_{LmB} = 2$, and the ratio of maximum to minimum values of available delay elements is 3?

6.48. Find optimum parameters for a one-phase clocking system with latches given the same parameters as in the previous problem, plus $D_{DQM} = 2.5$ and $D_{DQm} = 1.5$. Assume the maximum acceptable value of D_{LmB} is (a) 6; (b) 12.

6.49. Find optimum values of P and V for a two-phase clocking system with latches given: $D_{LM} = 16$, $T_{1L} = 2.5$, $T_{1T} = 2$, $T_{2L} = 1.5$, $T_{2T} = 2$, $D_{1DQM} = 2$, $D_{1DQm} = 1$, $D_{2DQM} = 2.5$, $D_{2DQm} = 1$, $D_{1CQM} = 3$, $D_{1CQm} = 1.5$, $D_{2CQM} = 2.5$, $D_{2CQm} = 1.5$, $S_1 = 2.5$, $H_1 = 1.5$, $S_2 = 3$, $H_2 = 2.5$, $C_{1Wm} = 2$, $C_{2Wm} = 3$. Assume the maximum acceptable value of D_{LmB} is (a) 10; (b) 4.

6.50. For a two-phase system with latches, assuming that the D_1-signals arrive early, derive a constraint to ensure that the signals get through the L_1-latches in time to be set up properly at the D_2-inputs. This will correspond to one of the bounds incorporated in expression (13), p. 231 of Subsection 6.6.3, bounding $W_1 + W_2$.

6.51. Given what has been shown here about how clock-pulse edge tolerances affect the speed of clocked systems, discuss the reasons why, as mentioned in connection with register design, "gating the clock" is generally considered undesirable in high-performance systems.

6.52. For the C-element described by the flow table of Figure 6.83, p. 249 list all of the total states from which transitions could lead to metastable behavior:
(a) As a result of a runt pulse at one input terminal. (Specify the input in each case.)
(b) As a result of a multiple input-change.

6.53. Show that the 00-column of the flow table for the arbiter (Figure 6.73, p. 236) can be specified in either of two other ways without altering the behavior of the device in any significant manner.

6.54. Refer to the candy machine controller (Example 6.33, p. 240). Show how to generate the logic for a display that indicates the customer's balance. (This is a very simple problem.)

6.55. As a result of field trials of the candy machine (Example 6.33, p. 240), it is found that the machine runs out of dimes far more often than it runs out of nickels.
(a) Without violating other aspects of the problem specifications, modify the flow table so as to reduce the number of dimes ejected as change.
(b) Generate logic expressions for the new table.

6.56. Implement the C-element with NOR-gates.

6.57. Using a 2-input C-element and some simple combinational logic, construct a 4-input C-element.

6.58. It is proposed that C-element C_2 in the buffer circuit of Figure 6.85, p. 251 be replaced by an AND-gate. Specify a scenario in which this could lead to a malfunction.

6.59. How might omitting the connections from the X-inputs to the OR-gate cause a malfunction of the Figure 6.85, p. 251 buffer circuit?

6.60. Specify in detail the steps to be performed in using the scan method to verify that the circuit of Figure 6.86, p. 254 performs properly when the X1 = 0 and $y_1 y_2 y_3 = 100$.

6.61. Write a reduced flow table describing the L_1-latch needed for the scan method used in the Figure 6.86, p. 254 system. (Omit input columns corresponding to forbidden states.)

*6.62. Refer to the Figure 6.87, p. 257 diagram of the linear feedback shift register. (a) Suppose the latches are all initially clear and that X is fixed at 0. Describe the output sequence. (b) Again with X fixed at 0, specify the sequence of y-states if the initial state is 110.

6.63. (a) Repeat Problem 6.62b, with X fixed at 1. (b) For Problem 6.62b, assume X = 1 only for the second clock cycle. How does this change the y-state after the seventh cycle?

6.64. For which, if any, of the following situations would aliasing occur if the circuit of Figure 6.88, p. 258 were used? Each pair refers to error signals produced during two consecutive tests: $[X_1, X_3]$, $[X_1, X_2]$, $[X_2, X_1]$, $[X_3, X_1, X_3]$, $[X_1 X_3, X_3]$, $[X_2, X_1, X_3]$.

7
Software Tools

It seems only fair that the systems that logic designers help produce should be used to assist them in their work. Given the complexity of the problems involved in logic design, there is ample scope for the application of the most powerful digital computers. Computers are indeed used in many phases of digital system design, documentation, manufacture, and maintenance. Computer aids include several categories: logic design, simulation (on various levels), timing analysis, and physical design. Another category, very-large-scale integration (VLSI) design tools, overlaps the others in several respects.

An important and difficult problem in digital systems design is what is sometimes referred to as *physical design*. This refers to the placement of logic elements on a chip so as to minimize wiring problems, and then the specification of wiring paths interconnecting these elements. Sometimes the placement is done manually and the wiring left to a program, but more commonly interactive systems are used for placement. The problem of automating this task for large systems is so complex that human intervention is considered necessary where truly tight designs are required.

The growing power and availability of workstation and desktop computers is stimulating the rapid development of many new software aids for designers. Only a brief survey of those topics most interesting to logic designers is included here. We begin with the subject of logic design.

7.1. LOGIC DESIGN

Most logic design problems are finite in nature, and so in principle are solvable by enumeration, i.e., by trying all feasible solutions and choosing the best one. In practice, however, the numbers grow at an enormous rate with the basic parameters. Consider, for example, the problem of finding a state assignment for a sequential function that results in minimal logic. One might attempt to find the best assignment by trying each possibility. But, as is the case with many combinatorial problems, the number of possible state assignments grows very rapidly with the size of the table. This approach is quite reasonable for four-row tables, for which there are only three significantly different assignments. For tables with between five and eight rows, where the number of possibilities ranges from 140 to 840, enumeration with the aid of a computer might be feasible. But for nine-row tables the number of trials increases to 10,810,800. Since, for each assignment, a significant amount of work is necessary to determine the complexity of the resulting combinational logic, the enumeration approach, even with the aid of a computer, does not seem useful for tables with more than eight rows. (These numbers are for the case where the minimum number of state variables is used. Allowing for more than the minimum number, which sometimes leads to better circuits, greatly inflates the number of cases.) Enumerative solutions for most other logic design problems are similarly unattractive.

For many important problems, there exist non-enumerative algorithms that permit us to find optimal solutions, provided that the problem sizes are not excessive. For example, finding minimal sum-of-product (SOP) expressions for functions of no more than six variables is well within the range of hand computation with Karnaugh maps. (More skilled practitioners can handle eight-variable problems, and many cases involving still more variables can be solved with variable entered maps.) With the aid of computers, optimum solutions can be found for larger problems, but complexity usually grows so fast with problem size that the extension possible with machine aids is often disappointingly small. Thus, most computer aids intended for application to large problems are designed to provide *near*-optimum solutions. These are generally quite acceptable.

Several approaches to the computer-aided design of combinational logic are introduced first. Then the less developed area of sequential logic circuit design is briefly surveyed.

7.1.1. Combinational Logic

A number of programs have been written to implement the Quine-McCluskey method for multiple-output logic functions (Subsection 3.5.2). Even with large, fast computers, problems with more than 15 or 20 inputs can exhaust available storage capacity or exceed reasonable computation times. Most such programs are therefore designed to fall back to incomplete procedures that yield good, but generally not optimal, results when time or memory are running short. For example, the programs might be designed to limit the number of prime implicants (pi's) generated, or might not explore all of the possible branches when seeking a minimal cover of the pi table. It is desirable, however, that the resulting circuits be, if not minimal, at least redundant. This is because redundancy in a circuit makes it more difficult to test.

Other computerized procedures for finding minimal SOP expressions rely on rather different algorithms that, even when executed to completion, do not guarantee minimal circuits. One such procedure is based on the Shannon expansion theorem (Theorem 15, Figure 2.3). It is illustrated below by means of a simple example.

The original minsum expression is

$$Z = \overline{A}\,BCDE + \overline{A}BC\overline{D}E + \overline{A}B\overline{C}DE + AB\,\overline{C}DE + A\overline{B}CDE + ABC\overline{D}E + AB\overline{C}DE. \tag{1}$$

Any common literals are first factored out and saved to be replaced at the end. In this case, the E is removed, leaving

$$Z_1 = \overline{A}\,\overline{B}CD + \overline{A}BC\overline{D} + \overline{A}B\overline{C}D + AB\,\overline{C}D + A\overline{B}CD + ABC\overline{D} + AB\overline{C}D. \tag{2}$$

If the resulting expression is unate (Section 3.6), it is simplified by deleting any *p*-terms that contain other *p*-terms, yielding a minimal expression and terminating the process. Else, as in the current example, the expression is expanded about a literal that appears most frequently in both complemented and uncomplemented form. Choosing A in this case yields

$$Z_1 = \overline{A}\left(\overline{B}CD + BC\overline{D} + B\overline{C}D\right) + A\left(\overline{B}\,\overline{C}D + \overline{B}CD + BC\overline{D} + B\overline{C}D\right). \qquad (3)$$

The same process is repeated for each subexpression within parentheses. For the cofactor of $\overline{A}$, we expand next about C, and for the cofactor of A, we expand about B, yielding

$$Z_1 = \overline{A}(\overline{C}(B\overline{D} + BD) + C(\overline{B}D)) + A\left(\overline{B}(\overline{C}D + CD) + B(C\overline{D} + \overline{C}D)\right). \qquad (4)$$

At each stage (including the first), the subexpression is checked, not only for unateness and literals that can be factored out but also for the applicability of the following transforms, corresponding to theorems: $X + \overline{X} = 1$, $X + \overline{X}Y = X + Y$, and $X + XY = X$ (where X and Y represent single literals). In the current example, this leads to simplifications of (4), resulting in

$$Z_1 = \overline{A}(\overline{C}(B) + C(\overline{B}D)) + A(\overline{B}(D) + B(\overline{C})). \qquad (5)$$

The expansion process ends for each subexpression when it is reduced to one p-term. This is the case for all subexpressions in (5). Now, the expansions are reversed through multiplications by one literal at a time. At each step, we look for opportunities to apply $\overline{X}Q + XQ = Q$, or $\overline{X}Q + XPQ = \overline{X}Q + PQ$. In the current example, the first backward step yields

$$Z_1 = \overline{A}(B\overline{C} + \overline{B}CD) + A(\overline{B}D + B\overline{C}). \qquad (6)$$

For the next backward step, both of the simplifications apply, resulting in

$$Z_1 = B\overline{C} + \overline{B}CD + A\overline{B}D. \qquad (7)$$

Restoring the common factor E, yields the final result:

$$Z_1 = B\overline{C}E + \overline{B}CDE + A\overline{B}DE. \qquad (8)$$

In this case, the result is indeed a minimal SOP expression. More generally, the method seems to yield *non*minimal expressions that are not too far from the minimum. This algorithm is not very difficult to program, and it is reported to be capable of handling substantial problems in reasonable amounts of time. More elaborate algorithms employing similar tactics have been successfully used for multiple-output functions. This approach, and methods based on the Quine-McCluskey method, generate SOP expressions that can be used directly to personalize programmable array logic (PALs) and programmed logic array (PLAs) (Subsections 4.2.2 and 4.2.3). Where multistage logic is to be used, they carry out only the first step.

The *logic synthesis system* (*LSS*), the next computer aid to logic design discussed, is of a rather different nature. It operates incrementally on representations of multistage logic circuits. LSS takes an initial circuit that realizes the desired function (usually multi-output), and searches for ways to simplify subcircuits with no more than about five gates. The restriction to local transformations makes it possible to deal with large circuits without requiring excessive computation time. While optimal solutions are not to be expected, LSS generally produces results that are quite satisfactory. Appropriate sets of transformations are programmed into the system for the particular technology employed. Both simple and complex transformations are included. Since the process is iterated, it is often the case that the results of one transformation, perhaps a subtle one, set up conditions for another transformation that looks trivial. Let us see what these transformations look like.

Examples of simple transforms are shown in Figure 7.1. The transform shown in Figure 7.1a follows at once from the canceling effect of two cascaded inverters. Transformation b is a consequence of the identity $B + \overline{A}\overline{B} = B + \overline{A}$. The justification of transformation c is that the inverter converts the NOR-gate to an OR-gate so that, prior to the final inverting bubble, we can utilize the associativity of Boolean addition. That is, $(A + B) + C = A + B + C$. This is a particularly useful transformation that designers should familiarize themselves with. It is often applicable, and can save chip area, power, and delay.

The transformation shown in Figure 7.2a exploits the wired-OR capability of the technology. It saves two gates while shortening the maximum signal path length by two gates. The transformation of Figure 7.2b simplifies one gate, eliminates an inverter, reduces the maximum signal path length, and, as a bonus, eliminates a logic hazard.

Note that LSS is not restricted to two-stage logic. It is particularly suited for the design of the kind of complex, irregular logic characteristic of the control circuitry in central processors. But there is more to LSS than a system for simplifying logic circuits via local transformations. It has been developed into a system for developing designs that can be fed into other software for producing and assembling logic chips to create large digital systems. Such factors as loads on off-chip drivers, power consumption, and both long- and short-path delays are taken into account. The system has been developed at IBM by means of close cooperation between researchers and those engaged directly

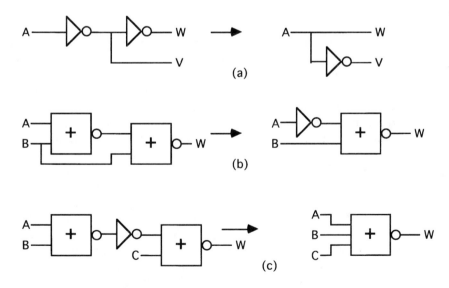

Fig. 7.1. Two simple and one not so simple logic transforms.

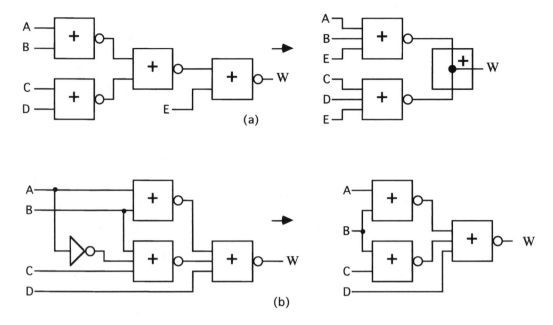

Fig. 7.2. More complex LSS transforms.

in the design of mainframe computers. It has proved to be a practical tool and is being used to an increasing extent. It is designed to be used interactively by designers, who can control when, and to what extent various operations are performed. LSS continues to evolve, and now includes features of a more global nature than were contained in earlier versions.

7.1.2. Sequential Logic

Less work has been done in providing tools for the design of sequential logic circuits. Various algorithms for particular aspects of the synthesis process have been implemented. For example, programs have been written implementing approximate procedures for reducing incompletely specified flow tables, a very difficult problem (Subsection 6.1.2). The easier problem of reducing fully specified tables has been implemented many times. Programs have also been written for finding good state assignments for synchronous sequential circuits, and many people have written programs for finding various types of state assignments for asynchronous sequential circuits.

There is increasing activity in this area, with packages becoming available that carry through the process from flow table to circuit, including programs for designing the combinational logic circuits. Some of these run on small personal computers. Many software design aids developed by companies for their own use are kept confidential. Analysis programs are also of interest, since, regardless of how circuits are designed, it may be necessary to check to see that they meet certain constraints. This is one important application of the simulation programs discussed next.

Design tools are currently under development for dealing with self-timed systems. These focus on such issues as race-free state assignments and hazard-free logic.

7.2. SIMULATION

An integral part of any design process is validating what has been done. The ultimate test, of course, is whether the finished product performs properly. But in the case of digital systems, the expense and delay involved in producing hardware is too high to permit design by trial and error. Prior to converting designs to silicon, every effort must be made to find and eliminate design errors and to improve the design as much as possible.

The most effective way to minimize errors is not to make them in the first place. Careful, systematic design practices, employing a modular, top-down approach, attention to detail, checking and rechecking of calculations, and reviews by colleagues are basic to good design. But where complex systems are involved, it is helpful to have additional verification in the form of step-by-step tracing of the processing of particular sequences of data. Computer simulations are very useful for this purpose.

It is not feasible, however, to simulate in complete detail the functioning of a complex digital system. An attempt to track the precise behavior of every transistor in a digital computer, as the machine executed even a simple program, would require so many operations on the part of the machine running the simulation that the time required would be truly astronomical. Thus, it is necessary to partition the simulation process into a number of phases, each dealing with a different level of detail. At the highest level is a simulator concerned only with the command structure of the subject machine. It makes no attempt to verify that the commands were properly implemented in hardware, but only checks out the consequences of the correct executions of a series of commands. At the other extreme is a circuit simulator, capable of modeling, in detail, the behavior of the electrical components of a logic circuit consisting of perhaps a few dozen gates. This circuit simulation is used to ensure that the electrical designs of the gates are valid, and to determine key parameters of these elements. The lower the level of simulation, the smaller the portion of the system tested in each simulation run must become in order to keep the running times reasonable. In the following subsections, the roles of simulators at four such levels are briefly outlined. Besides their role in design verification, other uses of simulators are pointed out.

Related to, but different than simulation, is the area of design verification. Here the object is to produce computer algorithms that examine a given design and determine whether it satisfies specifications defined in some formal language. Such techniques are useful aids in dealing with the ever increasing complexity of modern digital systems. We need all the help we can get. However, formal verification methods are fundamentally limited by the fact that they can only compare one formal specification with another. Many problems arise precisely in the process of producing the formal specifications used as the "answer sheet" by the verifier. There is no way to determine rigorously if a formal specification conforms to real-world requirements.

7.2.1. High Level

Before plunging into the details of designing a system, it is a good idea to test the overall concept to see if it is likely to meet the goals of the project. Finding and correcting flaws in the design concept is always easier at a higher level. Once the machine has been designed in detail, changing the word size, adding new kinds of registers, etc., are usually expensive, time-consuming processes. It is always better to make changes at the earliest possible stage of design.

For example, if the system being designed is a digital computer, it would be wise to simulate the proposed order structure on another machine. Programs typical of those expected to run on the new machine can then be written and executed. Various measurements can be made, for example, on the extent to which various features of the new machine are actually used. Run times can be estimated.

At a more detailed level, the register-transfer sequences (Subsection 6.5.1, p. 204) that specify how various subunits are to be controlled and how instructions are to be executed, can also be simulated to check them for correctness. General-purpose simulation programs exist to facilitate this process. High-level (sometimes called *behavioral* level) simulations are also used to begin the testing of microcode even before a laboratory model of the machine is available. If the simulation is sufficiently refined, interactions with peripheral units can be explored.

After the high-level design has been checked out, it is appropriate to proceed with the logic design phase. Quite possibly this may uncover failings in the high-level design, or perhaps inspire some useful elaborations. For example, it might be noted that, given the logic for implementing a particular instruction, some useful variations on this instruction might be implemented at very little added cost. When the logic design stage is essentially complete, a more detailed level of simulation is appropriate to verify that this work has been properly done. This process is discussed next.

7.2.2. Logic Level

Logic-level (also sometimes referred to as *functional* level) simulation is a fairly straightforward process. Descriptions of the circuits to be tested may be entered in one of two ways. Some systems require a list (sometimes called a *net list*) of input sources, gates, and outputs that specifies for each logic element and output terminal where its inputs come from. Other systems allow one to describe circuits graphically (e.g., using a puck, mouse, or other such graphics device). Assuming that the circuit is sequential, it is also necessary to specify various timing parameters. In addition to indicating the connections, a set of general declarations must be made specifying the relevant properties of the logic elements. Finally, the simulator must be initialized by giving it the initial states of the memory elements and a sequence of external inputs.

The simulation program then proceeds to trace through the combinational logic block to determine the outputs of each first-stage gate, then each second-stage gate, etc., until all block outputs have been determined for the first clock cycle. In many systems, timing details are ignored (actually, left to a different program: see Section 7.3). At the end of each cycle, the program sets the state devices, i.e., latches, etc., to new values, depending on the outputs just calculated, and then goes on to derive the logic circuit outputs for the next cycle. The results are suitably displayed. Note that when a simulator runs on a conventional computer, it must process sequentially events that are occurring simultaneously in different elements of the subject circuit. This is an important factor in slowing down simulation. In order to ease this problem, special-purpose, highly parallel simulation machines have been built. Although expensive, they make simulations of much larger systems feasible.

Many logic-level simulators have a *hierarchical* capability in that they permit the specification of complex logic blocks in terms of simpler blocks. This permits users to check out the logic for the implementation of particular blocks of logic, and then to define these blocks as components for testing in conjunction with other logic circuits.

Alternatively, a *top-down design* philosophy may be followed: the input-output relations of the complex blocks are defined first, and the interactions among such blocks simulated even before the blocks have been implemented in terms of simpler elements. Later, the logic circuits implementing the blocks are simulated.

In addition to verifying the correctness of logic designs, logic simulators are also used to measure the efficacy of test procedures. For example, a sequence of randomly chosen faults might be simulated and a set of tests applied to determine what proportion of the faults are detected. In order to model faults more accurately, it is usually necessary for logic simulators to allow signals to take on more than two values. Along with the usual 0- and 1-values, a third value, X, might be assigned where the signal is invalid, with a voltage value somewhere between the ranges assigned to 0 or 1. A fourth value might be assigned to signals that are valid 0s and 1s, but where the value is undetermined (for example, when power is first turned on, the states of some FFs might not be known). Problems arising in the modeling of gate circuits with certain types of faults and in the testing of logic circuits using pass networks (Figure 1.3) have stimulated the development of simulators that operate on individual switches (transistors) as the basic elements. These are introduced next.

7.2.3. Switch Level

A *switch-level* simulator operates at a level just below that of a logic simulator operating at the gate level. It accepts descriptions of circuits with individual metal-oxide-semiconductor (MOS) transistors as the basic elements, treating them as switches. The discussion here refers principally to MOSSIM II, the best known switch level simulator. Some of the basic concepts are outlined here to convey the general flavor of such systems. Let us see what type of situations motivated the development of such systems.

Consider the circuit shown in Figure 7.3. It consists of a pass-transistor network feeding an inverter from input sources A and B. Let us focus our attention on the signal P. If C = 1 and D = 0, then the signal at A is passed through transistor f to P. If C = 0 and D = 1, then we have P = B, since transistor g acts as a closed switch. What happens if C = D = 0? In this event, node-P is isolated from all other signals by extremely large resistances. But there is always some stray capacitance (i.e., a capacitance inherently present, as opposed to being deliberately inserted: see Appendix A.2.3) from P to

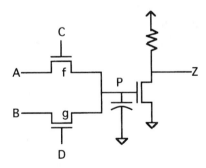

Fig. 7.3. A pass-transistor network, illustrating transistor strength and capacitive storage.

ground, as shown in the figure. This is characteristic of all real systems. The time constant for the discharge of this capacitance under these conditions may be assumed to be very large. Therefore, whatever signal was on P prior to its being isolated, will remain there for a considerable time. To be more specific, suppose that, for some time prior to $t = 0$, $A = C = 1$, and $D = 0$. Then $P = 1$. Now, assume that at $t = 0$, C is turned off and that D remains off. Then the voltage at P remains at a logic-1 value, "remembering" the value it was last given. Assuming the circuit is embedded in a clocked system, then this value can be expected to persist for many clock cycles. (This storage effect is utilized in what are called *dynamic logic* technologies, and is also the basis of the most commonly used type of random-access memory, known as *dynamic* RAM.) Switch-level simulators must take this type of behavior into account. Consider now the fourth possible state of CD in Figure 7.3.

What happens if $C = D = 1$? If $A = B$, there is clearly no problem: $P = A = B$. But suppose $A = 1$ and $B = 0$. Now the situation resembles that discussed in our discussion of wired logic (Subsection 1.3.1). The value of P depends on a voltage divider effect, i.e., it is determined by a ratio of the resistances of transistors f and g. In switch-level simulators such as MOSSIM II, rather than specify actual resistance values and require the simulator to make arithmetic computations, each transistor is assigned what is called a *strength* value, s_i. This is a measure of how strongly a transistor can connect a node to another node. The lower the resistance, the greater the strength. Strength values are a very coarse measure in that if transistors f and g, respectively, have different strengths, s_1 and s_2, where $s_2 > s_1$, then we assume that the resistance of g is smaller by an order of magnitude. If that is the case, then in the situation under discussion, the A-value dominates, so $P = 1$. If the two transistors were of roughly equal strength, then P would be assigned the undetermined value, X. The rule in general then is that if a node is connected to several different nodes via transistors of different strengths, the connection through the strongest transistor prevails. Sometimes it is necessary to consider the relative magnitudes of capacitances at different nodes.

Look at the pass-transistor circuit of Figure 7.4. Assume that B goes on for a clock cycle while D is off. Signal A is then passed to the node labeled C. During the next cycle, suppose B is off and D goes on. What are the subsequent signal values at C and E? Since signals can flow equally well in either direction through a MOS transistor, it is possible that E is copied by C, or that C is copied by E, or that they both assume some intermediate value that is not defined as a valid 1 or 0. The actual outcome depends on the relative values of the stray capacitances linking C and E to ground. If the C-capacitance is many times that of E, then charge can flow out of C and charge E to a value close to that of C without appreciably changing the charge (and therefore the voltage) of C. Conversely, if the E-capacitance is dominant, then both nodes assume the value of E.

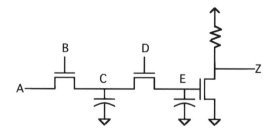

Fig. 7.4. Pass-transistor network illustrating node size.

If the capacitances are similar in value, and if E and D have different logic values, then both nodes will assume some intermediate value. One approach to this problem (used in MOSSIM II) is along lines similar to those followed in coping with the previous problem. The simulator is given a *size* value for each node that reflects the order of magnitude of the capacitance at the node. In this example, if C and E have sizes k_2 and k_1, respectively, where $k_2 > k_1$, then when D goes on, the C-signal is passed on to E. If a node is an *input* (able to supply as much charge as needed), then its size is ω, which is considered as dominating all other sizes. Thus, if in Figure 7.4, A is an input and B = D = 1, then the signals at C and E are both driven to 1, regardless of their previous values or of the sizes of C and E (which cannot match ω).

Using the concepts of transistor strength and node size, and signals, restricted to values 0, 1, or X, networks of transistors can be modeled on the basis of "ranking" rules (not presented here). That is, no attempt is made to make fine distinctions among signal values or among paths that feed signals to a given node. Suppose, for example, that activated transistors in the circuit at some time interconnect nodes P, Q, and R. If the size of P exceeds the sizes of both Q and R, then, regardless of the relative transistor strengths, the simulator assigns the present value of P as the next-state values of all three nodes. On the other hand, if P and Q are both input nodes, then the transistor strengths determine the next-state signal values.

The MOSSIM II switch level simulator is reported to be capable of handling circuits with over 10,000 transistors, simulating thousands of clock cycles in acceptable run times. It is capable of dealing with more subtle effects than can be handled with a gate-level simulator. But it is not intended to simulate detailed electrical behavior, which is necessary for such matters as determining logic circuit propagation delays. This is the task of circuit-level simulators, which are discussed next.

7.2.4. Circuit Level

Prior to the days of integrated circuits (ICs), when electronic circuits were constructed out of individual, discrete components, i.e., resistors, transistors, capacitors, etc., it was a common practice to construct laboratory models of new designs. These could then be tested, and the designs refined on the basis of actual measurements, prior to the construction of the production models. This is not possible with ICs. The effects of the actual wiring on the chip and of the detailed placement of the transistors is so important that very little would be learned about the electrical characteristics of the circuits by attempting to wire up off-the-shelf components to "breadboard" a particular chip design. Yet the design of a large-scale integration (LSI) or VLSI chip is so complex, that the need for design verification is much greater than ever. The solution to this dilemma is the detailed circuit-level simulator.

This is a program that incorporates a great deal of quantitative knowledge about ICs, including many second-order effects, that has been built up as a result of both theoretical studies and laboratory as well as industrial experience. The simulators execute iterative numerical routines to track the detailed changes in current, charge, and voltage at various points in the subject circuits. Because of the large amount of computation necessary, these programs can only deal with low-level circuits.

Thus, the operation of a latch, or perhaps an edge-triggered D-flip-flop (ETDFF), would be a suitable subject for simulation by such programs as SPICE or ASTAP (an IBM tool). It would not be feasible to simulate the behavior of a cache memory system at this level. In addition to acquiring timing information, it is important to determine how vulnerable the circuit is to noise, how its behavior is affected by variations in temperature and power supply voltage, and how its parameters are affected by expected manufacturing tolerances. It would not be unusual to expend hundreds or even thousands of hours of computer time to check out the circuitry incorporated in an LSI chip. Circuit simulators are also valuable research and learning tools.

We turn now to the problem of determining propagation delays through combinational logic networks, an important issue that most logic- and switch-level simulators handle only superficially, in order to keep their run times within reasonable bounds.

7.3. TIMING ANALYSIS

As is clear from the discussion of clocking schemes (Section 6.6), both long-path and short-path delays are of great importance in the design of digital systems. For large systems with a great many circuits, varying considerably in path length, fan-in, and fanout, it is difficult to calculate and to keep track of the bounds on these parameters. The problem is exacerbated in large projects where many people are involved, and where circuit changes are being made even as the system is about to go into production. For these reasons, programs have been developed for computing long- and short-path delays, and for identifying the critical paths that determine the extreme values.

Suppose we knew the maximum propagation delay for each logic element. Then, after enumerating all of the actual signal paths, we could add up the delays along each path to find the maximum delays for the individual paths. The long-path delay would then be that corresponding to the maximum of these values. A corresponding procedure based on the minimum propagation delays could then be used to find the short-path delay. A difficult part of this procedure is in the very first step: identifying the actual signal paths. Consider the circuit shown in Figure 7.5. Signal flow is from left to right. The logic elements are not identified as to type, but are labeled in the upper right corners. The numbers in the middle of each block represent propagation delays. Now observe that there is a path from an input at the left end through boxes labeled B, F, and H to an output at the right end. But the existence of such a "topological" path does not mean that any signal can actually traverse it. It may be that there is no possible input that

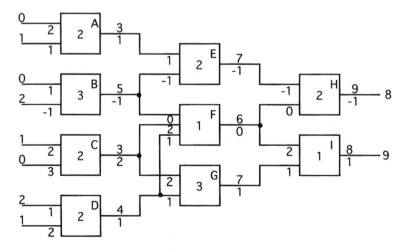

Fig. 7.5. Simple example for a timing analyzer.

sensitizes this particular path. In this case, the delay along the path is of no interest. But filtering out such "pseudo" paths would entail a very substantial amount of computation. Hence, timing analyzers designed to operate on large systems do not make this distinction. Rather, they assume all topological paths are genuine, and leave it to the user to determine which paths can be ignored. Now let us see how a timing analyzer might work.

The numbers at the left end of Figure 7.5 represent arrival times of signals at the eight inputs. The two numbers at the right end represent the *required* arrival times at the two outputs. Actual arrival times are computed by a simple computation performed on each block, starting at the input end. Since the latest arrival time of an input to block A is 1, and since the delay through that block is 2, the arrival time (under the worst-case assumption) of its output signal is clearly 3, indicated by the 3 above its output line. In a similar manner, the output arrival times for blocks B, C, and D are computed. Next, the system proceeds to the next level, starting with block E. The latest arrival time of an input to E is 5 (the signal from B), and so, adding that to the delay in E, yields the arrival time 7 at its output. The results of the complete process are shown in the figure by the numbers above the signal lines. Now observe that the arrival time at output I is 8, which is early by one unit of time. This signal could have arrived one unit of time later and still been on time. Such a margin is referred to as *slack*. The number 1 below the signal line at the I-output indicates this slack. The H-output is *late* by 1, and so a slack of -1 is indicated under its output. Now a second pass is made back through the network to calculate the slacks at each point. This process is just a bit more complex than the first pass.

Adding the arrival time and slack of a block output and subtracting the block delay gives the latest acceptable arrival times for its inputs. Subtracting the actual arrival time of an input from this number gives the slack for that input. Thus, for block H, the acceptable arrival time at its inputs is 6 (if they arrived at $t = 6$, then the output would be at $t = 8$, just in time). Input slacks are shown in the figure just below the input lines to each block. The slack for a block *output* is the *minimum* of the slacks for the inputs fed by that output. Hence, the slack for the output of F is the minimum of the slacks for the F-inputs to the H- and I-blocks, or 0. The results of this backward pass are shown in Figure 7.5; input and output slacks are shown below the signal lines.

The critical path is clearly indicated by the signals with negative slack. If the input signal at the left end feeding the lower input to B (which has -1 slack) were made to arrive one unit earlier, or if the sum of the propagation delays for blocks B, E, and H (the blocks lying along the critical path) could be reduced by 1, then the arrival time requirements would all be met. There is another application of the results obtained by this analysis. Consider block C, which has output slack of 2. If the propagation delay for that block were increased by 1, this would merely reduce the slack by 1, causing no problems anywhere. In many technologies, designers are given the option to trade speed for power at individual gates. Thus, increasing block delays where less speed is tolerable can result in reduced power dissipation, often an important design parameter.

The preceding example was simplified in that only one delay value was shown for each block. In most technologies, we would have to specify two propagation delays, one for outputs changing from 0 to 1, and another for outputs changing from 1 to 0. In order to utilize these parameters, it is necessary to indicate for each input whether it drives the output up or down. It is not difficult to include this elaboration, which entails carrying out two separate calculations of the type just shown. Computing minimum delays through the network involves using lower bounds on the propagation delays and modifying the process by assuming that block outputs are generated upon the arrival of *any* (as opposed to *all*) inputs. A more basic elaboration incorporated in some timing analysis systems is to use a statistical approach. Instead of using maximum and minimum values for propagation delays, nominal values and standard deviations are used. These are processed to produce results specifying ranges of arrival times expressed statistically. This may make possible tighter designs than can be achieved via worst-case analyses.

The block delays are usually found from circuit-level simulations and take into account wiring delays and the consequences of fanout. Suitable safety margins are generally incorporated in these estimates. Clearly, the results of such an analysis are critically dependent on the validity of the input data. Running times for timing analyzers of the type described here are roughly linear in the number of logic blocks, and such systems have been used successfully in the design of large computers.

OVERVIEW AND SUMMARY

There are several interesting methods for generating economical logic circuits. The Quine-McCluskey method is the basis for some, but it is limited to finding SOP expressions and cannot handle very large problems. Another approach is based on the Shannon expansion. It too produces SOP expressions, but it seems to yield good, though generally not optimal, results in cases involving substantial numbers of variables. The logic synthesis system (LSS) is based on the use of local transformations to incrementally optimize large circuits not confined to two-stage logic. It has been used successfully to aid in the design of large computers, and is integrated into a family of design-automation programs at IBM. (More recent versions of LSS incorporate features of a global nature.) Problems specific to sequential logic have received much less attention, although some early work was done on computer programs for reducing flow tables and for encoding them efficiently.

Simulation is important for design verification at all levels and for generating and measuring the efficacy of testing sequences. High-level simulation is used to evaluate overall design decisions, such as choice of the instruction set, to estimate the speed of the system, and to check out register transfer sequences. Logic-level simulation is used to find logic design errors and to perform various functions related to testing, such as generating test sequences and searching for faults that they would not catch. For MOS-implemented systems, a somewhat more detailed level of simulation is useful, particularly in connection with test generation and evaluation. These are called switch-level simulators, as they treat individual transistors as the basic building blocks. They categorize them as switches, falling into one of a few categories ranked by the order of magnitude of resistance when turned on. Capacitance in the circuits is also taken into account via a similar ranking. Such simulators are able to cope with capacitive signal storage. Circuit-level simulators deal in great detail with the electrical behavior of the basic gate and storage elements. They are indispensable in the design of ICs, as there is no other practical way to verify their designs before chips are actually produced. These simulators also produce information about such matters as propagation delays and sensitivity to noise.

Timing-analysis programs identify critical paths through logic circuits that determine long- and short-path delays. They can deal with very large systems because they ignore the logic performed and simply make one forward pass and one backward pass through the logic to find arrival times and slacks for signals at all points.

Sources

An early computer implementation of the Quine-McCluskey method is reported on in [Bartee, 1961]. The discussion here of logic minimization using the Shannon expansion is based on the work described in [Brayton, 1982, 1984]. More can be learned about the logic synthesis system in [Darringer, 1984], and in [Trevillyen, 1986]. An example of current work in automated logic circuit design is in [Nowick, 1992]. A program for minimizing incompletely specified flow tables is presented in [Grasselli, 1966], and an early program for finding efficient state assignments is the subject of [Armstrong, 1962]. As an example of software now available, a Columbus, Ohio, company called SOFCAD Electronics is selling an IBM PC program called CALCAD that it claims, among other things, implements the Quine-McCluskey method for 8-variable functions and synthesizes clocked sequential circuits. Valid Logic Systems of Sunnyvale, California, has for years been marketing the SCALD system, a hardware–software graphics package for generating hierarchical sets of logic-circuit diagrams, doing logic-level simulation, and doing timing analysis. Logic-level simulators are discussed in [Szygenda, 1976], and the MOSSIM II switch-level simulator is described in [Bryant, 1984]. The treatment of timing analysis is based on [Hitchcock, 1982]. A discussion of VLSI design tools, including an introduction to the placement and wiring problem, is in [Mukherjee, 1986]. PSpice, a version of the circuit level simulator SPICE that runs on various personal computers, is described in [Tuinenga, 1988]. Good general sources for ongoing work in the general field of software aids for digital design are the *IEEE Transactions on Computer Aided Design of Integrated Circuits and Systems*, and the IEEE magazine, *Design and Test of Computers*.

Problems

7.1. In the example worked out to illustrate the expansion method for logic minimization (see p. 269), the expression reached at step (2) was expanded about A. Rework the same problem from this point, but expand about B instead of A. Note that this path leads to more opportunities for early application of simplifying steps. Compare results.

7.2. Prove that the transformations of *(a) Figure 7.1 and (b) Figure 7.2 are valid.

7.3. What logic hazard is eliminated by the Figure 7.2*b* transformation?

7.4. For the circuit of Figure 7.3, write a minimal SOP expression for Z, assuming that, at all times $C + D = 1$, and that the strength of f exceeds that of g.

7.5. Assume that for the circuit of Figure 7.4, the sizes of nodes C and E are, respectively, k_1 and k_2 where $K_1 > K_2$. Suppose, with $A = D = 0$ and $E = 1$, B is pulsed, after which D is pulsed. What will be the resulting values of C, E, and Z?

7.6. What class of simulator would be most suitable for calculating the signature of a test set p. 257?

*7.7. For the network shown in Figure 7.5, change the delays in boxes E, F, and G, to 3, 2, and 1, respectively. Using the same input arrival times and required output arrival times, carry out the timing analysis for long paths to find the two output arrival times and the eight input slacks.

7.8. For the Problem 7.7 network, assume the minimum delay for each element is half the given maximum delay, and that the earliest allowable arrival times are 2 and 4, respectively, for the upper and lower outputs. Compute the earliest arrival times and the short-path input slacks.

A Postscript
on Professionalism

This is addressed to those who are now, or who intend to become, applied scientists or engineers. As suggested by the heading, the subject matter of this note is rather broader than the specific area of technology treated in the body of this book. Unfortunately, it is absent from most curricula.

Challenger, Chernobyl, Bhopal are words that suggest the grave consequences that can result from the incompetent application of sophisticated technology. These disasters resulted from more than simple errors. They are direct consequences of the roles played by engineers and scientists in decision making related to the products of their work. Subsequent to the flaming end of the Challenger, we learned that the engineers (including engineering managers) who worked on the subsystem that failed had urged that the flight be postponed. But their individual and collective professional judgments were overruled by upper level executives for reasons apparently stemming from financial and political pressures. The engineers did not appeal to any higher authority. Nor did they attempt to warn the astronauts whose lives were at risk. As a result they must live with the knowledge that their acquiescence contributed to a terrible tragedy.

But for them to have acted otherwise would have required a degree of moral courage that not many possess. There have been numerous analogous episodes in which engineers who resisted such pressures were vilified and suffered severe career and financial damage. In 1972, three engineers working on the development of the high-technology Bay Area Rapid Transit System (BART) were fired for exposing bad engineering on the project. The results of independent investigations subsequently validated their concerns. Even more convincing was a train crash caused by a faulty circuit component. This incident caused the Institute of Electrical and Electronics Engineers (IEEE) to set up procedures to support members whose careers were jeopardized as a result of their adherence to ethical practices. Seeing to it that such procedures are properly implemented by their professional societies and that supporting federal and state legislation is enacted, should be a concern of every engineer. It should not be necessary for responsible professionals to stand alone against large organizations.

A distinguishing characteristic of professionals is that they consider their work responsibilities in the broadest sense. For example, a professional does not blindly accept the specifications for a logic circuit and proceed to implement them without understanding the function of the circuit within the larger system. Does it make sense to do it this way? Are there better alternatives? Of course, one must not carry this to absurd extremes. Normally, this process need not take very much time, and it is likely to benefit the project in every respect.

The preceding discussion concerned doing the job right. Perhaps even more important is that the right job be done. Engineers should not behave as hired hacks, blindly applying their talents without regard to the end use of the product. It is not within the

282

purview of engineering ethics to define what particular end uses of technology are proper. But there *is* a professional obligation to consider seriously whether proposed assignments conflict with one's own moral values, and to act accordingly. It is quite possible that after thoughtful consideration two individuals may come to opposite conclusions about whether a given project is or is not harmful to the public safety or welfare, so that one agrees to work on the project and the other declines. Both may be acting as ethical professionals. The validity of their respective conclusions, while possibly an issue of great importance, is a matter outside the realm of engineering ethics.

Given the ever increasing role of technology in shaping human existence, it is crucial that it be used with wisdom. By virtue of their knowledge and occupations, engineers and applied scientists have a special responsibility to help ensure that this powerful force is used for the benefit of humanity. They are usually the first to know about important developments and are often in position to foresee possible dangers as well as opportunities before these become evident to others. They should speak out on such issues in a responsible manner, both within their workplaces and, where appropriate, outside.

Appendix

A.1. NUMBER SYSTEMS AND CODES

Since ten-fingered humans are addicted to the decimal system, and since computers operate most efficiently in binary, it is necessary to understand how both integers and fractions can be translated from one system to the other. Certainly those involved in the design of computers must understand how to do arithmetic in the binary system. But computers operate on text as well as on numbers. This motivated the development of American Standard Code for Information Interchange (ASCII) and other standard coding schemes that can specify alphabetic characters and other nonnumerical symbols in terms of 1s and 0s.

In addition to binary and decimal, we are sometimes interested in other number bases. For example, octal and hexadecimal are often used to represent strings of 0s and 1s compactly. We begin with a discussion of number systems in general.

A.1.1. Number Systems

When an integer is expressed in positional notation with base b, each numeral is weighted by a power of b, depending on its position. The rightmost, or *least significant*, digit, is given weight $b^0 = 1$, and in general the ith digit from the right has weight b^{i-1}. Thus, in base b the number x would be expressed as $a_{n-1}a_{n-2} \cdots a_0$ (where the a_i-digits range from 0 through $b - 1$) and we have the equation:

$$x = a_{n-1}b^{n-1} + a_{n-2}b^{n-2} + \cdots + a_0. \tag{1}$$

Consider, for example, the number 7542 expressed in our usual decimal, or base-10, notation. In the notation just presented, we would express it as: $7 \times 10^3 + 5 \times 10^2 + 4 \times 10^1 + 2 \times 10^0$. The value of the numeral 4 is $4 \times 10^1 = 40$, the value of the 7 is $7 \times 10^3 = 7000$, etc.

For engineering reasons related to reliability and other factors, it has been found most effective to use 2-valued signals within computers. Hence, it is best to represent numbers in base-2 notation; we refer to these as binary numbers.

Because so many problems in logic design, computer programming, and other aspects of the computer field involve powers of 2, it is very useful for people working in the area to be able to compute mentally the value of any power of 2. Fortunately this is not a difficult skill to acquire. One must first memorize the first 10 powers of 2 $(1, 2, 4, 8, 16, 32, 64, 128, 256, 512, 1024)$. Note then that $2^{10} = 1024$ is within 2.5 percent of $1000 = 10^3$. Thus, to estimate the value of a fairly large power of 2, we can replace each multiple of 10 in the exponent of 2 by 10^3 in the result, and then multiply by the value of 2 raised to the power of the exponent modulo-10. This is easier done than said. For example, 2^{25} can be found quickly by noting, in effect, that it is equal to $2^{10} \times 2^{10} \times 2^5 =$ (approximately) $10^3 \times 10^3 \times 32 = 32$ million. In the computer field the abbreviation K is commonly used to mean 2^{10}, so that a 64K-memory chip contains 64×2^{10} or 65,536 bits.

A.1.2. Arithmetic in Binary

Arithmetic operations in the binary number system are carried out the same way as in the more familiar decimal system. But the individual steps are easier to execute since there is only one nonzero numeral. Hence, only one operation each for addition and multiplication need be memorized. Each of these operations ($1 \times 1 = 1$ and $1 + 1 = 10$) involve two nonzero operands. One example of each of the four basic arithmetic operations is shown below.

1. Addition

$$
\begin{array}{r}
110100110 \\
+\,101111111 \\
\hline
1100100101
\end{array}
$$

2. Subtraction

$$
\begin{array}{r}
1100011001 \\
-\,1011111011 \\
\hline
11110
\end{array}
$$

3. Multiplication

$$
\begin{array}{r}
1011 \\
\times\,1101 \\
\hline
1011 \\
1011 \\
1011 \\
\hline
10001111
\end{array}
$$

4. Division

$$
\begin{array}{r}
101 \\
110\overline{)100010} \\
110 \\
\hline
1010 \\
110 \\
\hline
100\ \text{(the remainder)}
\end{array}
$$

A.1.3. Negative Numbers

Since digital computers do a great deal of arithmetic computation, negative numbers will often be encountered and we must have a convenient way of dealing with them. That is, we need a simple way to represent numbers as being either positive or negative.

The most obvious approach is to reserve one bit of a word representing a number as the sign bit. (Although real computers represent numbers with anywhere from 16 to 64 bits, the principles involved are easier to illustrate with a much smaller word—4-bit words will be used here.) For example, if our computer had 4-bit words to represent numbers, then we might specify that all positive numbers have a 0 in the leftmost bit and that this bit be 1 for all negative numbers. The rightmost three bits would specify the magnitude of the number in binary, as discussed earlier. Thus, the number 5 would be specified as 0101 while -4 would be written as 1100. There are several drawbacks to this apparently simple scheme, commonly referred to as the sign-magnitude method. One is that if two arbitrary numbers are to be added, it is necessary to examine the signs of each of them to determine whether to generate the sum or difference of the magnitudes, and a further logical operation is required to decide what sign to attach to the result. A second difficulty is that there are two representations of 0, since $+0 = -0$. In our 4-bit example, both 0000 and 1000 represent 0. This complicates comparison operations, since if two numbers are being tested for equality, special provision must be made for the possibility that one is $+0$ and the other -0. For these reasons, most computers use different approaches, and by far the most common technique is the *two's complement* method presented below.

In order to understand this method, it is essential to have a firm grasp of a few simple points about binary number representations.

1. The most significant (leftmost) bit of an n-bit binary number has weight 2^{n-1}. Thus, the binary number 1000 represents the number $2^{4-1} = 2^3 = 8$.
2. With n bits, we can represent magnitudes ranging from 0 to $2^n - 1$. Thus, with four bits we can specify numbers from 0 (0000) through 15 (1111).
3. If the result of an arithmetic operation (say addition) exceeds the number of bits in the register provided for, it then is easy to arrange for the lower bits to go into the register with the excess upper bits simply being discarded. Thus, what appears in the register is the original number modulo-2^n. For example, if we try to stuff the binary number representing 37 (whose binary representation is 100101) into our 4-bit register, what actually goes in is the number 0101 (5 in decimal notation), which is 37 modulo-16.

Finally, a word about notation. In our discussion, symbols such as k or k_i, will represent absolute values, i.e., positive numbers. Negative numbers will be written as $(-k)$.

Now we can introduce two's complement numbers. For an n-bit system, the two's complement of a number k is defined as $2^n - k$. Thus, for $n = 4$, the two's complement of 5 is $2^4 - 5 = 16 - 5 = 11$, which would be written as 1011. The basic idea is to represent negative numbers as the two's complements of their absolute values. Positive numbers are represented as the simple binary equivalents, so 5 would simply be 0101.

But then how do we distinguish between positive and negative numbers? In the preceding example, how do we know that 1011 stands for -5 as opposed to 11? This is accomplished by recognizing that, given n bits, a total of 2^n different numbers (positive and negative) can be specified. We must partition this set between the non-negative numbers (positive or 0) and the negative numbers. This is done by reserving all of the codes with 0 for the most significant bit for the non-negative numbers, and all of the codes with 1 in the first position for the negative numbers, the latter being represented by the two's complements of their magnitudes. Thus, as in sign-magnitude notation, the leftmost bit serves as a sign bit. This works out very nicely because the binary representations of all of the non-negative numbers from 0 through $2^{n-1} - 1$ have 0 for the most significant bit, and the two's complements of the numbers from -1 through -2^{n-1} all have 1 for the most significant bits. This is illustrated in the table of Figure A.1 for the case of $n = 4$.

Decimal Number	Two's Complement Representation
7	0111
6	0110
5	0101
4	0100
3	0011
2	0010
1	0001
0	0000
-1	1111
-2	1110
-3	1101
-4	1100
-5	1011
-6	1010
-7	1001
-8	1000

Fig. A.1. Two's complement system for 4-bit numbers.

Readers should examine this table carefully to verify the various statements made earlier. Note that all 16 codes have been allocated. There is only one representation of 0. However, since 0 takes up one code, this leaves an odd number of codes to be allocated between the positive and negative numbers. In this system, the extra code goes to the negative numbers in that (-2^{n-1}) is represented, but 2^{n-1} is not represented. In our 4-bit example, -8 appears, but 8 does not appear. This asymmetry is an inevitable price of insisting on a single representation of 0. More about this later.

Let us now see how the system works out with respect to arithmetic operations. First we note that a number, whether positive or negative can be negated (i.e., its sign changed) by taking its two's complement. This is obviously true by definition for positive numbers. Suppose now we have the two's complement representation of $(-k)$, which would be $2^n - k$. If we take the two's complement of this, we obtain $2^n - (2^n - k) = k$, which is as it should be. The case of 0 merits special consideration. What is the two's complement of 0? Our procedure specifies this as $2^n - 0 = 2^n$. At first, this looks like trouble. But then we note that 2^n exceeds the maximum number that can be represented by n bits, and so, as noted at the outset of this discussion (item 3), the actual number that will be stored is the calculated number modulo-2^n, which, in this case is 2^n modulo-2^n, or simply 0. Thus, 0 is its own two's complement, and all is well.

Now consider what happens when two's complement numbers are added. Adding two non-negative numbers produces another non-negative number, provided that the sum does not exceed $2^{n-1} - 1$ (which we have specified as the largest allowable non-negative number). For example, adding 0100 to 0010 yields 0110. (The situation in which the sum is too large, known as an overflow, is treated at a later point.)

Suppose two negative numbers $(-k_1)$, and $(-k_2)$ expressed in two's complement are added. This is equivalent to finding $2^n - k_1 + 2^n - k_2 = 2^n + 2^n - (k_1 + k_2)$. Since what actually appears in an n-bit register is the result modulo-2^n, this is equivalent to $2^n - (k_1 + k_2)$, which is the two's complement of $(-k_1) + (-k_2)$, assuming $k_1 + k_2$ does not exceed 2^n (which would be another instance of overflow). As an example, consider, for our 4-bit word size, adding -2 and -3. In two's complement form this would be equivalent to $(16 - 2) + (16 - 3) = 16 + 16 - (+2 + 3)$, which, modulo-$2^4$, is equal to $16 - (2 + 3) = 16 - (5) = 11$, the two's complement version of -5.

If a non-negative number k_1 is added to a negative number $(-k_2)$, where $k_1 < k_2$, the result would be $k_1 + (2^n - k_2) = 2^n - (k_2 - k_1)$, which is the two's complement representation of the correct result, a negative number. For example, $1 + (-6)$ would correspond to $1 + (2^4 - 6) = 2^4 - (6 - 1) = 2^4 - 5$, the two's complement form of -5. If $k_1 \geq k_2$, then the result can be expressed as $2^n + (k_1 - k_2)$, which is the correct answer, a positive number, modulo-2^n. An instance of this situation is the sum $5 + (-3)$, which in two's complement form, corresponds to $5 + (2^4 - 3) = 2^4 + (5 - 3) = 2^4 + 2 = 2$ modulo-2^4. No overflow is possible when numbers of opposite signs are added.

Thus, in all cases, correct results are obtained. Let us now consider how to compute the two's complement of an n-bit binary number k. If we have an arithmetic unit that can perform subtraction, the simplest approach is to generate $0 - k$. This is clearly equal, modulo-2^n, to $2^n - k$. For example, in binary, with $n = 4$, $0000 - 0101 = 1011$. The carry out from the leftmost position is simply ignored.

An approach that does not require subtraction is based on the fact that the binary representation of $2^n - 1$ consists of a string of n 1s. For example, $2^4 - 1 = 15$ is written in binary as 1111. If we subtract k from $2^n - 1$, the effect is simply to complement each bit of the binary form of k. In other words, all that is necessary to convert the binary representation of k to the representation of $2^n - 1 - k$ is to complement each bit. Thus, $1111 - 0101 = 1010$. But this number is just 1 less than the two's complement. Hence, we can obtain the two's complement of a binary number by complementing each bit of the number and then incrementing the result by 1. Thus, to obtain the two's complement of 6 in our 4-bit system, we go from 0110 to 1001 to 1010. We will sometimes use the notation $TC(k)$ to represent the two's complement of k. Thus, for the preceding example, we might write $TC(0110) = 1010$. Of course, this works both ways, i.e., $TC(1010) = 0110$.

Another method is based on the rule for incrementing by 1 (see the discussion of counting on p. 201). Starting at the least significant end, all bits up to and including the first 0 encountered are complemented, and all of the more significant bits are left unchanged. Suppose we apply this to the complement–increment technique. Because of the complementing step, the incrementing operation restores to the original values all of the bits starting with the least significant (rightmost) 1 of the original string. It inverts all of the bits to the left of this bit. So the rule is simply to complement all bits to the left of the rightmost 1 in the number. For example, applying this idea to 0110 means that we should complement the two leftmost bits, yielding 1010. This technique allows us to generate two's complements very easily by inspection. It also lends itself to implementation by relatively simple iterative circuits (see Sections 5.2 and 5.3 and Problem A.9).

If a string of 1s and 0s represents an unsigned binary number, the magnitude of this number can be calculated, as pointed out in Section A.1.1, by adding the weights corresponding to each position containing a 1, where these weights are 2^{i-1} for the ith bit from the right. That is, the weight of the rightmost position is 1, the next position has weight 2, the next 4, etc. A careful examination will show that if the same string of bits is considered to represent a signed two's complement number, the only change that need be made is to change the sign of the weight of the leftmost bit (the sign bit) to a minus sign. This does not change the value assigned to positive numbers, since the sign bit for a positive number is 0. But the value of a negative number is changed, negatively, by twice the weight of the leftmost bit (which is 2^{n-1}). Twice this number is 2^n, so in effect we are converting $2^n - k$ to $(2^n - k) - 2^n = -k$, the desired result. Thus, for the 4-bit system, the number represented in the two's complement realm by the sequence 1010 can be calculated as $-2^3 + 2^1 = -6$.

As indicated earlier, the magnitude of the largest positive integer in an n-bit system is $2^{n-1} - 1$, and the largest magnitude of a negative number is 2^{n-1}. What happens if, as a result of an addition of two valid numbers, the result is out of the permissible range? In our 4-bit system, this would occur as a result of the operation $5 + 4$. In virtually all computers, such an event, called an overflow, is detected by the hardware, so that programmers have the opportunity to specify what, if any, remedial action should be taken. In the two's complement system, there is a relatively simple solution to the overflow detection problem. It is based on carries in and out of the sign bit (the leftmost bit). Let us begin with the addition of positive integers.

Since the sign bit of a positive number is 0, when we add two such numbers, there can never be a carry out of this position. If, however, the result of the addition exceeds the limit of $2^{n-1} - 1$, there will be a carry out of the $n - 1$th bit into the sign-bit position. For example, elaborating on the $5 + 4$ example, we have $0101 + 0100 = 1001$. (The value of the sign bit was changed to 1 by the carry.) Thus, for the addition of two positive numbers, there is overflow if and only if there is a carry *into* the sign-bit position, but no carry *out* of it. Consider next the adding of two negative numbers.

Since the sign bit of a negative number is 1, adding two such numbers always results in a carry out of the sign-bit position. If the result of the addition is a valid number, then the sign bit of the result will also be 1. This can occur only if there was a carry into this position. Thus, for the valid addition $(-3) + (-5) = (-8)$, we have $1101 + 1011 = 1000$. An example of such addition with overflow is $(-4) + (-5)$, which corresponds to $1100 + 1011 = 10111$. In the overflow case, there is no overflow into the sign-bit position. So for the case of addition of negative numbers, there is overflow if and only if there is a carry *out* of the sign-bit position, but no carry *into* it. Now consider adding a positive and a negative number. Since one of the sign bits is a 1 and the other a 0, there will be a carry out of the sign position if and only if there is a carry into it. For the example $(-3) + (5)$, corresponding to $1101 + 0101 = 0010$, there is a carry in and a carry out of the sign position. On the other hand, for the case of $(3) + (-5)$, corresponding to $0011 + 1011 = 1110$, there is no carry in and no carry out of the sign position. For the sum of a positive and negative number, there can never be any overflow.

Summarizing, then, we see that all cases are covered by the following rule: For the addition of any two valid numbers expressed in the two's complement system, there is overflow if and only if the carry out is not equal to the carry in to the sign-bit position. This condition can be detected by an XOR-gate with inputs corresponding to the carries in and out of the leftmost bit.

One more situation must be considered. In our initial discussion of two's complements, it was pointed out that there is a valid negative number without a matching valid positive number with the same absolute value. In general, this is (-2^{n-1}). For our $n = 4$ example, this number is -8 (See Figure A.1). This is a perfectly valid number, behaving the same way as all other negative numbers in addition operations. What happens if we attempt to negate this number, that is, execute the operation $-(-2^{n-1})$? The result would be a positive number out of range, since 2^{n-1} is not expressible in an n-bit two's complement system. If the subtraction method is used, then it is easy to see that there will be a carry out of the sign position, but no carry in to it. If the two's complementing operation, which is how negation is accomplished, is carried out using the complement–increment method, the result of the incrementing part is a carry in to the sign-bit position, but no carry out. If the iterative circuit method is used, it is not hard to see that a similar simple detection method exists. In all cases, the behavior of 2^{n-1} is unique with respect to the carries involving the sign bit, so that detection is easily achieved.

When a computer has 32-bit words, it is extremely unlikely that the number 2^{n-1} will be encountered in any particular operation. (What are the odds?) But this number *will* occasionally appear and be negated. Unless this rare event is provided for, it may result in unexpected and perhaps catastrophic consequences. This is precisely the kind of situation that causes embarrassing, costly, and sometimes tragic computer failures.

A.1.4. Integer Conversions to Base 10

Since computers usually operate in binary and people operate in decimal, it is necessary to be able to convert numbers between decimal and binary forms. Before discussing these specific conversions, it will be useful to consider the more general case of conversions between decimal and base-b numbers for arbitrary values of b.

Consider first conversions from base b to base 10. Working in base-10 arithmetic, this is a straightforward process. We simply apply Equation (1). For example, to convert the base-3 (ternary) number 2102 (the base is often indicated by a subscript, e.g., 2102_3) to decimal, we obtain $2 \times 3^3 + 1 \times 3^2 + 2 \times 3^0 = 65$. For an n-digit number, this process requires a maximum of $n(n-1)/2$ multiplications. A more efficient procedure, requiring no more than $n - 1$ multiplications, is derived by repeated factoring of the right side of Equation (1). For the case of 2102, this yields

$$((2 \times 3 + 1) \times 3 + 0) \times 3 + 2 = 65.$$

In the general case we have

$$x = (\ldots(((a_{n-1}b + a_{n-2})b + a_{n-3})b + a_{n-4})b + \cdots + a_1)b + a_0. \qquad (2)$$

This procedure is easily executed starting from within the innermost parentheses and working out, i.e., multiply the leftmost digit of x by b, then add the next digit of x, then multiply the result by b, and continue alternatively adding a digit of x and multiplying by b. The process terminates with the addition of a_0.

A.1.5. Conversions from Base 10

Consider now the inverse process, the conversion of a number from base 10 to base b. If we divide both sides of Equation (1) by b, an examination of the right side indicates that the remainder is a_0, the least significant digit of x written in base-b notation. The quotient is in the same general form as the original equation, but now a_1 is the rightmost term of the left side. Dividing the quotient by b therefore yields a_1 as the remainder.

Repeating this process generates all of the digits of the base-b representation of x, from right to left. As an example, consider finding the base-8 (octal) representation of 971_{10}. We have the following sequences of operations:

$$971/8 = 121 + 3/8; \quad 121/8 = 15 + 1/8; \quad 15/8 = 1 + 7/8; \quad 1/8 = 0 + 1/8.$$

The octal representation of 971 is therefore 1713. (Verify this by using the method of Equation (2) to convert back to decimal.)

Note that all operations in this process are carried out in the decimal system, which corresponds to the base of the *original* number. The previous algorithm, for converting from base b to base 10 is also executed using decimal arithmetic—in this case, the system of the *final* number. Thus, we now have methods generally applicable to any number base conversion, and we can choose to operate in the arithmetic of either of the two systems involved. Suppose it is essential (or very convenient) always to use one particular base, b^*, for all such arithmetic, even where b^* is not the base of either of the systems involved (e.g., b^* might be 10 if this is to be a hand computation, or it might be 2 if it is to be executed by a computer). Then we might convert the original number into a base-b^* number, using the division method, and then use the multiply and add method to convert from b^* to the base of the final number. In both cases, we would be operating with base-b^* arithmetic.

A.1.6. Fractions

In a base-b system, we express "radix" fraction (generalization of decimal fraction) x as $0.a_1 a_2 \cdots a_n$, where the a_i are numbers between 0 and $b - 1$, and where

$$x = a_1 b_+^{-1} a_2 b^{-2} + \cdots + a_n b^{-n}. \tag{3}$$

In the decimal system, multiplication by 10 is equivalent to moving the decimal point one position to the right, and division by 10 is equivalent to moving the decimal point one position to the left. Analogously, it is evident from (1) and (3) that for a number written in base b, multiplication or division by b has the effect of moving the radix point (the generalization of the decimal point) one step to the right or left, respectively.

To convert the decimal fraction x to base b, we must identify the a_i coefficients in (3). The key point is to note that if the right side of (3) is multiplied by b, the result is a_1, an integer, added to a sum of fractions totalling less than 1. Hence, the *integer part* of the product of x and b is a_1. If we multiply the fractional part of this initial product by b, the result will be a_2, and if we continue this process, we generate digits of the base-b equivalent of x in decreasing order of significance (i.e., from left to right). The procedure just described is illustrated below by the generation of the binary equivalent of 0.324_{10} to three significant figures:

$$0.324 \times 2 = 0.648; \quad 0.648 \times 2 = 1.296; \quad 0.296 \times 2 = 0.592; \quad 0.592 \times 2 = 1.184.$$

Hence, we have 0.0101_2 as the desired result. (Strictly speaking, we should generate one extra digit and use it for rounding purposes. In this case, it is clear that the value of the next digit is 0, so our result is unchanged.)

Conversion of radix fraction y from base b to base 10 can be accomplished efficiently by factoring the right side of Equation (3) to obtain Equation (4) below (which is analogous to Equation (2)):

$$y = (\ldots((a_k/b + a_{k-1})/b + a_{k-2})/b + \cdots + a_1)/b. \tag{4}$$

Note that, in contrast to the analogous procedure for integers based on (2), we start with the *rightmost* digit and terminate with *division*. In the example below we convert the ternary fraction 0.1202 to decimal form:

$$y = (((2/3 + 0)/3 + 2)/3 + 1)/3.$$

We evaluate the expression by starting with the innermost term and working out. We divide the rightmost digit by 3, add the next digit, divide the result by 3, add the next digit, etc., terminating with a final division by 3. The result in this case is 0.580. (Precision for this procedure depends on the precision of our arithmetic.)

A.1.7. Conversions Between Base b and Base b^k

Consider now the case of converting a number from base b to base b^k. Using the division method, we would be operating in base-b arithmetic, repeatedly dividing by the base of the "destination" system, namely b^k. Each such division in equivalent to k divisions by b. But each of these amounts to moving the radix point one place to the left. Hence, each division by the destination base consists of moving the radix point k places to the left. Since the number being divided (the dividend) each time is an integer, the remainder, which corresponds to a digit of the desired number, is the part of the quotient to the right of the radix point, i.e., the rightmost k digits of the dividend.

It follows then that to convert a base-b number x to base b^k, all we need do is partition the digits of x into subsequences of length k, starting from the right, and convert each of these subsequences into the corresponding base-b^k digit. For example, to convert x = 1001010111_2 to octal ($8 = 2^3$—so $k = 3$) we group the digits of x as follows: {001}{001}{010}{111}. Then we convert each group to an octal digit yielding 1127_8. Conversions in the reverse direction are correspondingly simple.

To go from base b^k to base b, we need only convert each digit of the original number to a k-digit (*don't forget to add leading 0s as necessary to fill out the required length*) base-b number. Thus, to convert 6207_9 to base 3, we replace 6 by 20, 2 by 02, 0 by 00, and 7 by 21 to obtain 20020021_3.

A.1.8. Binary, Octal, and Hexadecimal

In return for the simplicity of the binary number system (very simple arithmetic operations and only two symbols), we must pay the price of writing relatively long sequences of digits to represent a given number —more than triple the number of digits required for a decimal representation. Here is where the techniques just described come in handy. Since $8 = 2^3$, it is very easy to convert back and forth between binary and octal numbers. Thus, we can use octal numbers to express binary numbers in compact form. For example, 1000111010110110_2 can be abbreviated by 107266_8. (Note that the same technique can be used to abbreviate *any* sequence of 0s and 1s, even if it does not represent a number.)

Although converting from decimal to binary is a straightforward matter using the division method, it is a lengthy computation to perform by hand when more than a few digits are involved. The reader might wish to verify this assertion by converting 97,603 to binary form (the answer is 10111110101000011). It is much easier to convert first to octal (obtaining 276,503) and *then* to binary—try it! A similar advantage exists when octal is used as an intermediate stage for conversions from binary to decimal.

In most computers, information is usually in multiples of four bits (16, 32, and 64 are common word sizes, and the 8-bit *byte* is almost universally used). It is therefore often convenient to abbreviate bit strings in groups of four, using the base-16, or *hexadecimal* (generally abbreviated as *hex*) number system. Then each byte is represented by two hex numerals. Since there are 16 numerals in hex, as opposed to only 10 in our everyday decimal system, 6 additional numerals are needed. Rather than invert new symbols, the common practice is to use the letters A through F to represent the numerals with values from 10 through 15, respectively. Thus, 506_{10} corresponds to $1FA_{16}$ $(1 \times 16^2 + 15 \times$ '16 + 10).

A.1.9. Binary-Coded Decimal

It is important that computers be able to represent numbers internally in decimal form. In some cases, where internal arithmetic operations are relatively simple, as is often the case for commercial data processing, conversions back and forth between decimal and binary would be more time-consuming than the arithmetic intrinsic to the problem. Hence, it might be advantageous to have the machine operate internally in decimal. Even where the internal operations are to be in binary, it is necessary to represent the decimal numbers that constitute the input to the machine, and it is necessary to generate decimal numbers as the output. How can this be done, given the assertion made earlier in this appendix that 2-valued signals are much preferred? The answer is to encode each decimal digit independently as a sequence of binary digits (or bits). Several different encodings have been used for this purpose, but we confine ourselves here to the most common, and most straightforward, scheme, which is simply to convert each decimal digit to its binary equivalent. This scheme, called *binary-coded decimal* (*BCD*), requires four bits per digit. Leading 0s are inserted as necessary to fill out the 4-bit slots for digits less than eight in value. Thus, 2089_{10} is represented in BCD as the concatenation of the four sequences 0010, 0000, 1000, 1001, or 0010000010001001. Note that the *binary* representation of 2089 is 100000101001, a 12-bit as opposed to a 16-bit number. This less efficient use of bits reflects the fact that only 10 of the 16 possible sequences of the four bits used to represent a decimal digit are utilized to represent a decimal digit.

A.1.10. ASCII Code

Although digital computers were originally developed to perform complex numerical computations, they are currently used for a wide range of other purposes, many of which involve the processing of symbols other than numbers, particularly alphanumeric characters. (Indeed, the very words you are now reading were typed into a computer controlled by a word processing program.) It is therefore necessary to have a means for representing, within a computer, the sort of characters found on a standard typewriter. Since such data is often circulated among different computers, and may be transmitted over communications networks, it is useful to standardize the codes used.

The most popular such standard is the American Standard Code for Information Interchange ASCII (pronounced as'kee). Since seven bits are used per character, a total of 2^7 or 128 different characters can be represented. The first three bits of an ASCII code word specify a class of characters, and the last four bits indicate the precise character within that class. For example, the numerals are all prefixed by 011, the 4-bit suffix indicating the particular numeral in BCD. Other characters with prefix 011 include the colon, semicolon, and question mark. Uppercase letters are prefixed by 100 and 101, and lowercase letters are prefixed by 110 or 111. For example, 5 is encoded as 0110101, C as 1000011, D as 1000100, c as 1100011. Various other symbols are prefixed by 010, for example, 0100000 is a space and 0101110 is a period. A number of ASCII codes are used for control signals such as "end of transmission," or "carriage return." A major competitor of ASCII is Extended BCD Interchange Code (EBCDIC), the 8-bit code originated by IBM.

A.1.11. Gray Code

There are situations requiring a sequence of binary code words in which adjacent pairs of elements in the sequence differ in only one bit position. Such a sequence, called a *reflected Gray code*, is shown below (let the four bit-positions be designated as a_1, a_2, a_3, a_4):

$$0000, 0001, 0011, 0010, 0110, 0111, 0101, 0100,$$
$$1100, 1101, 1111, 1110, 1010, 1011, 1001, 1000.$$

This sequence can be constructed as follows: Begin with 0000, then keep a_1, a_2, a_3 the same and change a_4 to 1 to obtain the second word 0001. The next two words are obtained by "reflecting" the first two words, in the sense that a_1 and a_2 are kept the same, a_3 is changed to 1 for both of them, and the value of a_4 is the same in word 3 as in word 2, and the same in word 4 as in word 1. Now we have 0000, 0001, 0011, 0010. To generate the next four words, we retain a_1 at its previous value of 0, change a_2 to 1 for all of them, and "reflect" the values of a_3, and a_4 by setting their values in words 5, 6, 7, and 8 to equal their values in words 4, 3, 2, and 1 respectively. The last eight words can be generated by changing the value of a_1 and reflecting the values of a_2, a_3, and a_4 about their values in the first eight words. This general scheme is applicable to Gray code sequences of other sizes. Another approach to generating sequences of this type is pointed out in connection with Karnaugh maps, introduced in Chapter 3.

A.1.12. Error-Correcting Codes

Bits of data words are sometimes corrupted in the course of transmission or memory operations.

A common technique for *detecting* that a bit has been changed, say in transmission, is to add to each word a *parity-check bit*, which is generated by taking the modulo-2 sum of the other bits in the word. The total number of 1s in the enhanced word is clearly *even*. At the receiving end, the complete word, including the check bit, is fed into another modulo-2 adder. If the output is a 1 (meaning that the number of 1s in the received word in *odd*), then this indicates that a bit has been changed. This is called a *parity-check failure*. There is no indication as to *which* bit is in error. This *parity check* scheme gives no error indication, of course, if 2, 4, or any other even number of bits are corrupted.

When parity checks fail, systems are designed to take appropriate action, such as halting, setting off alarms, or retrying the process that led to the failure. As logic elements on integrated-circuit (IC) chips continue to shrink, memory and logic circuits are becoming increasingly vulnerable both to hard and soft faults (Section 3.9). It is therefore very desirable to go beyond the mere *detection* of faulty bits to provide means for automatically *correcting* errors of this type. This would enable systems to operate without interruption despite occasional transient errors. Furthermore, memory chips with a few isolated faulty cells would still be usable. A very simple error-correcting method is presented next.

Suppose each bit is individually triplicated. That is, from information bit b we generate three bits, $b^1 = b^2 = b^3 = b$. The three b's are then stored or transmitted in place of b.

Whenever we wish to use b in some logic function, we regenerate in from the b's with a majority gate (Subsection 5.1.1). That is we set $b = M(b^1, b^2, b^3)$. Now, even if one of the b's has been corrupted, the resulting b-value is still correct. For example, suppose $b = 1$. Then, as originally generated, $b^1 = b^2 = b^3 = 1$. Suppose b is to be stored in memory. Then all three of the b's are stored. If b^2, for example, should be changed to a 0 by noise in the memory, then when b is regenerated after readout, it will be set to $M(1, 0, 1) = 1$, the correct value. This is an effective technique, but it is very costly in that storage requirements or transmission capacity must be tripled in order to use it. A widely used scheme for achieving the same general result more economically is described below. It is based on an elaboration of the concept of simple parity checks.

The idea is to assign each bit of a word to one or more overlapping subsets called *check sets*. Each check set includes one parity-check bit, whose value is chosen to make the overall parity of the check-set even. If a single bit of a word is altered, then the parity of each check set containing it changes. The check sets are so chosen that each bit belongs to a different collection. Therefore, by determining which check sets have odd parity (fail a parity check), the altered bit can be uniquely identified. It is then a simple matter to restore it to the correct value. Let us see now how the check sets can be chosen.

Consider a word with bits b_i, where i ranges from 1 to 7. Bits b_1, b_2, and b_4 are the check bits, and the other four are the information bits. Designate the overlapping check sets as S_1 S_2, and S_3. Include b_i in S_k if the kth bit from the right of the binary representation of i is a 1. Thus, for $i = 3 = 011_2$, b_i is assigned to S_1 and to S_2. Similarly, b_6 is assigned to S_2 and to S_3, and b_7 is assigned to S_1, to S_2, and to S_3. Note that each of the check bits, b_1, b_2, and b_4 is assigned to exactly *one* check set. Thus, their values can be assigned as unique functions of the other bits to make the parity of each check set even, as follows: $b_1 = b_3 \oplus b_5 \oplus b_7$, $b_2 = b_3 \oplus b_6 \oplus b_7$, and $b_4 = b_5 \oplus b_6 \oplus b_7$. Remember that b_3, b_5, b_6 and b_7 are the information bits, i.e., the bits that constitute the data. The 7-bit word is generated as shown in Figure A.2.

The parity of check set S_i is P_i, computed as the modulo-2 sum of all of the bits in S_i. Thus, $P_1 = b_1 \oplus b_3 \oplus b_5 \oplus b_7$, etc. Let us see what happens if a bit of a word encoded

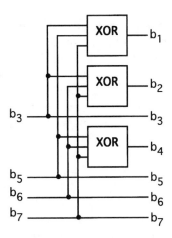

Fig. A.2. Generating the parity bits to complete the 7-bit word.

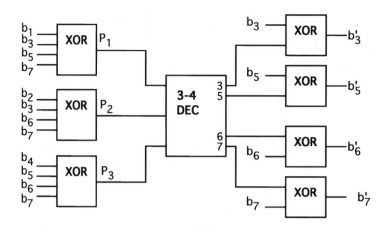

Fig. A.3. Logic circuitry for correcting a single faulty bit in a word.

as described here is corrupted. The correction circuitry is shown in Figure A.3. Suppose, for example, that b_5 is altered by noise in the system. Then the parities of each check set containing b_5 will be 1, and the others (assuming no other errors) will be 0. Thus we have the parity vector $P_3 P_2 P_1 = 101$. This indicates that the faulty bit is in sets S_1 and S_3. From the way the sets were constructed, this uniquely identifies the faulty bit as b_5. In fact, the subscript of the faulty bit is obtained by regarding the vector $P_3 P_2 P_1 = 101$ as a binary number. Note that the scheme makes no distinction between check bits and information bits. A fault in b_2, for example, would produce the parity vector 010, which singles out b_2 in precisely the same manner that b_5 was identified. Thus, any single fault is pinpointed and can therefore be corrected (although in the circuit shown, errors in the parity bits are not corrected). The P_i-signals, produced by the parity-check circuits, control the 3-4 decoder, the outputs of which select which, if any, of the information bits is to be corrected. The corrected information bits are labeled with primes. Note that if the data at this point is to be transmitted or stored (as opposed to being processed), then a 3-7 decoder could be used to make the check bits as well as the information bits subject to correction. Protection against a subsequent error would thus be retained.

Rather than tripling the number of bits in the code word, as was done with the simple scheme mentioned earlier, only three extra bits were needed to accommodate four information bits. It should be obvious how the same scheme can be extended to provide single-error correction for a 15-bit word, consisting of 4 check bits and 11 information bits. This general scheme is called a *Hamming single-error-correcting code* (*SECC*). By adding one more overall check bit, the scheme can be extended to *detect*, but not correct, all double errors.

This is achieved by setting $b_8 = b_1 \oplus b_2 \oplus b_3 \oplus b_4 \oplus b_5 \oplus b_6 \oplus b_7$ and adding the overall parity check $P = b_1 \oplus b_2 \oplus b_3 \oplus b_4 \oplus b_5 \oplus b_6 \oplus b_7 \oplus b_8$. If there is no error, then P and all of the P_i are 0. Should a single bit be in error, then $P = 1$, and the P_i-checks identify the faulty bit. If a double error occurs, then at least one of the P_i will be 1, but $P = 0$. This indicates that two bits are faulty, but there is no way to determine *which* bits are incorrect. More elaborate codes exist that can detect *and* correct multiple errors. But these are rather costly and so are less commonly used.

A.2. SIMPLE ELECTRICAL CIRCUITS

Mastering the material of this section will *not* qualify anyone as an electrical engineer. The subject matter is limited to what is absolutely necessary for an elementary understanding of the operation of logic-circuit elements. This understanding should be sufficient for an appreciation of such basic issues as the relation between speed and power.

We begin by showing how to calculate voltages and currents in circuits for which time is not a factor. Next, the simplest type of transistor is introduced and modeled as a switch. Finally, capacitance, the enemy of speed, is brought into the picture and its basic role outlined.

A.2.1. Resistors and Voltage Sources

Electric current is the flow of electric charge. Current flow through a wire, for example, is measured in terms of the number of unit charges that pass a given point per unit time. Charge is measured in *coulombs*. The basic unit of current is the *ampere*, defined as the number of coulombs flowing per second. The symbols Q and I, respectively, are commonly used for charge and current, so we would write this as:

$$I = Q/t. \tag{5}$$

A *conductor* is a material, such as copper and most other metals, that permits the flow of current through it. At the other extreme is an *insulator*, such as glass, that completely blocks the passage of current. In between are *semiconductors*, such as silicon, that, depending on their exact composition and on the temperature, can conduct *or* insulate. Current flow through a conductor is analogous to the flow of water through a pipe. Even if the pipe is perfectly level, there is friction to be overcome. That is, there is *opposition* to the flow of the water. This must be overcome by the application of *pressure*. The analogous opposition to the flow of electric current is called *resistance*, measured in *ohms*. Corresponding to pressure, current is pushed along by *voltage*, measured in *volts*. Current flow through any path is directly proportional to the applied voltage, and is inversely proportional to the resistance of the path, as expressed by *Ohm's law*:

$$I = V/R. \tag{6}$$

These ideas are illustrated in the following example.

Example A.1

The diagram in Figure A.4*a* depicts three basic elements connected together. At the left, labeled V_S, is a *voltage source*. This represents a device, such as a battery or electrical generator, that produces a voltage, V_S, across its terminals. Voltage V_S remains fixed regardless of the current flowing through the source, Next is a switch, S, which when open represents an infinite resistance, so that no current can pass through it. When closed, S has 0 resistance, so that it allows current to pass through without opposition. Finally, at the right there is a *resistor*, labeled R. It is a device that has a resistance of value R. The lines joining these elements are *wires*, assumed to offer no resistance to the flow of current. When the switch S is open, the circuit is broken, and no current can flow. When S is closed, current flows *out* of the *positive* terminal of the voltage source in the direction of the arrow labeled I (corresponding to the current

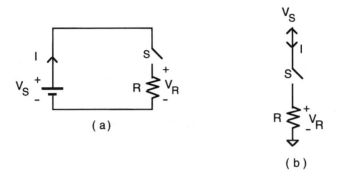

Fig. A.4. A circuit with a switch, voltage source, and resistor drawn in two equivalent ways.

magnitude). The current flows down through the switch and the resistor back into the *negative* terminal of the voltage source. When the switch is closed, the total resistance in the path between the voltage source terminals is R, and so, according to Ohm's law, the current I is V_S/R. When S is closed, the voltage across the resistor is the same as that across the source. When S is open $V_R = 0$.

The diagram shown in Figure A.4b represents *exactly* the same circuit depicted in Figure A.4a. The little triangle at the bottom represents *ground*, a point in the circuit used as a reference for voltage measurements. If several ground symbols appear in a diagram, it is assumed that they are all connected together by 0-resistance wires. The arrow at the top, labeled V_S, indicates that a voltage source of magnitude V_S is connected from that point to ground. Where symbols representing voltages, such as V_R are shown, it is assumed that they stand for voltages with respect to the ground voltage (assumed to be 0). This form is very common and is used here in subsequent circuit diagrams.

Ohm's law is used in several different ways. In the previous example, the current through a resistor was determined by dividing the voltage across it by the resistance. In other cases, we may know the value of the resistance and the value of the current flowing through it. We can then apply Ohm's law in the reverse direction to determine the voltage across the resistor by multiplying the resistance value by the current. It is important to understand that a voltage source pushes current *out* of its positive terminal, and takes it back in through the negative terminal. Current flows *into* the terminal of a resistor that is most positive, and *out* of the terminal that is negative. An examination of Figure A.4a should make this clear. Circuits with two resistors are considered next.

Example A.2

Resistors R_1 and R_2 in the Figure A.5a circuit are connected in *series*, i.e., in such a way that the *same* current flows through both of them. No current flows through the line connected to the junction of R_1 and R_2. It is there simply to indicate that V_2, the voltage (or *potential difference*, as it is sometimes called) between this junction and ground, may be considered as an output of the circuit. Since the current must be forced through *both* resistors, their effect is additive. That is, the equivalent resistance of a group of resistors connected in series is equal to the sum of the individual resistance values. This is

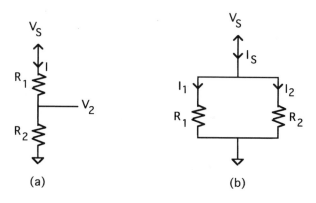

Fig. A.5. Resistors connected in series and in parallel.

intuitively evident, and can be justified formally by application of Kirchhoff's voltage law, which is: "The algebraic sum of the voltages around any closed path in a circuit is always 0." The current flowing out of the source and through the resistors is therefore $I = V_S/(R_1 + R_2)$. This particular circuit configuration is of prime interest from the point of view of logic circuits. The key variable is V_2, the voltage across R_2. Having already calculated the current through R_2, we can find V_2 simply by multiplying the current by R_2. This gives us

$$V_2 = V_S R_2/(R_1 + R_2). \tag{7}$$

Note that the proportion of the total voltage V_S that appears across R_2 is the ratio of R_2 to the total resistance. The series combination thus acts as a voltage *divider*, in that it divides V_S between the two resistors in proportion to their values.

In the circuit of Figure A.5*b*, resistors R_1 and R_2 are connected in *parallel* in the sense that their corresponding terminals are connected together so that the same voltage is across both of them. The currents through the resistors are $I_1 = V_S/R_1$ and $I_2 = V_S/R_2$. At any node in a circuit, the algebraic sum of the currents entering that node is 0. (This is Kirchhoff's *current* law.) Thus, we have $I_S = I_1 + I_2$, or $I_S = V_S(1/R_1 + 1/R_2)$. Now we can use Ohm's law to solve for R, the resistance equivalent to the parallel combination:

$$R = \frac{V_S}{I_S} = \frac{1}{(1/R_1) + (1/R_2)} = \frac{R_1 R_2}{R_1 + R_2}.$$

Connecting resistors in parallel results in an equivalent resistance *lower* than any of the individual resistances. The next example involves both serial and parallel connections.

Example A.3

In Figure A.6, current I_T flows through the source and R_3, and then divides, part of it going though R_1 and the other part through R_2. Suppose the problem is to find the output voltage V_o. A good approach is to think of the circuit as consisting of R_3 in series with the parallel combination of R_1 and R_2. Then we can find V_o with the aid of the voltage divider concept embodied in equation (7). From the analysis of Figure A.5*b*, the

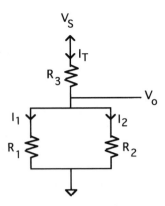

Fig. A.6. A series-parallel circuit.

parallel combination has resistance $R_1R_2/(R_1 + R_2)$. The total resistance is found by adding this to R_3 to obtain $R_3 + R_1R_2/(R_1 + R_2)$. Applying (7) gives us the desired result:

$$V_o = V_S \frac{(R_1R_2/(R_1 + R_2))}{R_3 + (R_1R_2/(R_1 + R_2))} = V_S \frac{R_1R_2}{R_3(R_1 + R_2) + R_1R_2}.$$

An important issue in the design of electric circuits, particularly logic circuits, is that of power. Power supplies for computers are very expensive and, as explained below, power is associated with heat, a waste product of computing that is often very costly to dispose of. In general, power is defined as the rate of energy production or consumption per unit time. It is measured in watts. The power output of a voltage source is given by

$$P = IV. \tag{8}$$

Power dissipation in a resistor is given by the same equation. In the latter case, by applying Ohm's law to replace V in (8) by IR yields the very useful equivalent equation:

$$P = I^2R. \tag{9}$$

The energy dissipated in a resistor is converted to heat. This heat must somehow be removed, or else the temperature of the resistor will increase until something unpleasant happens. Therefore, for any given resistor, the heat-disposal problem increases with the square of the current flowing through it. Heat removal may be simply by the natural circulation of air, as is the case for such devices as hand-held calculators. It may require forced-air circulation by means of an electric fan (as is the case with the small computer that is processing these words as I type them). Or, in the case of large, high-performance computers, the circuit boards may be embedded in housings fitted with elaborate plumbing and refrigerating arrangements to conduct away the heat in circulating fluids.

The preceding discussion dealt with ideal elements of several types. It should be understood that such elements do not exist in the real world. What does exist are elements that approximate ideal behavior over certain ranges of the relevant parameters. There are no ideal voltage sources, since any real source is limited as to how much current it can deliver before its output voltage falls off significantly. But we can, for example, get a power supply that maintains its output voltage within 5 percent of 3 volts provided that the current drain does not exceed 50 mA (milliamperes). Similarly, resistors behave as specified only if they are not permitted to overheat (usually this results from excessive current flow). Except for the special case of superconductors, all connecting wires have some resistance. In many cases, particularly on integrated-circuit chips, this resistance is a major concern.

The ideas presented here in relation to simple circuits composed of resistors, switches, and voltage sources are directly applicable to circuits employing transistors. It is possible to approximate the behavior of transistors used in logic circuits in terms of these simpler elements. This is the subject of the next section.

A.2.2. How an MOS Transistor Works

There are two principal types of transistors, metal-oxide silicon (MOS) and bipolar. Because MOS transistors are easier to understand, and are the most widely used in logic circuits, only this type is treated here. The explanation presented here is on a very rudimentary level, but it is sufficient to convey a rough idea as to how these devices work. The particular device to be described is a negative carrier-type metal-oxide-silicon field-effect transistor (nMOSFET). It will be referred to here simply as an MOS transistor.

MOS transistors are constructed on silicon chips. Millions can be placed on the surface of a chip 10 mm (millimeters) square. A few terms pertaining to semiconductor physics must be defined at the onset. We are interested here in two types of semiconductors: *n*-type and *p*-type. A *p*-type semiconductor has a preponderance of *positive* charges (the *majority carriers*) available to carry current. But, and this is important, it does have a small number of *minority* carriers consisting of *negative* charges. The reverse is true of *n*-type semiconductors. The nMOSFETs under discussion are constructed on a block (or *substrate*) of *p*-type silicon.

In the cross-sectional diagram, Figure A.7*a*, a region of the substrate has been converted to *n*-type silicon. This region (marked with dots) is called a *channel*. Metal contacts (indicated by diagonal stripes) are attached to the substrate at the ends of the channel. The leftmost terminal is called the *source* and the rightmost terminal is called

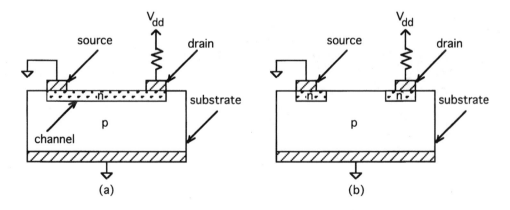

Fig. A.7. Conducting and nonconducting semiconductor devices.

the *drain*. These are connected together externally by a resistor in series with a voltage source. The substrate as a whole is connected to ground via a metallic contact attached to its base. In this situation, negative charges would flow out of the source, through the channel, and out of the drain. This, of course, corresponds to a positive current flowing through the channel in the opposite direction. Note that the device is perfectly symmetrical. If we were to reverse the polarity of the voltage source, then the flow of charges would also be reversed. In that event we would call the right-side terminal the source. In general, the source is defined as the terminal *from* which the charges flow into the channel, and the drain is the terminal that receives these charges. This semiconductor device could serve as a resistor with a value determined by the density of negative charges in the *n*-type channel.

Now suppose the channel is eliminated, but that two separate *n*-type islands in the *p*-type substrate remain under the contacts as shown in Figure A.7*b*. Virtually no current would flow. The negative charges would be attracted out of the drain island by the positive side of the voltage source. These carriers could not be replenished because the positive carriers in the region between the islands would be attracted away from the boundary with the drain island by the negative terminal of the voltage source (which is connected to the source since both are grounded). Now we are in a position to discuss the nMOSFET shown in Figure A.8*a*.

The device shown in Figure A.7*b* is enhanced to produce the Figure A.8*a* device by the addition of a third terminal, the *gate*. This is constructed by first coating the surface of the device above the area between the source and drain islands with a thin insulating glass layer. Then the contact, a layer of metal, is placed over the glass. It is important that the gate is completely insulated from the substrate. A voltage source V_G is connected from the gate to ground. The substrate region under the gate is, as in Figure A.7*a*, called the channel.

When $V_G = 0$, the gate has no effect at all and the device behaves exactly as did the one in Figure A.7*b*, i.e., as an open circuit. Now suppose that V_G has a substantial positive value. Then the gate, being positive, attracts into the channel area under it significant numbers of minority negative carriers from the rest of the substrate. For a sufficiently positive value of V_G, called the *threshold voltage*, V_{th}, this converts the channel from a *p*-type region to an *n*-type region, thereby uniting the source and drain

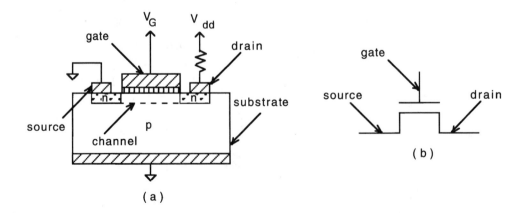

Fig. A.8. Cross sectional diagram and circuit symbol for an nMOSFET.

islands into a single *n*-type region. Now we have a situation that is substantially the same as that shown in Figure A.7a, in which the device acts as a conductor.

Thus, V_G, defined as the voltage of the gate with respect to the source, controls the conductivity of the channel, so that the device can be used as a switch. Typically, V_{dd} is in the neighborhood of 3 volts. If V_G is below V_{th}, typically about 1 volt, the transistor acts virtually as an open circuit. For V_G significantly greater than V_{th}, the transistor conducts, acting as a closed switch (although its resistance is not negligible). A common circuit symbol for a FET is shown in Figure A.8b. When used to construct logic-circuit devices, the transistor is often modeled as a simple switch, controlled by a voltage. When decisions must be made about currents, voltages, and the physical sizes of the transistors, it is necessary to use a more accurate model, in which a resistor, R_T, is placed in series with the switch.

It is also possible to construct transistors of the same type just discussed with the roles of *p* and *n* reversed. These are referred to as pMOS transistors because they utilize *positive* charge carriers. In a pMOS transistor, the source is the terminal that feeds the positive charges into the channel. Because the positive charge carriers in semiconductors (called *holes*) are more sluggish than their negative cousins (electrons), p-MOSFETs are inferior to nMOSFETs and so are seldom used by themselves. However, substantial advantages result when nMOS and pMOS transistors are used together on the same chip. This technology, known as *complementary MOS* or *CMOS*, is discussed in Subsection A.3.4.

The other major type of transistor is the *bipolar* transistor. It is much more difficult to explain its operation in simple terms, and so no attempt to do so is made here. Bipolar transistors are also used to construct logic gates and are the elements found in such logic technologies as emitter-coupled logic (ECL) and transistor–transistor logic (TTL) (Appendix A.4).

A key factor in determining power requirements and speed is the inevitable but usually unwanted presence of capacitance in electrical circuits. This is the subject of the next subsection.

A.2.3. Capacitors and Time Constants

A capacitor can be thought of as a sheet of insulating material sandwiched between two conducting plates. If a voltage is connected across the plates, there is a sufficient transient current flow to charge one side positively and the other negatively. The unit of capacitance is the *farad*. The capacitance of an element is defined as the charge acquired per unit volt, that is,

$$C = Q/V. \tag{10}$$

Once a capacitor has been charged by a voltage source, it retains the charge after the voltage source has been disconnected until it "leaks" off through imperfect insulation. The capacitance of a device can be computed by the formula $C = \epsilon A/d$, where ϵ is a constant pertaining to the nature of the insulator, A is the area of the sandwich, and *d* is the distance between the conducting plates. If capacitors with values C_1 and C_2 are connected in parallel, the combination behaves as a capacitor with value $C_1 + C_2$. How do capacitors behave as circuit components?

Example A.4

Consider the circuit shown in Figure A.9, consisting of a series connection of a voltage source, a switch, a resistor, and capacitor. Assume the capacitor is initially discharged, so that the voltage across it is initially 0 (this can be seen by solving (10) for V and substituting 0 for Q). As long as the switch remains open, no current flows and no voltage appears across the capacitor. Assume the switch is closed at time t = 0. With no voltage across the capacitor or the switch, the full source voltage appears across the resistor, producing an initial current $i = V_S/R$ flowing around the circuit. (It is customary to use lowercase i for current when we are interested in its variation over time.) Thus, there is a flow of charge *into* the upper plate of the capacitor and out of the lower plate. Then, with the passage of time, the capacitor begins charging, and a growing voltage V_C appears across it, polarized as shown in the figure. Since V_C is opposing V_S in the sense that it is trying to push current in the opposite direction, the net voltage across R, $V_G = V_S - V_C$ steadily decreases, thus decreasing the current. Thus, both the charge and V_C continue to increase, but at ever decreasing rates. It is not possible for V_C ever to exceed V_S, because if $V_C = V_S$, i falls to 0 and the process ceases. Therefore what happens is that V_C asymptotically approaches V_R from below, as shown in the graph of Figure A.9*b*.

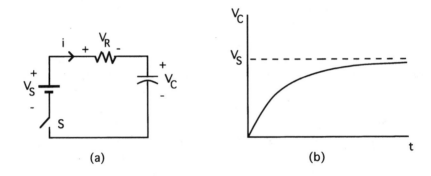

(a)   (b)

Fig. A.9. A capacitor in a series circuit.

An important reason for introducing capacitors here is to make possible an understanding of timing factors in logic circuits. It is therefore necessary to do some mathematics. Assuming S closed, we begin by applying Kirchhoff's voltage law (see Example A.1 in Subsection A.2.1) to obtain $V_S - V_R - V_C = 0$, or $V_R + V_C = V_S$. Applying Ohm's law to replace V_R by Ri, and (10) to replace V_C by Q/C, yields $Ri + Q/C = V_S$. Differentiating both sides by t gives us

$$R\frac{di}{dt} + \left(\frac{1}{C}\right)\frac{dQ}{dt} = \frac{dV_S}{dt} = 0$$

(since V_S is a constant). Now, we note that (dQ/dt) is charge per unit time, or current, so we have the simple differential equation

$$R\frac{di}{dt} + \left(\frac{1}{C}\right)i = 0. \tag{11}$$

The solution to this equation is $i = i_0 e^{-t/RC}$, where i_0 is the value of i immediately after S is closed. In our informal analysis, we assumed that the capacitor was initially discharged. Now let us take a more general view and assume some nonzero initial charge on the capacitor. Suppose this charge makes V_{C0} the initial value of V_C. Then the initial value of i becomes: $i_0 = (V_S - V_{CO})/R$. The general solution for i is thus: $i = ((V_S - V_{C0})/R)e^{-t/RC}$. Given i, it is easy to find V_C by using the expression $V_C = V_S - Ri$. Let us also generalize a bit further by referring to the value V_C is approaching as its *final* value V_{Cf} (i.e., its asymptotic value). (In this case we have found that $V_{Cf} = V_S$.) Then we can write

$$V_C = V_{Cf} - (V_{Cf} - V_{C0})e^{-t/RC}. \tag{12}$$

Let us now check out (12) against the results of the informal analysis. If $V_{C0} = 0$, as assumed, then at $t = 0$, (12) yields $V_C = V_{C0} = 0$, confirming the previous calculation. As t approaches infinity the exponential term in (12) approaches 0, indicating that V_C approaches V_{Cf}, as shown in the graph of Figure A.9b. Now let us consider the question of time.

The exponent in the solution is $-t/RC$. Thus, each time t increases by the amount RC, the magnitude of the exponent increases by one. This means that $V_{Cf} - V_C$ is divided by the constant e (which is about 2.718) every RC units of time. It is usual to refer to *RC* as the *time constant* of the circuit, represented by the symbol τ. During each interval of length τ, the difference between V_C and its final value is reduced by 63 percent (i.e., to about 37 percent of its value at the start of the interval). After 2τ it is down to about 13 percent of its original value, after 3τ it is down to about 5 percent, etc. The concepts developed thus far are sufficient for timing analyses of RC circuits with one capacitor. Equation (12) applies to *all* of the time-varying voltages in any such circuit. (The form of the equation for *currents* in such circuits is the same, with the time constant unchanged.) It is only necessary to determine the time constant, the initial value of the variable of interest, and its asymptotic (final) value. The procedure is as follows:

1. The time constant is calculated by first finding the equivalent resistance across the terminals of the capacitor, assuming all voltage sources are replaced by short circuits (i.e., that the values of the voltage sources are all set to 0). This value is then multiplied by the capacitance to get τ.
2. The initial value of the variable is calculated by assuming that the voltage across the capacitor is fixed at its initial value (reasonable because the voltage across the capacitor depends on the charge stored in it, and this cannot change instantly).
3. The final value is calculated by treating the capacitor as an open circuit (reasonable because the flow of charge into it approaches 0 in the long run).
4. Inserting these parameters into (12) gives us the expressions for the variable.

Now let us apply this approach to a realistic problem.

Example A.5

It would be wrong to malign capacitance in general. Capacitors are very useful elements in many situations. But unwanted *stray* capacitance is a pernicious factor in all digital circuitry. Every transistor has capacitance associated with it, as does every wire both on and off our chips. The values are very small, measured in picofarads (pf, meaning 10^{-12} farads), but the effects are major.

Consider the circuit shown in Figure A.10a consisting of a resistor, transistor, and voltage source. It represents a particular type of logic gate (an inverter), and typifies a wide class of circuits. An equivalent circuit for it is shown in Figure A.10b, based, in part on the discussion in the previous subsection. This model is a crude one in that it fails to take into account such factors as the nonlinearity of all of the components involved. For example, the *pullup* resistor R_U is usually implemented by a *depletion-mode* transistor, which is a variation of the type of transistor previously discussed. It can be made to behave like a resistor whose value increases when the current through it increases. Such behavior results in improved performance of the gates in which it is used. We will not discuss this effect here.

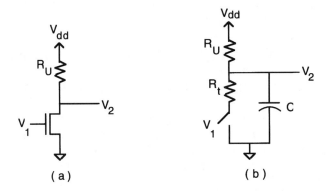

Fig. A.10. An nMOS inverter and an equivalent circuit for it.

The switch is open if the input voltage V_1 is below V_{th} for the transistor, and it is closed if the input voltage V_1 is above V_{th}. The values of resistor R_t corresponds to the resistance of the transistor channel when it is in the conducting state. The capacitor C represents the stray capacitance associated with the transistor *and* with any other devices and wiring connected to its drain. Our goal is to estimate how much time it takes for the output voltage V_2 to respond to changes in input voltage V_1 (in both directions). Reasonable values for the circuit parameters are $V_{dd} = 3$, $R_t = 3K$ (3000 ohms), $R_U = 18K$, $C = 1$ pf.

Assume V_1 has been close to 0, cutting off the transistor, for many time constants. This means that the switch has been open for a long time; no current at all is flowing in the circuit. Then there is no voltage across R_U, and hence $V_2 = V_{dd}$. Now assume V_1 changes to nearly 3 volts at $t = 0$, thus causing the switch in our model to close at that time. The initial value of $V_2 = 3$; it does not change instantly. To compute the final value of V_2, we treat C as an open circuit, i.e., as though it were not there at all, and calculate the voltage that would appear across its terminals. This is precisely the voltage divider problem of Example A.2 (see Figure A.5a.) We find $V_{2f} = V_{dd}R_t/(R_t + R_U) = 3(3)/(3 + 18) = 0.43$. The time constant is found by noting that, if we short circuit the power supply (a constant voltage is like a short circuit when we are dealing with transient effects), then R_U and R_t are effectively connected in parallel across C. Therefore, we find $t = R_UR_tC/(R_U + R_t) = 2.6 \times 10^{-9} = 2.6$ ns. Thus, we see that V_2 falls from 3 toward 0.43 with a time constant of 2.6 ns. In two time constants it falls to about $0.43 + 0.37^2 \times (3 - 0.43) = 0.78$ volts. If we assume a threshold voltage of 1, then this is certainly low enough to be satisfactory as a logic 0. Hence, the propagation delay in this gate for an output change from the high to the low value is roughly $2 \times 2.6 = 5.2$ ns. Now let us consider the delay for a change in the reverse direction.

At $t = 0$, the input signal V_1 changes from 3 volts to a value less than the threshold voltage, say 0.75 volts, opening the switch in the equivalent circuit. Now V_2 is changing from $V_{20} = 0.43$ to $V_{2f} = 3$, the reverse of the previous case. With the switch open, the equivalent resistance across C is R_U. Hence, $t = R_U C = 18K \times 10^{-12} = 18$ ns. Again accepting 2t as enough time to bring V_2 close enough to its final value (within about 14 percent), we have a delay of about $2 \times 10 = 36$ ns for this transition.

Although the preceding analysis is based on a greatly simplified version of reality, it does give us a good idea as to the magnitudes of the delays. Perhaps more important are several general principles that it illustrates. One striking point is the large disparity in the propagation delay through the gate between the case when the output is rising and the case when it is falling. This is due to the difference in the resistive components of the time constants. In one case the capacitor is *charged* through the relatively large resistor R_U, while in the second case it *discharges* through the relatively small resistance R_t (with R_U in parallel). The fact is that, in this example, $R_U = 6R_t$ is not the result of an arbitrary choice, but is the result of an engineering compromise. On the one hand, it is necessary to ensure that the low value of V_2 is sufficiently small—that is, significantly less than the threshold voltage, V_{th}. This can be achieved by making R_U sufficiently large. On the other hand, we want to keep R_U as small as possible, so as to hold down the value of the time constant during the period when the output signal is increasing. A technology such as nMOS logic, in which a compromise must be made in determining the ratio between the values of pullup and pulldown resistances, is called *ratioed logic*.

How could we make this circuit go faster? Clearly, t must be minimized. This can be done by reducing C and/or by reducing R_U and R_t. The capacitance is affected by the number of devices V_2 fans out to, the length and nature of the wiring attached to it, and various physical factors having to do with the design, layout, and sizing of the devices on the chip. There are limits to what can be done. Reducing the resistances is a matter well within the control of the designer. For a given manufacturing technology, halving R_t, the transistor resistance, would entail doubling the channel width, and hence, the chip area occupied by the transistor. But perhaps even more important, Ohm's law comes into play. Suppose, for example, we halved both R_U and R_t, thereby halving t and doubling the speed. Then, assuming V_{dd} remains unchanged, it is clear from Figure A.8b that the steady-state current flow is thereby doubled. This, in turn, means that the *power* requirement is doubled. The point illustrated is that, within limits, speed and power are directly proportional. This is virtually a universal principle applicable within any technology.

A.3. MOS-Based Technologies

A.3.1. nMOS

During most of the 1980s, the most widely used technology was nMOS (*negative* carrier *metal-oxide-silicon*). It is the lowest cost technology, and allows the highest density of devices on a chip. Its power requirements are modest, but it is also the slowest of the families discussed here. While still important, nMOS has been displaced by complementary MOS (CMOS) in leading edge technology. Since nMOS transistors,

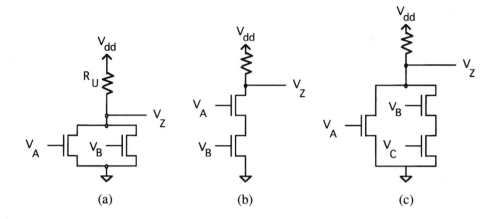

Fig. A.11. nMOS gates.

together with pMOS transistors constitute the building blocks of CMOS, an understanding of NMOS is still important.

An nMOS inverter is depicted in Figure A.10. A NOR-gate, a NAND-gate, and a complex gate realizing the function $\overline{A + BC}$ are shown in Figure A.11. When current in the pulldown circuit must pass through a long series of transistors, circuit operation is delayed. Therefore, the preferred gate for nMOS is the NOR-gate, since it entails minimal-length (one) pulldown circuits. Circuits with longer chains are sometimes used, but the delay penalty is generally considered to be very significant for circuits with more than about four transistors in series (such as a NAND-gate with fan-in exceeding four).

A related problem causes difficulties with pass-transistor circuits. Consider the circuit of Figure A.12, which realizes the XOR function $(A\overline{B} + \overline{A}B)$ with only two transistors (provided that $\overline{A}$ and $\overline{B}$ are available). Suppose $A = 0$ and $B = 1$. Then the upper transistor is cut off and the lower transistor conducts, passing the $\overline{A} = 1$ signal to the output Z. Let us examine the electrical behavior of this circuit in this situation. Assume that a logic-1 signal corresponds to a voltage of V_{dd}, and that a logic-0 signal corresponds to a voltage of 0 (the positive logic assumption of Section 2.8). Suppose that initially $A = B = Z = 0$ and that B then changes to 1. Since the left terminal of the lower transistor is at a higher voltage than the right side (V_{dd} as opposed to 0), current flow through the channel will be from left to right. Since this current is carried by negative

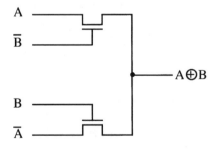

Fig. A.12. nMOS pass-transistor network.

charges, the flow of *charges* is from right to left, making the right terminal the source and the left terminal the drain. Initially, the voltage difference between gate and source is V_{dd}, so that the channel conducts charge well. But the effect of this current flow is to increase the source voltage (by charging the various stray capacitors between the source and ground). The effect of increasing the source voltage is to *decrease* the voltage between gate and source, which has the effect of *increasing* the channel resistance. This slows down the rate of increase of the output voltage of this circuit and, when the gate-to-source voltage approaches the threshold voltage, the channel ceases to conduct. Hence, the output voltage of this circuit cannot rise above $V_{dd} - V_{th}$.

Actually, the situation is worse than this. When the source voltage is well above the potential of the substrate (which is always grounded, i.e., at 0 voltage) the threshold voltage is increased as a consequence of a phenomenon known as the "body effect." This increase can easily be as much as 50 percent, significantly inflating the consequences of the effect described earlier. The result is to degrade the quality of the output of the circuit in the sense that the voltage for a 1-signal is substantially less than V_{dd}, which is what an ordinary gate produces. It also slows down the gate when the output must change to a 1. Note that when the same pass transistor must transmit a 0 from left to right, the flow of current is from right to left. Hence, the negative charge flow is from left to right and the source is now the *left* end of the transistor. The potential between gate and source remains fixed at V_{dd}, and so there is no problem in driving the output voltage all the way down to zero. Because of the degraded-1 problem, pass-transistor circuits are used only in special situations where their drawbacks are judged to be not too serious. The same problems are the reasons why nMOS transistors are seldom used in pullup circuits (with pulldown resistors). Such circuits might at first seem to be good ways to realize AND-gates and OR-gates. In CMOS technology, introduced in the next subsection, there are nice solutions to the problems of nMOS, particularly that of the asymmetry between rise times and fall times of output signals, and the problem of constructing pass networks without degraded outputs.

A.3.2. CMOS

Since CMOS is a blend of nMOS and pMOS, it is necessary to point out some important aspects of pMOS circuits. Consider Figure A.13, which shows a pMOS inverter, analagous to the nMOS inverter of Figure A.10. (Note that pMOS transistors are drawn with bubbled gates as shown here to indicate that they are activated by *low* gate-to-source voltages as opposed to the *high* gate voltages that activate nMOS transistors.) Assume positive logic, with V_{dd} at a value of, say, 3 volts. When V_1 is high

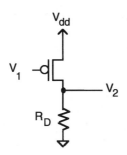

Fig. A.13. A pMOS inverter.

(1-input), the transistor is off, so that the pulldown resistor, R_D, brings output V_2 down to 0. When V_1 is low (0-input), the transistor conducts and, if the transistor parameters and the value of R_D are properly chosen, V_2 will be reasonably close to V_{dd}. When $V_1 = 0$ and the transistor is conducting, current flows, in the form of positive charges, from the top to the bottom in the diagram. Thus, the transistor source is the upper terminal, and so the voltage between gate and source remains constant at $-V_{dd}$. Therefore V_2 rapidly increases to its maximum value ($V_{dd}R_D/(R_D + R_t)$). It is therefore possible for V_2 to approach V_{dd} as R_D gets arbitrarily large. When V_1 is raised to within V_{th} of V_{dd}, the transistor is cut off and V_2 goes to 0.

Now we are in position to see how nMOS and pMOS transistors can be combined to form a CMOS inverter with some very desirable characteristics. The basic idea is to use a pMOS transistor for the pullup element, and an nMOS transistor for the pulldown element, as shown in Figure A.14. For the circuit shown there, when $V_1 = V_{dd}$, then the pullup transistor (pMOS) acts as an open switch and the pulldown transistor (nMOS) acts as a relatively low resistance. This causes V_2 to fall to 0 (or to remain at 0 if it is already at this value). If V_2 must change from the high value to the low value, then the charge stored in the capacitance between V_2 and ground discharges through the conducting nMOS transistor at a relatively fast rate.

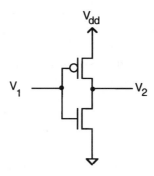

Fig. A.14. A CMOS inverter.

Now suppose that V_1 is switched to 0. Then the pullup transistor is activated and the pulldown transistor is cut off. This connects V_2 to the V_{dd} terminal through the conducting transistor and drives it up to V_{dd}. If V_2 had initially been a 0, then the output capacitance would be charged though the pullup transistor. Note the symmetry here. If the transistors are designed to have equal resistance values when activated, then turning on the output would take the same time as turning it off. There are no resistance ratios to determine the output voltage as in the case of nMOS or pMOS inverters. Even if the two transistors are not balanced with respect to conduction resistance, the output will still swing between 0 and V_{dd}. Thus, this circuit is not "ratioed"—it is termed *ratioless*. (Because electron mobility is roughly triple that of hole mobility, nMOS transistors conduct current more easily than do their pMOS cousins. To redress this balance and even out the switching times of CMOS gates in both directions, pMOS transistors are usually made two or three times wider than the nMOS transistors in the same CMOS gates.)

An important feature of the CMOS inverter is that while the input signal is constant, whether at logic 0 or logic 1, exactly one of the two transistors is conducting with the other behaving essentially as an open circuit. Thus, in both situations, there is no conducting path from the V_{dd} terminal to ground. Therefore, there is no current flowing and so no power is consumed. During very brief transient intervals immediately following an input change, both transistors may simultaneously act as closed switches so that current will flow through them. A second, and usually more important transient effect, is current flow from the V_{dd} terminal to charge the various stray output capacitances when the output is changing from 0 to 1. Contrast this with the nMOS and pMOS inverters. In the first case, a steady-state current flows when the output signal is 0, and in the second case, a steady-state current flows when the output signal is 1. Now let us see how the inverter circuit can be generalized to realize other logic functions.

In nMOS, a 2-input NAND-gate (See Figure A.11*b*) is realized by constructing an active pulldown network consisting of two transistors in series, each controlled by one of the inputs. The pullup circuit is *passive*, consisting simply of a resistor (actually a depletion-mode transistor). Such a circuit has characteristics similar to those of the nMOS inverter discussed earlier. It is a ratioed circuit, and output changes from 0 to 1, which entail the charging of capacitances through the relatively large pullup resistor, are much slower than are changes from 1 to 0. The CMOS version of a NAND-gate is depicted in Figure A.15*a*. This version differs from the nMOS NAND-gate in that the pullup circuit is now *active*. It consists of the parallel combination of two pMOS transistors, controlled by the same input signals that feed the nMOS pulldown transistors.

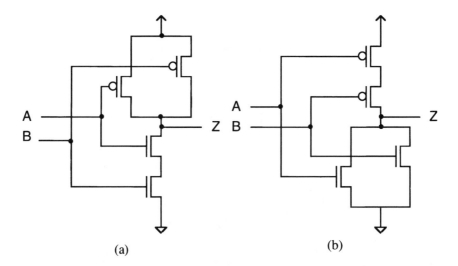

Fig. A.15. CMOS gates.

Observe that the pullup circuit realizes the complement of the function realized by the pulldown circuit. When there is a path through the pulldown circuit (i.e., when *both* nMOS transistors are conducting), there is *no* path through the pullup circuit (both pMOS transistors are open circuited). Conversely, when either pulldown transistor is acting as an open switch, thereby blocking the pulldown path, the pullup transistor with the same input is conducting, thus creating a path through the pullup circuit from the V_{dd} terminal to the output. As in the case of the CMOS inverter, the output voltage swings all the way between 0 and V_{dd}, no power is consumed in the steady state, and delays are relatively small for output changes in either direction.

The dual circuit, realizing a NOR-gate, is shown in Figure A.15b. Readers may verify that this circuit is valid and that it has the same advantageous features as the NAND-gate circuit. Gates with fan-in exceeding two can be obtained by simply adding transistors in both the series and the parallel circuits. As was pointed out in the discussion of nMOS, path delays grow rapidly with the number of transistors in a path. Furthermore, pMOS transistors are inherently slower than nMOS transistors. Therefore, since the pMOS transistors are in series in CMOS NOR-gates, the preferred CMOS gate is the NAND-gate.

Complex CMOS gates realizing functions other than NAND and NOR can be designed as exemplified by the function described by $Z = A\overline{B} + \overline{C}$ (see Figure A.16):

1. Construct the nMOS pulldown circuit with transmission (see Subsection 1.2.1, p. 2) equal to the *complement* of the function to be realized.

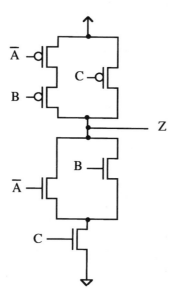

Fig. A.16. Complex CMOS gate.

2. Construct the pMOS pullup circuit with transmission equal to the function to be realized. The topological dual of a series-parallel switching circuit is the circuit obtained by replacing parallel branches with series branches and vice versa. (In the example, a series connection of a single-element and the parallel connection of two elements, is replaced by the parallel connection of a corresponding single element and the corresponding two elements connected in parallel.) Such a transformation results in a circuit realizing the dual transmission of the original circuit. Note that the pMOS transistors behave as the logical complements of nMOS transistors (e.g., a

gate voltage that causes an nMOS transistor to act as a closed switch, makes a pMOS transistor act as an open switch). Thus the pullup circuit constructed as above realizes the complementary transmission of the pulldown circuit.

Note that, in the example, it is necessary to generate $\overline{A}$ outside of the complex gate, perhaps by using an inverter. The function realized by any complex gate (this applies to nMOS as well as CMOS gates) is always a negative function (see Section 3.6) of the inputs to the transistors. The subfamily of circuits that includes both the standard (e.g., NAND) gates and the complex gates just described is generally referred to as *standard* or *complementary* CMOS. (The term *complementary* here refers to the fact that there are active pullup and pulldown circuits with complementary transmission.)

Now let us see how pass transistors can be used in the CMOS world. The problem with nMOS pass transistors, as described above, is that 1-signals are degraded. Similarly, pMOS pass transistors, degrade 0-signals. But if *both* nMOS and pMOS transistors are available, as they are in CMOS, we can construct a switch that does not degrade *any* signal. Consider the circuit shown in Figure A.17 (called a *transmission gate*), in which an nMOS transistor is connected in parallel with a pMOS transistor, and where the former is controlled by the signal S and the latter by $\overline{S}$. The signal to be passed through the switch is X.

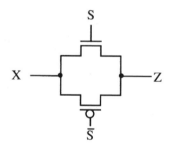

Fig. A.17. CMOS transmission gate.

If the control signal S = 0, then, regardless of the value of input X, both transistors will be off. If S = 1 and if X = 0, then the source for the nMOS transistor is on the left and the transistor is fully on, allowing the voltage at the right end (corresponding to signal Z) to go all the way to 0. The pMOS transistor will be on and help transmit the 0-signal only as long as the Z-voltage exceeds V_{th} by a significant amount. The nMOS transistor ensures that Z will go all the way to 0. If S = 1 and X = 1, then the nMOS transistor assists in transmitting the signal until the Z-voltage gets within V_{th} of the final value, while the pMOS transistor stays on to take Z all the way to V_{dd}. (Note that in both cases, the body effect inflates V_{th} significantly.) But, in any event, the nMOS transistor transmits the 0-signals and the pMOS transistor passes along the 1-signals, both without the degradation discussed earlier in connection with simple one-transistor switches.

Note that a price must be paid in that two transistors are necessary for the switch itself, and two more may be needed for an inverter to generate the complement of the control signal. Problems can also arise if long chains of transmission gates are cascaded. This is equivalent to having a series connection of many transistors in a conventional gate. Delays increase more than linearly with the number of elements in the chain. The solution is to insert inverters at intervals sufficiently small to keep the delays within satisfactory bounds. Examples of applications of the CMOS transmission gate are to designs of an XOR-gate (Section 2.9), a 2-1 MUX (Subsection 4.1.2), and a latch (Subsection 6.4.2).

Because CMOS circuits include two kinds of transistors, the fabrication process is complicated, making it more expensive than simple nMOS. For many functions, CMOS implementations require more transistors, thus increasing area requirements. But, in return, very substantial advantages are obtained:

1. For a given level of manufacturing capability, CMOS circuits can be made faster than nMOS circuits. This is the result of having low time constants for output changes in *both* directions.
2. Power requirements for CMOS are dramatically lower than for nMOS because there is essentially no current drawn from the power supply when signals are static. (For this reason, the earlier uses of CMOS were in digital watches, where battery life is a major issue.)
3. CMOS systems are less sensitive to noise, because the 0- and 1-signals are maximally separated—0 signals at 0 volts, and 1-signals at V_{dd}.
4. Designs for CMOS are easier to carry out and are and less sensitive to process variation because of the ratioless feature.

It is because of these advantages that CMOS became the dominant technology in the 1990s. There are many variations of CMOS, such as various forms of dynamic CMOS (which includes several kinds of "domino" logic) and cascode logic. These are usually more difficult to design than is standard CMOS (even when extended to include transmission gates), and are used in special situations requiring very low power and/or very high speed. Incorporating bipolar transistors on CMOS chips (BiCMOS logic) is a technique that is gaining favor, as the bipolar transistors can provide more punch to drive off-chip circuits.

A.3.3. Galium Arsenide

Galium arsenide has been used for many years for optical devices, such as solid-state lasers, and for microwave amplifiers. Electron mobility, an important measure of performance in semiconductors, is more than four times higher in galium arsenide than in silicon. Hence, there are important advantages to using it as the raw material for constructing digital logic circuits.

Because of the gap between hole mobility and electron mobility is more than three times greater for galium arsenide than for silicon, complementary circuits (i.e., CMOS) are not used in galium arsenide devices. Instead devices corresponding to nMOS are favored, and so galium arsenide logic gates are ratioed. For this reason, the gate of choice in galium arsenide circuits is the NOR-gate, as it is in silicon nMOS families. Finally, it should be noted that the gates in galium arsenide transistors are not insulated as they are in silicon transistors. This means that some current (though not much) has to be supplied to control the channel conductivity. It is reported that power consumption in galium arsenide circuits is essentially independent of switching frequencies, which is quite different from the situation in CMOS. There were many obstacles to be overcome, but it appears (as of 1996) that galium arsenide is finding a place in the development of high-performance digital devices. Apparently, both silicon and galium arsenide will be found in digital systems over the next decade.

A.4. Bipolar Families

In general, bipolar transistors are faster and are more capable with respect to transmitting signals to remote units, a task requiring inherently higher power output than on-chip operations.

TTL, an old, established family with many branches, has for many years dominated the SSI and MSI fields. The most natural TTL-gate is the NAND-gate, but a great variety of TTL chips, featuring all kinds of logic gates, latches, decoders, register segments, arithmetic logic unit (ALU) segments, etc., can be ordered through catalogs. However, the basic circuitry is rather complex and does not lend itself well to large-scale integration. It is easily outclassed in this respect by MOS-based families, which also require less power. On the other hand, it has a potent rival in the high-performance field, as indicated below. Nevertheless, TTL is likely to remain as a significant technology, largely because of its wide availability off the shelf in so many forms, and because it is very familiar to users.

The fastest silicon-based technology is ECL, sometimes referred to as current-switched logic. It is the clear choice at present for supercomputers that rely on high-performance circuitry (as opposed to depending on massive parallelism). Its power consumption is also highest, and while it is being used for LSI, it is not a serious contender in the very-large-scale integration (VLSI) field, due to the complexity of its circuitry. Almost all ECL logic is based on NOR-gates, sometimes with an OR-output also available, and often with wired-OR capabilities. Complex gates can be implemented in some versions of ECL.

A bipolar technology that seems to have the potential to compete with MOS technologies with respect to scale of integration and low power consumption is *integrated injection logic* (I^2L), also called *merged transistor logic* (*MTL*). It is relatively fast, particularly with respect to its low power requirements. However, perhaps for reasons related to manufacturing difficulties, I^2L has not yet found its way into many systems.

Overview and Summary

Converting numbers between different number bases is an important operation. Two basic conversion schemes for integers are the division-remainder method, in which the arithmetic is done in the original system, and the multiply-and-add scheme, in which the arithmetic is done in the "destination" system. Negative numbers are most often represented by two's complements, which allows the uniform handling of addition and subtraction of positive and negative numbers. For fractions, the corresponding schemes involve the multiplication-integer-part method and the divide-and-add method, respectively. Conversions between binary and either octal or hexadecimal are very simple. Where it is necessary to represent decimal numbers within a digital system, each digit is separately encoded. The most common scheme is binary-coded decimal or BCD. The 7-bit ASCII code is used for non-numerical symbols. Where it is important that adjacent members of a sequence of binary-code words differ in only one bit, a Gray code is used. With Hamming single-error-correcting codes (SECC) it is possible to detect and correct errors in single bits of words that occur during the transmission or storage of data.

Simple circuits with constant voltage sources, switches, and resistors can be analyzed using Ohm's law and the Kirchhoff voltage and current laws. Series and parallel circuits are duals in the sense that the same current passes through all elements in series, whereas the same voltage is across all elements in parallel. Current flowing through a resistor dissipates energy in the form of heat, which must be disposed of. An nMOS transistor can be operated as a voltage-controlled switch, in that a positive voltage on the gate pulls negative charge carriers into the channel, thereby converting it into a conducting path between source and drain. Capacitors store charge as a result of the flow of current into them. The voltage across a capacitor grows linearly with the amount of charge stored. Changing the voltage across a capacitor takes time, which is proportional to the product of the capacitance and the resistance through which the charge flowing in or out must pass. Capacitors act as open circuits over long periods of time when there are only static voltage sources in the circuit. For the purpose of calculating very short-term effects, capacitors look like constant voltage sources. Delays in logic circuits are largely produced by the need to charge and discharge stray capacitances associated with the transistors and with the wiring. Faster operation results if resistances are reduced, but this increases the power supply current and hence the power consumed.

The dominant VLSI technology as of the mid-1990s is CMOS. It is faster than the previous leader, nMOS, because it employs an active pullup network. Power dissipation is lower because essentially no current flows under steady-state conditions. CMOS also features higher noise immunity, in that the signal levels swing all the way from 0 to V_{DD}. The price paid is the need for more transistors for most functions, and the increased complexity of the fabrication process. Technologies, such as ECL, based on bipolar transistors, offer higher speed, but cost more per gate and require more power. Still more costly are BiCMOS chips that add bipolar transistors to CMOS technology. The bipolar transistors are used where high speed is necessary despite heavy capacitive loads, as is the case when driving off-chip devices. Another contender in the high-performance arena is the galium arsenide nMOS family.

SOURCES

An in-depth treatment of arithmetic methods can be found in [Waser, 1982]. Error correcting codes are treated in [Peterson, 1972]. There are many good introductory texts on electrical circuits and electronics, including microelectronics; for example, [Millman, 1979], [Mukherjee, 1986], and [Weste, 1992]. The last two books include good treatments of CMOS. Galium arsenide devices are treated in [Deyhimy, 1995].

PROBLEMS

A.1.   Compute the following sums in binary arithmetic:
    (a) 101101 + 0111
    (b) 1011101 + 100011
    (c) 1011 + 1011 + 10011

A.2.   Compute the products in binary arithmetic:
    (a) 1010 × 1101
    (b) 1110 × 1001

A.3.   Compute the quotients and remainders in binary arithmetic for:
    (a) 100110 ÷ 1011
    (b) 101110 ÷ 1001

A.4.   Carry out the following division problem, in binary arithmetic, to five significant bits: 101110.1 ÷ 1101 (round off in binary as in the following examples: .110 → .11; .101 → .11; .111 → 1.0).

A.5. Convert each of the following decimally specified numbers to two's-complement-form 6-bit numbers, if this is possible. If any are out of range, state this.

   *(a) 29

   *(b) -14

   (c) -2

   (d) -40

   (e) 32

   (f) -32

   (g) 0

A.6. Convert each of the following two's complement numbers to their decimal equivalents.

   *(a) 100001

   *(b) 010101

   (c) 000000

   (d) 111111

A.7. Find the two's complements of each of the following two's complement numbers. Indicate if there are problems with any of these:

   *(a) 100100

   *(b) 001100

   (c) 100000

   (d) 111000

A.8. Show how each of the following arithmetic operations would be carried out in a system using two's complement numbers. Point out any overflow situations (assuming 6-bit words):

   (a) $010101 - 011001$

   (b) $011010 + 001001$

   (c) $101010 - 110011$

   (d) $100010 + 100100$

   (e) $100000 + 011111$

A.9. Construct a flow table for an iterative function that will generate the two's complement of any two's complement number fed into it. Assume the signal flow is from the least to the most significant bits. Show how the circuit can be modified to indicate that the input is the one number that has no two's complement.

A.10. Convert to decimal form the binary numbers:

   (a) 100000

   (b) 10000000001

   (c) 1011101011011

   (d) 11010011001110

A.11. Convert the following decimal numbers to binary:

   (a) 572

   (b) 8765

   (c) 10351

A.12. Convert as specified:

   (a) 1011_3 to binary

   (b) 1212201_3 to base 9

   (c) 7580_9 to ternary

A.13. Convert as stated:
  (a) 0.7531_{10} to binary (12 places)
  (b) 0.011011101_2 to decimal (3 places)
  (c) 753.24_{10} to binary (6 places)
  (d) 369.28_{10} to ternary (4 places)

A.14. Express each of the following decimal numbers in binary, octal, hexadecimal, and BCD:
  *(a) 60
  (b) 307
  (c) 500371

A.15. Given that the ASCII codes for d and D are, respectively, 1100100 and 1000100, what are the ASCII codes for e and E?

A.16. Construct a 12-word code, each word composed of four bits, that is analogous to Gray code. Start with 0000 and make each word differ from the previous word in exactly one bit position. The last word in the sequence should also differ from 0000 in exactly one bit position.

A.17. *(a) Suppose the word 1011 is to be encoded using a SECC. Specify all of the bits of the code word.
  (b) Repeat for the word 010011.

A.18. Modify the solutions to the previous problem to allow double-error-correction.

A.19. Suppose that a Hamming SECC is used for a 7-bit word size. What happens if there is a double fault involving b_3 and b_5? If a SECC is used, and a double error occurs, is it possible that it will appear as though *no* error occurred? What if there is a triple error?

A.20. (a) What is the equivalent resistance of a string of three series-connected resistors with values 3, 6, and 8 ohms?
  (b) What is the equivalent resistance of the same set of resistors connected in parallel?

*A.21. In the circuit of Figure A.18 find
  (a) I_3
  (b) V_1

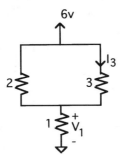

Fig. A.18.

A.22. Sketch a cross-sectional diagram of a pMOS FET. How should V_{dd} be polarized?

A.23. In an nMOSFET with $V_{th} = 1.3$, $V_{dd} = 4$, $V_G = 2$, and a 2K resistor connected from drain to V_{dd}, how much current flows from the gate to the source?

A.24. What happens to the channel resistance of an nMOSFET as V_G increases from 0 to 3, when $V_{th} = 1.2$? Do not give numerical answers. Just sketch a rough graph.

*A.25. For the circuit of Figure A.19, assume that A has been and remains closed. If B has been open for a long time prior to $t = 0$ when it closes, find:

    (a) The initial value of V_C.

    (b) The asymptotic value of V_C.

    (c) The time constant.

    (d) An expression for V_C as a function of time starting at $t = 0$.

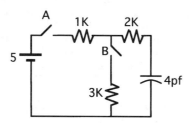

Fig. A.19.

A.26. Repeat the previous problem, but now assume that after A and B have both been closed for a long time, A *opens* at $t = 0$.

A.27. (a) What would be the effect of decreasing the value of the pullup resistor, R_U, in the inverter circuit of Figure A.10 on the propagation delays?

    (b) What *other* effect would this have?

    (c) Would it be helpful if V_{th} were decreased?

    (d) Increased?

A.28. In Section 2.6, there is a description of a NOR-gate.

    (a) Draw a diagram of a 2-input nMOS NOR-gate, using the transistor symbol as in Figure A.10*a*.

    (b) Then draw an equivalent circuit analogous to Figure A.10*b*.

    (c) For the worst-case situation when the output *falls*, how would the delay compare with that found for the inverter, assuming the transistors, all have the same parameters?

    (d) What about the case where the output *rises*?

A.29. Repeat the previous problem, but with a NAND-gate. Add part (e): How would the speed of the NAND-gate compare with that of the NOR-gate for rising and falling outputs?

A.30. In nMOS technology, how does the power required by a 2-input NOR-gate compare with that required by an inverter? What about a 3-input NOR-gate? A 2-input NAND-gate?

A.31. Assuming double-rail inputs, draw a circuit diagram showing how to implement directly the expression $Z = \overline{AB}(C + \overline{DE})$ with a complex CMOS gate.

References

Abramovici, Miron, Melvin A. Breuer, and Arthur D. Friedman, *Digital Systems Testing and Testable Design*, IEEE Press, 1994.

Armstrong, Douglas B., "On the Efficient Assignment of Internal Codes to Sequential Machines," *IRE Trans. Electron. Comput.*, Vol. EC-11, No. 5 (Oct. 1962), pp. 611–622.

Armstrong, Douglas B., Arthur D. Friedman, and P. R. Menon, "Design of Asynchronous Circuits Assuming Unbounded Gate Delays," *IEEE Trans. Comput.*, Vol. C-18, No. 12 (Dec. 1969), pp. 110–120.

Bartee, Thomas C., "Computer Design of Multiple-Output Logical Networks," *IRE Trans. Electron. Comput.*, Vol. EC-10, No. 2 (1961), pp. 21–30.

Birtwistle, G., and A. Davis, Eds., *Asynchronous Digital Circuit Design*, Springer-Verlag, New York, 1995.

Brand, Daniel, "Redundancy and Don't Cares in Logic Synthesis," *IEEE Trans. Comput.*, Vol. C-32, No. 10 (Oct. 1983), pp. 947–952.

Brand, Daniel, "Detecting Sneak Paths in Transistor Networks," *IEEE Trans. Comput.*, Vol. C-35, No. 3 (Mar. 1986), pp. 274–278.

Brayton, Robert K., et al., "Fast Recursive Boolean Function Manipulation," *Proc. ISCAS* (Apr. 1982), pp. 58–62.

Brayton, R. K., G. D. Hachtel, C. T. McMullen, and A. L. Sangiovanni-Vincentelli, *Logic Minimization Algorithms for VLSI Synthesis*, Kluwer, Boston, 1984.

Breuer, Melvin A., and Arthur D. Friedman, *Diagnosis and Reliability of Digital Systems*, Computer Science Press, Rockville, Md., 1976.

Bryant, Randal E., "A Switch-Level Model and Simulator for MOS Digital Systems," *IEEE Trans. Comput.*, Vol. C-33, No. 2 (Feb. 1984), pp. 160–177.

Caldwell, Samuel, *Switching Circuits and Logical Design*, Wiley, New York, 1958.

Chan, A. Y., "Easy-to-Use Signature Analyzer Accurately Troubleshoots Complex Logic Circuits," *Hewlitt-Packard J.* (May 1977), pp. 9–14.

Chaney, Thomas J., and Charles E. Molnar, "Anomalous Behavior of Synchronizer and Arbiter Circuits," *IEEE Trans. Comput.*, Vol. C-22, No. 4 (Apr. 1973), pp. 421–422.

Clare, C. R., *Designing Logic Systems Using State Machines*, McGraw-Hill, New York, 1973.

Darringer, John A., Daniel Brand, John V. Gerbi, William H. Joyner, Jr., and Louise Trevillyan, "LSS: A System for Production Logic Synthesis," *IBM J. Res. Dev.*, Vol. 28, No. 5 (Sept. 1984), pp. 537–545.

Deyhimy, Ira, "Galium Arsenide Joins the Giants," *IEEE Spectrum*, Vol. 32, No. 2 (Feb. 1995), pp. 33–40.

Dolotta, Theodore A., and Edward J. McCluskey, "The Coding of Internal States of Sequential Circuits," *IEEE Trans. Comput.*, Vol. EC-13 (Oct. 1964), pp. 549–562.

Eichelberger, Edward B., and T. W. Williams, "A Logic Design Structure for LSI Testability," *Proc. 14th Design Automation Conf.*, June 20–22, 1977, pp. 462–468.

Ercegovac, Milos D., *Digital Systems and Hardware/Firmware Algorithms*, Wiley, New York, 1985.

Fletcher, William I., *An Engineering Approach to Digital Design*, Prentice-Hall, Englewood Cliffs, N.J., 1980.

Flores, Ivan, *The Logic of Computer Arithmetic*, Prentice-Hall, Englewood Cliffs, N.J., 1963.

Friedman, Arthur D., *Fundamentals of Logic Design and Switching Theory*, Computer Science Press, Rockville, Md., 1986.

Friedman, Arthur D., and Premachandran R. Menon, *Theory and Design of Switching Circuits*, Computer Science Press, Rockville, Md., 1975.

Fujiwara, Hideo, *Logic Testing and Design for Testability*, MIT Press, Cambridge, 1985.

Gilchrist, Bruce, J. H. Pomerene, and S. Y. Wong, "Fast Carry Logic for Digital Computers," *IRE Trans. Comput.*, Vol. EC-4 (Dec. 1955), pp. 133–136.

Gill, Arthur, *Linear Sequential Circuits*, McGraw-Hall, New York, 1966.

Grasselli, Antonio, "Mimimal Closed Partitions for Incompletely Specified Flow Tables," *IEEE Trans. Electron. Comput.*, Vol. EC-15, No. 2 (Apr. 1966), pp. 245–249.

Hachtel, G. D., et al., "An Algorithm for PLA Folding," *IEEE Trans. Comput. Aided Design*, Vol. CAD-1, No. 2 (1982), pp. 63–67.

Hartmanis, Juris, and Richard E. Stearns, *Algebraic Structure Theory of Sequential Machines*, Prentice-Hall, Englewood Cliffs, N.J., 1966.

Hayes, John P., "A Unified Switching Theory with Applications to VLSI Design," *Proc. IEEE*, Vol. 70 (Oct. 1982), pp. 1140–1151.

Hayes, John P., *Introduction to Digital Logic Design*, Addison-Wesley, Reading, Mass., 1993.

Hennessy, John L., and David A. Patterson, *Computer Architecture: A Quantitative Approach*, Morgan Kaufmann, San Francisco, Cal., 1990.

Hennessy, John L., and David A. Patterson, *Computer Organization & Design: The Hardware / Software Interface*, Morgan Kaufmann, San Francisco, Cal., 1994.

Hennie, Frederick C., III, *Iterative Arrays of Logical Circuits*, MIT Press, Cambridge, Mass., 1961. (Also, Wiley, New York.)

Hill, F. J., and G. R. Peterson, *Introduction to Switching Theory and Logic Design*, 3rd ed., Wiley, New York, 1981.

Hitchcock, Robert B., Sr., Gordon L. Smith, and David D. Cheng, "Timing Analysis of Computer Hardware," *IBM J. Res. Dev.*, Vol. 26, No. 1 (Jan. 1982), pp. 100–105.

Hodges, D. A., and H. G. Jackson, *Analysis and Design of Digital Integrated Circuits*, McGraw-Hill, New York, 1983.

Huffman, David A., "The Synthesis of Sequential Switching Circuits," *J. Franklin Inst.*, Vol. 257, No. 3 (Mar. 1954), pp. 161–190; and No. 4 (Apr. 1954), pp. 275–303.

Huffman, David A., "The Synthesis of Iterative Switching Circuits," MIT RLE *Quarterly Progress Reports* (Jan. 15, 1955), pp. 63–67.

Huffman, David A., "The Synthesis of Linear Sequential Coding Networks," in *Information Theory*, C. Cherry, Ed., Academic Press, New York, 1956, pp. 77–95.

Hurtado, M., and D. L. Elliott, "Ambiguous Behavior of Logic Bistable Systems," *Proc. 13th Annu. Allerton Conf. on Circuit and System Theory*, Oct. 1975.

Kacprzak, Tomasz, "Analysis of Oscillatory Metastable Operation of an *RS* Flip-Flop," *IEEE J. Solid-State Circuits*, Vol. SC-23, No. 1 (Feb. 1988), pp. 260–266.

Kambayashi, Yahiko, and Saburo Muroga, "Properties of Wired Logic," *IEEE Trans. Comput.*, Vol. C-35, No. 6 (June 1986), pp. 550–563.

Karnaugh, Maurice, "The Map Method for Synthesis of Combinational Logic Circuits," *Trans. AIEE*, Part I, Vol. 72, No. 9 (1953), pp. 593–598.

Katz, Randy H., *Contemporary Logic Design*, Benjamin/Cummings, Redwood City, Cal., 1994.

Keister, William, A. E. Ritchie, and Seth H. Washburn, *The Design of Switching Circuits*, Van Nostrand, New York, 1951.

Kohavi, Zvi, *Switching Theory and Finite Automata*, 2nd ed., McGraw-Hill, New York, 1978.

Koren, I., *Computer Arithmetic: Algorithms*, Prentice-Hall, Englewood Cliffs, N.J., 1993.

Lee, Samuel C., *Digital Circuits and Logic Design*, Prentice-Hall, Englewood Cliffs, N.J., 1976.

Liu, C. N., "A State Variable Assignment Method for Asynchronous Sequential Switching Circuits," *J. ACM*, Vol. 10 (Apr. 1963), pp. 209–216.

Maley, Gerald, *A Manual of Logic Circuits*, Prentice-Hall, Englewood Cliffs, N.J., 1971.

Mano, Morris, *Digital Design*, 2nd ed., Prentice-Hall, Englewood Cliffs, N.J., 1991.

Mano, Morris, *Computer System Architecture*, 3rd ed., Prentice-Hall, Englewood Cliffs, N.J., 1993.

Marino, Leonard R., "The Effect of Asynchronous Inputs on Sequential Network Reliability," *IEEE Trans. Comput.*, Vol. C-26, No. 11 (Nov. 1977), pp. 1082–1090.

Martin, Alain, "From Communicating Processes to Delay-Insensitive Circuits," *UT Year of Programming Institute on Concurrent Programming*, C. A. R. Hoare, Ed., Addision-Wesley, Reading, Mass., 1989.

McCluskey, Edward J., "Minimization of Boolean Functions," *Bell Syst. Tech. J.*, Vol. 35, No. 5 (1956), pp. 1417–1444.

McCluskey, Edward J., "Logical Design Theory of NOR Gate Networks with No Complemented Inputs," *Proc. 4th Annu. Symp. Switching Circuit Theory & Logic Design*, Oct. 28–30, 1963.

McCluskey, Edward J., *Logic Design Principles*, Prentice-Hall, Englewood Cliffs, N.J., 1986.

Mead, Carver, and Lynn Conway, *Introduction to VLSI Systems*, Addison-Wesley, Reading, Mass., 1980.

Miczo, Alexander, *Digital Logic Testing and Simulation*, Wiley, New York, 1986.

Millman, Jacob, *Microelectronics*, McGraw-Hill, New York, 1979.

Morgan, C. P., and D. B. Jarvis, "Transistor Logic Using Current Switching and Routing Techniques and Its Application to a Fast-Carry Propagation Adder," *Proc. IRE*, Part B, Vol. 106 (Sept. 1959), pp. 467–468.

Mukherjee, Amar, *Introduction to nMOS and CMOS VLSI Systems Design*, Prentice-Hall, Englewood Cliffs, N.J., 1986.

Muller, David E., "The General Synthesis Problem for Asynchronous Digital Networks," *IEEE Conf. Record of 8th Annu. Symp. on Switching Theory and Automata*, Oct. 1967, pp. 71–82.

Nowick, Steven M., and David L. Dill, "Exact Two-Level Minimization of Hazard-Free Logic with Multiple Input Changes," *Proc. Int. Conf. Comput. Aided Design*, IEEE Comput. Soc. Press, 1992, pp. 626–630.

Paull, Marvin C., and Stephen H. Unger, "Minimizing the Number of States in Incompletely specified Sequential Switching Functions," *IRE Trans. Electronic Comput.*, Vol. EC-8, No. 3 (Sept. 1959), pp. 356–367.

Peterson, Wesley W., and E. J. Weldon, Jr., *Error Correcting Codes*, 2nd ed., Wiley, New York, 1972.

Reyneri, Leonardo M., and Dante Del Corso, "Oscillatory Metastability in Homogeneous and Inhomogeneous Flip-Flops," *IEEE J. Solid-State Circuits*, Vol. SC-25, No. 1 (Feb. 1990), pp. 254–264.

Rosenberger, Fred U., Charles E. Molnar, Thomas Chaney, and Ting-Pien Fang, "Q-Modules: Internally Clocked Delay-Insensitive Modules," *IEEE Trans. Comput.*, Vol. C-37, No. 9 (Sept. 1988), pp. 1005–1018.

Roth, Charles H., Jr., *Fundamentals of Logic Design*, 4th ed., West, St. Paul, Minn., 1992.

Sandige, Richard S., *Modern Digital Design*, McGraw-Hill, New York, 1990.

Seidensticker, Robert B., *The Well-Tempered Digital Design*, Addison-Wesley, Reading, Mass., 1986.

Seitz, Charles, "System Timing," in *Introduction to VLSI Systems*, C. Mead and L. Conway, Chapt. 7, Addison-Wesley, Reading, Mass., 1980.

Shannon, Claude, "A Symbolic Analysis of Relay and Switching Circuits," *Trans. AIEE*, Vol. 57 (1938), pp. 713–723.

Sims, J. C., Jr., and H. J. Gray, "Design Criteria for Autosynchronous Circuits," *Proc. Eastern Joint Computer Conf.*, Dec. 3–5, 1958, pp. 94–99.

Sutherland, Ivan, "Micropipelines," *Commun. ACM*, Vol. 32, No. 6 (June 1989), pp. 720–738.

Szygenda, S. A., and E. W. Thompson, "Modelling and Digital Simulation for Design Verification and Diagnosis," *IEEE Trans. Comput.*, Vol. C-25, No. 12 (Dec. 1976), pp. 1242–1253.

Tannenbaum, Andrew S., *Structured Computer Organization*, 3rd ed., Prentice-Hall, Englewood Cliffs, N.J., 1990.

Texas Instruments, Inc., *The TTL Data Book for Design Engineers*, 2nd ed., Texas Instruments, Inc., Dallas, 1976.

Tracy, James H., "Internal State Assignments for Asynchronous Sequential Machines," *IEEE Trans. Electron. Comput.*, Vol. EC-15, No. 4 (Aug. 1966), pp. 551–560.

Trevillyan, Louise, William Joyner, and Leonard Berman, "Global Flow Analysis in Automatic Logic Design," *IEEE Trans. Comput.*, Vol. C-35, No. 1 (Jan 1986), pp. 77–81.

Tuinenga, Paul W., *Spice: A Guide to Circuit Simulation and Analysis Using PSpice*, Prentice Hall, Englewood Cliffs, NJ, 1988.

Unger, Stephen H., *Applications of Hypercube Diagrams to the Design of Switching Circuits*, MS thesis, MIT, Cambridge, Mass., June 1953.

Unger, Stephen H., *Asynchronous Sequential Switching Circuits*, Wiley-Interscience, New York, 1969. (Reissued by Krieger, Malabar, Florida, 1983.)

Unger, Stephen H., "The Generation of Completion Signals in Iterative Combinational Circuits," *IEEE Trans. Comput.*, Vol. C-26, No. 1 (Jan. 1977a), pp. 13–18.

Unger, Stephen H., "Tree Realizations of Iterative Circuits," *IEEE Trans. Comput.*, Vol. C-26, No. 4 (Apr. 1977b), pp. 365–383.

Unger, Stephen H., "A Building Block Approach to Unclocked Systems," *Proc 26th Annu. Hawaii Int. Conf. on System Sciences*, IEEE Computer Society Press, New York, 1993, pp. 339–348.

Unger, Stephen H., and Chung-Jen Tan, "Clocking Schemes for High-Speed Digital Systems," *IEEE Trans. Comput.*, Vol. C-35, No. 10 (Oct. 1986), pp. 880–895.

Unger, Stephen H., "Hazards, Critical Races, and Metastability," *IEEE Trans. Comput.*, Vol. 44, No. 6 (June 1995), pp. 754–768.

Waite, William M., "The Production of Completion Signals by Asynchronous, Iterative Circuits," *IEEE Trans. Comput.*, Vol. EC-13 (Apr. 1964), pp. 83–86.

Wakerly, J. F., *Digital Design Principles and Practices*, 2nd ed., Prentice-Hall, Englewood Cliffs, N.J., 1994.

Waser, Shlomo, and Michael J. Flynn, *Introduction to Arithmetic for Digital Systems Designers*, Holt, Rinehart & Winston, New York, 1982.

Weste, Neil H. E., and Kamran Eshraghian, *Principles of CMOS VLSI Design: A Systems Perspective*, 2nd ed., Addision-Wesley, Reading, Mass., 1992.

Williams, Michael J., and J. B. Angel, "Enhancing Testability of Large Scale Integrated Circuits via Test Points and Additional Logic," *IEEE Trans. Comput.*, Vol. C-22, No. 1 (Jan. 1973), pp. 46–60.

Winkel, David, and Franklin Prosser, *The Art of Digital Design*, Prentice-Hall, Englewood Cliffs, N.J., 1980.

Wood, R., "A High Density Programmable Logic Array Chip," *IEEE Trans. Comput.*, Vol. C-28 (Sept. 1979), pp. 602–608.

Solutions to Selected Problems

Chapter 1
1.3.

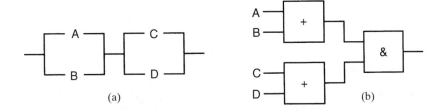

(a)

(b)

1.9.

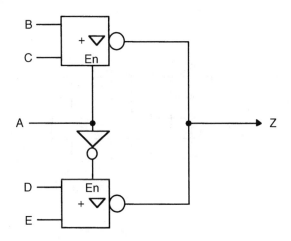

Chapter 2
2.7. (a) $(AB + \overline{C}D)(AB + \overline{C}E) = AB + \overline{C}DE$ (dual form of distributive law)

2.12. (a) $AB + CD + A\bar{C} + D\bar{E} = A(B + \bar{C}) + D(C + \bar{E}) = (A(B + \bar{C}) + D)(A(B + \bar{C}) + C + \bar{E}) =$
$(A + D)(B + \bar{C} + D)(A + C + \bar{E})(B + \bar{C} + C + \bar{E}) = (A + D)(B + \bar{C} + D)(A + C + \bar{E})$

2.13.(a)

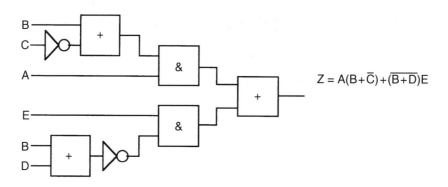

$$Z = A(B+\bar{C})+\overline{(B+D)}E$$

Chapter 3

3.1. (a) Two inverters, a 2-input AND-gate, a 3-input AND-gate, a 3-input OR-gate. Thus we have g = 5, gi = 10. The longest path is through an inverter, an AND-gate, and an OR-gate; mpl = 3.

3.7. (a) Within a 5-cube, $5 - 1 = 4$ literals specify a 1-cube, and $5 - 3 = 2$ literals specify a 3-cube. A 3-cube contains a 1-cube if the literals specifying it are a subset of the literals specifying the 1-cube. Thus the number of 3-cubes containing a 1-cube is equal to the number of ways we can choose two literals out of four, i.e. 4(choose)2 = $4!/(2!(4 - 2)!) = 6$.

3.8.

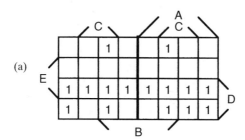

(a)

3.14a. The maps of the function and its complement are shown below in maps a1 and a2. Each function can be covered as shown by *essential* pi's (we are not usually this lucky.) This leads directly to the SOP expression for function-a: $f = BCD + AB\bar{D} + \bar{B}C\bar{D} + A\bar{B}D$. From the map for the complement of function-a (a2) we find the minimal SOP expression for the complement to be: $\bar{f} = B\,\bar{C}\,D + \bar{B}\,\bar{C}\,\bar{D} + \overline{AB}\overline{D} + \bar{A}\,\bar{B}\,D$. Applying a DeMorgan theorem yields: $f = (\bar{B} + C + \bar{D})(B + C + D)(A + \bar{B} + D)(A + B + \bar{D})$.

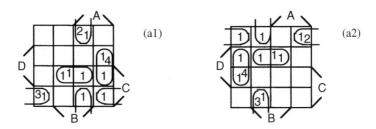

(a1)   (a2)

3.19.

(a)
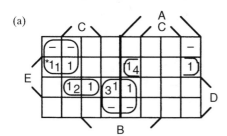

(b)
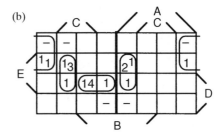

Branching required at first step. Note that for the b-branch, a small amount of lookahead was needed at step-3, to avoid a second branch. Solution (a) is better in that it includes 2 2-cubes. $Z = \bar{A}\,\bar{B}\,\bar{D} + \bar{A}CDE + B\bar{C}D + AC\bar{D}E$.

3.31. (a) There are only two columns required in the process for finding the pi's, which are -01, -10, 1-1, and 11- (*both* the 1-points *and* the don't cares are used). This leads to a covering table with three columns (corresponding to the 1-points 010, 101, and 111) and four rows, one for each pi. It is at once evident that -10 is essential to cover 010. It then follows that 1-1 covers the remaining two 1-points. Thus the solution is: $B\bar{C} + AC$.

3.34.

(a)
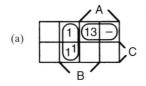
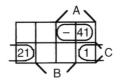

$$Z_1 = \bar{A}B + A\bar{C}, \quad Z_2 = \bar{B}C + \underline{A\bar{C}}.$$

3.39. Map the function, as below.

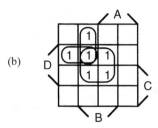

(b)

(b) All crossings of the A-boundary from $A = 0$ to $A = 1$ either leave Z unchanged, or turn it off. Hence the function is negative in A. Similar arguments show that it is also negative in C and positive in both B and D. Thus this is a unate function. Note that all of the pi's are essential, and intersect at $ABCD = 0101$.

3.43. (a) For $AB + AC + D$, $gi = 7$, $g = 3$, $mfi = 3$, $mpl = 2$. Factoring yields $A(B + C) + D$, for which $gi = 6$, $g = 3$, $mfi = 2$, $mpl = 3$.

Chapter 4
4.9.

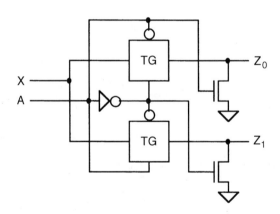

4.12. $Z = 1(A\overline{B}\overline{C}) + D(\overline{A}B\overline{C}) + E(\overline{A}\,\overline{B}C) + \overline{E}(A\overline{B}C)$. All other minterms of ABC have, or (due to don't care entries) can be assigned, coefficients of 0. Connect A, B, and C respectively to control inputs A_2, A_1, and A_0 of the MUX. In accordance with the above expression, Connect a 1-signal to MUX input X_4, connect D to input X_2, E to X_1, $\overline{E}$ to X_5, and connect 0-signals to the other four X-inputs.

Chapter 5
5.3. $M(A, B, C, D, E) = S_{3, 4, 5}(A, B, C, D, E)$
   (a) $M(A, B, C, D, E) + \overline{S}_{2, 3}(A, B, C, D, E) = S_{3, 4, 5}(A, B, C, D, E) + S_{0, 1, 4, 5}(A, B, C, D, E) = S_{0, 1, 3, 4, 5}(A, B, C, D, E)$.

5.11. $Z_1 = Z_2 = 0$ if $A = B$, $Z_1 = 1$ if $A > B$, $Z_2 = 1$ if $B > A$. (We could also have added a third input to be set to 1 if $A = B$).

(a)

AB

	00	01	11	10	
1	1,00	2,01	1,00	3,10	Most significant bits equal so far
2	2,01	2,01	2,01	2,01	B is greater
3	3,10	3,10	3,10	3,10	A is greater

Chapter 6
6.4.

AB

	00	01	11	10	
1	1,0	2,0	1,0	1,1	no borrow from right
2	2,1	2,0	2,1	1,0	borrow from right

6.8. The completed pair chart is shown in (a). It leads to equivalence classes: {1, 245, 36}. These are used to produce the reduced table shown in (b).

1

XX	2					(a)
XX	16 46 56 XX	3				
XX	36 45	16 56 35 XX	4			
XX	24	16 26 24 56 XX	24 36 25	5		
XX	16 34 25 56 XX	25	16 35 25 XX	16 45 23 XX	6	

(b)

	A	B	C	D	
1	2,0	3,0	2,1	1,1	(1)
2	2,0	3,0	2,1	1,0	(245)
3	2,0	2,0	3,1	3,0	(36)

6.10. Starting with the K-maps of Fig 5.13, find minimal SOP expressions for the *complements* of Y_1, Y_2, and

Z by treating the 0s of the map as ones and vice versa. This leads to the choice of subcubes shown below and the corresponding SOP expressions: $\overline{Y}_1 = \overline{X} + y_1$, $\overline{Y}_2 = \overline{X}$, and $\overline{Z} = X + y_1 + \overline{y}_2$.

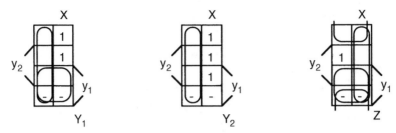

Applying the DeMorgan laws gives us: $Y_1 = X\overline{y}_1$, $Y_2 = X$, $Z = \overline{X}\overline{y}_1 y_2$. (A peculiarity of this problem is that the SOP expressions for the complemented variables are degenerate in the sense that they can *also* be considered as POS expressions. But we still must complement them to obtain expressions—also degenerate—for the uncomplemented variables.) The resulting circuit is shown below.

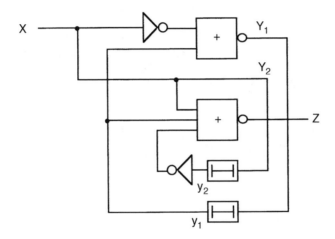

6.15. The waveforms are as shown below:

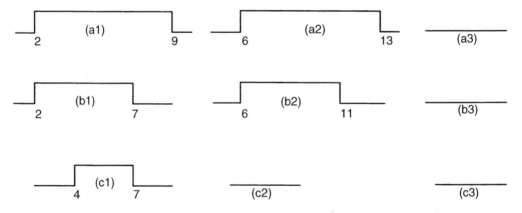

Clearly, the ordering of the delays in the circuit can make a difference. The difference between circuits (2) and (3) is the ordering, and this has a major effect on the processing of waveforms (a) and (b), as evidenced by the difference between output (a2) and (a3) and between (b2) and (b3). Reversing the sequence of input pulses can also do more than reverse the sequence of output pulses, as shown by the differences between (b1) and (c1) and between (b2) and (c2). Replacing the 2-unit inertial delay in circuit (2) by a 2-unit pure delay will cause the output to be reduced to a steady 0 (as in all three outputs from circuit (3)). This is because the first delay no longer eliminates the negative pulse thereby presenting the second delay with a longer pulse that it cannot filter out. But if the second delay is *smaller* than the first, as is the case in circuit (3), then replacing it with a pure delay never has any effect.

6.18. The K-maps are shown below. In order to get the minimal POS forms needed for a NOR-gate realization, the 0-points are minimally covered by subcubes as shown on the maps. This leads to expressions for the complements: $\overline{Y}_1 = D\overline{y}_1 + \overline{y}_2 + C\overline{y}_1$, $\overline{Y}_2 = \overline{D}\overline{y}_2 + C\overline{y}_2 + CD y_1$, $\overline{Z} = \overline{y}_2 + C + D$. Using the DeMorgan laws gives us the desired POS forms for the uncomplemented Ys and Z: $Y_1 = (\overline{D} + y_1)(y_2)(\overline{C} + y_1)$, $Y_2 = (D + y_2)(\overline{C} + y_2)(\overline{C} + \overline{D} + \overline{y}_1)$, $Z = y_2\overline{C}\overline{D}$. These expressions are then implemented as discussed in subsection 2.6.

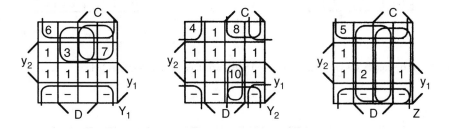

6.21. The circuit is shown below.

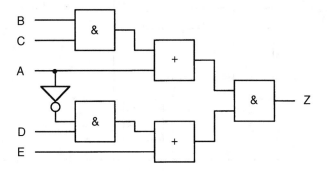

First apply hazard preserving transformations to obtain a SOP expression with the same static hazards: $(A + BC)(\overline{A}D + E) = A\overline{A}D + AE + \overline{A}BCD + BCE$. (We do *not* delete the $A\overline{A}$ D term since that would not be hazard preserving.) Next the p-terms thus obtained are mapped, (of course $A\overline{A}$ D does not appear on the map.) Since all adjacent pairs of 1-points on the map are included in subcubes, there are no 1-hazards. The $A\overline{A}$ D must now be checked to see if it corresponds to any 0-hazards. Examine the region where D = 1 to see if it includes pairs of 0-points differing only in the A-coordinate. There are indeed three such pairs of points: (00010, 10010), (00110, 10110), and (01010, 11010). Each corresponds to 0-hazard.

6.25. There is an essential hazard starting in state 1-11, with C changing, and another starting in 3-10 with C changing.

6.27. Replacing NORs and NANDs converts the combinational logic functions to their duals, and therefore the sequential function is also converted to its dual. Since the dual is also obtainable by complementing all inputs and outputs, the flow table for the new circuit can be obtained from the given flow table by simply complementing all inputs (column headings) and output entries. The first flow table below depicts the result. The second table is obtained from the first by re-ordering the columns in the usual way.

$$X_1 X_2$$

	11	10	00	01
1	1,1	1,1	2,0	3,0
2	2,0	2,1	2,0	3,0
3	1,1	2,1	3,1	3,0

$$X_1 X_2$$

	00	01	11	10
1	2,0	3,0	1,1	1,1
2	2,0	3,0	2,0	2,1
3	3,1	3,0	1,1	2,1

6.32. The two-stage AND-OR logic is replaced with two-stage NAND logic, as below:

(a)

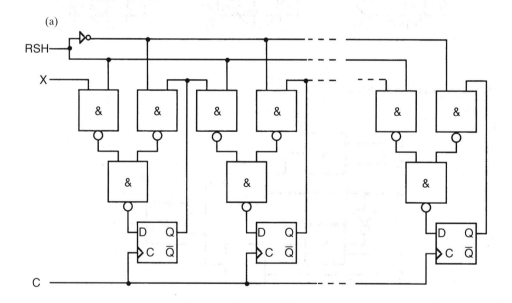

6.43. The logic expressions for the 2-stage logic design are: $Y_1 = \overline{A}B + By_1 + \overline{A}\,\overline{B}y_2$, $Y_2 = A\bar{y}_1 + A\overline{B} + \overline{A}\,\overline{B}y_2$, $Z_1 = A\bar{y}_1$, $Z_2 = \overline{A}B + By_1$.

Factoring yields: $\overline{Y_1} = B(\overline{A} + y_1) + \overline{A}\,\overline{B}y_2$, $Y_2 = A(\bar{y}_1 + \overline{B}) + \overline{A}\,\overline{B}y_2$, $Z_1 = A(\bar{y}_1 + \overline{B})$, $Z_2 = B(\overline{A} + y_1)$. Each component is shared. The corresponding 3-stage circuit has a total of 15 gate inputs as opposed to 21 for the 2-stage circuit.

6.62. (a) Initially, all inputs to the XOR-gate are 0s, so its output, Y_1, is 0. The situation after the first clock pulse is the same as before it, i.e. all y's are 0, so none of the y's ever changes state, and the output sequence is simply 000, 000, ...

(b) $Y_1 = X \oplus y_1 \oplus y_3$. Since X is fixed at 0, this becomes $Y_1 = y_1 \oplus y_3$. Also, $Y_2 = y_1$, and $Y_3 = y_2$. Then the following states after $y_1 y_2 y_3 = 110$ are: 111, 011, 101, 010, 001, 100, and back to 110.

Chapter 7

7.2(a)

(7.1a) $W = \overline{(\overline{A})} = A \qquad (\overline{\overline{X}} = X)$

(7.1b) $W = \overline{B + \overline{A}B} = \overline{B + \overline{A}} \qquad (X + \overline{X}Y = X + Y)$

(7.1c) $W = C + \overline{(A + B)} = \overline{\overline{C} + A + B} \qquad (\overline{\overline{X}} = X)$

7.7. The calculation is shown in the figure below, resulting in output arrival times of 10 and 8. The input slacks are 1, 0, 0, -2, 1, 2, 0, and 1.

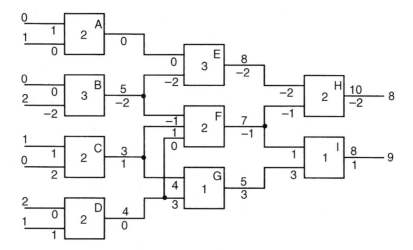

Appendix

A.5. (a) $29 = 011101_2$

(b) $14 = 001110_2$ The two's complement (TC) of this number is: 110010.

A.6. (a) 100001 (assuming a 6-bit word) is a negative number. TC(100001) = 011111_2 So 100001 represents -31_{10}

(b) 010101 is a positive number, corresponding to 21_{10}

A.7. (a) TC(100100) = 011100

(b) TC(001100) = 110100

A.14. (a) $60_{10} = 74_8 = 111100_2 = 3C_{16} = 0110\ 0000_{BCD}$

A.17. (a) A 7-bit word is necessary. Let the information bits be $b_3 b_5 b_6 b_7 = 1011$. Then the check bits are calculated as $b_1 = b_3 \oplus b_5 \oplus b_7 = 1 \oplus 0 \oplus 1 = 0$, $b_2 = b_3 \oplus b_6 \oplus b_7 = 1 \oplus 1 \oplus 1 = 1$, and $b_4 = b_5 \oplus b_6 \oplus b_7 = 0 \oplus 1 \oplus 1 = 0$. The code word is therefore: 0110011.

A.21. The total resistance in series with the voltage source is $1 + 2(3)/(2 + 3) = 2.2$. Then the current through the voltage source (and therefore through the 1 ohm resistor) is $6/2.2 = 2.73$. Then $V_1 = 1(2.73) = 2.73$ (answer to b). The voltage across the 3 ohm resistor is therefore $6 - 2.73 = 3.27$. Hence $i_3 = 3.27/3 = 1.09$ (answer to a).

A.25. With B open for a long time, there is no current flow in the circuit (C acts as an open circuit in the steady state), so that $V_{C0} = 5$ (answer to a). After B has been closed for a long time, again C acts as an open circuit, so no current flows through it (and the 2K resistor). The 5 volts are divided between the 1k and 3K resistors that are in series with the source. The voltage across the 3K resistor is $5[3/(3 + 1)] = 5(3)/4 = 3.75$. This is also V_C, since no voltage appears across the 2K resistor. So the answer to b is $V_{Cf} = 3.75$. The time constant is the product of the capacitance and the resistance seen across the capacitor. With B closed, this is 2K in series with the parallel combination of the 3K and 1K resistors (we treat the voltage source as a short circuit—i.e. as a 0 ohm resistor.) Thus the discharging resistance is $2 + 1 \times 3/(1 + 3) = 2.75$K. Then the time constant is $T = 4 \times 10^{-12} \times 2.75 \times 10^3 = 11 \times 10^{-9} = 11\text{nS}$ (answer to c). Now we have; $V_C = V_{Cf} + (V_{C0} - V_{Cf})e^{-t/T} = 3.75 + (5 - 3.75)e^{-t/T} = 3.75 + 1.25e^{-t/T}$.

Index

,